T0289852

Laser-Based Additive Manufacturing of Metal Parts

Advanced and Additive Manufacturing Series

SERIES EDITOR

Ali K. Kamrani

University of Houston, Texas, USA

PUBLISHED

Laser-Based Additive Manufacturing of Metal Parts: Modeling, Optimization, and Control of Mechanical Properties
Linkan Bian, Nima Shamsaei, and John M. Usher

Computer-Aided Inspection Planning: Theory and Practice
Abdulrahman Al-Ahmari, Emad Abouel Nasr, and Osama Abdulhameed

Laser-Based Additive Manufacturing of Metal Parts

Modeling, Optimization, and Control of Mechanical Properties

Edited by

Linkan Bian, Nima Shamsaei, and John M. Usher

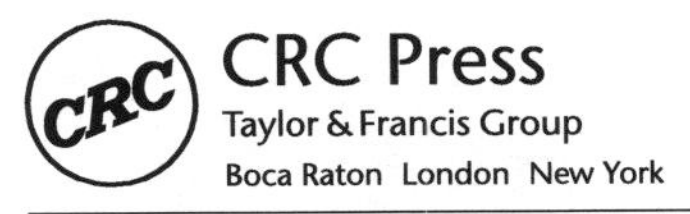

CRC Press
Taylor & Francis Group
Boca Raton London New York

CRC Press is an imprint of the
Taylor & Francis Group, an **informa** business

CRC Press
Taylor & Francis Group
6000 Broken Sound Parkway NW, Suite 300
Boca Raton, FL 33487-2742

Printed on acid-free paper

International Standard Book Number-13: 978-1-4987-3998-6 (Hardback)
International Standard Book Number-13: 978-1-4987-3999-3 (eBook)

Library of Congress Cataloging-in-Publication Data

Names: Bian, Linkan, editor. | Shamsaei, Nima, editor. | Usher, John (John M.), editor.
Title: Laser-based additive manufacturing of metal parts : modeling, optimization, and control of mechanical properties / edited by Linkan Bian, Nima Shamsaei, John M. Usher.
Description: Boca Raton: CRC Press, Taylor & Francis, 2018. | Includes bibliographical references.
Identifiers: LCCN 2017012442 | ISBN 9781498739986 (hardback : alk. paper)
Subjects: LCSH: Three-dimensional printing. | Machine parts--Design and construction. | Metal products--Design and construction.
Classification: LCC TS171.95 .L365 2018 | DDC 671.8--dc23
LC record available at https://lccn.loc.gov/2017012442

Visit the Taylor & Francis Web site at:
http://www.taylorandfrancis.com

and the CRC Press Web site at:
http://www.crcpress.com

Contents

Preface

Overview

Since the advent of laser-based additive manufacturing (LBAM) technologies for producing metallic parts in the late 1990s, their usage continues to be extended toward many applications. Metallic LBAM parts are now used as functional parts within many industrial sectors such as tooling, dental, medical, and aerospace. This popularity is driven by virtue of LBAM unique feature of generating complex-shaped, functionally graded, or custom-tailored parts that can be utilized for a variety of engineering and industrial applications. The recent flurry of LBAM research and development can be attributed to the manufacturing industry at large, seeking new "in-house" methods to fabricate custom structures and to the consumer/societal appeal of "three-dimensional (3D) printing." The idea of "printing" a 3D prototype or part, as opposed to purchasing or sending the computer-aided design (CAD) out-of-house for machining, has become alluring to technical and lay persons across the globe. A large number of companies have begun to use LBAM technologies to reduce time-to-market, increase product performance, and reduce manufacturing costs. Hailed by some as the "third industrial revolution," LBAM has the potential to revolutionize cost-effective mass customization and time-efficient mass production.

While LBAM processes have advanced greatly in recent years, many challenges remain to be addressed, such as relatively poor part accuracy (net shape), insufficient repeatability and consistency of parts, and lack of in-process control methodologies. To realize additive manufacturing (AM) potential for ushering in the "third industrial revolution," products must be fabricated rapidly, efficiently, and inexpensively while meeting all stringent functional requirements. The key to large-scale customization and mass production, two important LBAM business frontiers, is the capability of LBAM of a vast array of components with desired features. To compete with conventional mass production processes, AM technology needs to advance further in order to drastically reduce the cost of production, improve the performance of fabricated parts, and achieve consistency between parts. This book is designed to educate readers on the history/development of LBAM, help them grasp the fundamental process of physics and postfabrication mechanical properties, understand the issues of process optimization and control, and recognize the various applications of LBAM.

Target Readership

Advanced manufacturing education and training is critical to our economy, considering that 83% of US manufacturers recently reported an overall shortage of qualified employees. University–industry collaborative AM research in the United States has been initiated primarily through a variety of federally funded programs. To ensure and facilitate the transition of AM moving from the laboratory to a production environment, AM education and training has begun to play an important role in establishing a healthy engineering education ecosystem among US universities and colleges. Taking full advantage of AM will require educating the current workforce and recruiting a new generation of students. Identified focus areas for education and workforce development include AM foundational understanding, representative processes, design for AM, quality assurance for AM, and advanced AM research and education. *The purpose of this book is to fill an existing need in education materials for AM by providing background knowledge, reviewing recent research advances in mechanical/microstructural properties of LBAM built parts, and addressing topics on how to design, optimize, and control the thermal history to achieve desired part features.*

This book is not only designed to educate researchers, practicing engineers, and manufacturing industry professionals, who are interested in entering the arena of AM, but also to provide a *textbook* for senior undergraduate or graduate students. The book will include in-depth, detailed discussions of mechanical properties, materials science, and mathematical/numerical methods. Each chapter will begin by providing theoretical background information, terminology, definitions, etc., on the subject matter to aid nonsubject expert readers as they delve further into the chapter. To help readers understand the topics, examples, case studies, and problems/solutions will be presented at the end of each chapter. Problems will allow students/readers to test their comprehension of the topics requiring them to utilize mathematical models, engineering software, and external resources to solve/answer the problems effectively.

Contents

This book consists of a total of nine chapters, each contributed by invited authors. Chapter 1 introduces the history and development of LBAM as a process and how it has evolved from rapid prototyping to an automated technique for direct conversion of 3D CAD data into physical objects. The concepts and technology related to LBAM will be described and differentiated from other AM methods. A large number of LBAM processes, in conjunction with their corresponding support technologies such as software systems and postmanufacturing processing, will be reviewed. Chapter 2 focuses on methods and technologies related to the CAD of parts with the objective of achieving better part quality (e.g., mechanical properties, geometric accuracy, life cycles, etc.). Research issues related to designing parts with complex geometry (e.g., overhang structures, staircase effects, etc.) as well as the related research efforts will be reviewed. Chapter 3 compares the mechanical/microstructural properties of LBAM parts with those produced using traditional manufacturing processes. Advantages and limitations of LBAM will be discussed. Since the loading condition for many metallic parts in automobiles, airplanes, heavy machinery, bio-implants, and other industrial applications is cyclic, fatigue is the most prominent failure mode. Since fatigue failures account for up to 90% of all mechanical failures, Chapter 4 focuses on recent studies in the fatigue properties of LBAM parts. Studies related to the optimization of these parameters to achieve application-specific physical/mechanical properties are reviewed in Chapter 5. Chapter 6 reviews mainstream control methods that focus on maintaining constant melt pool morphology/temperature and deposited layer height. Due to the complex, coupled thermo physical dynamics associated with the LBAM process, process parameters need to be adjusted in real time to ensure that desired mechanical/microstructural properties can be obtained, which is in direct contrast to the traditional “set-it-and-forget-it” open-loop process that consists of time-invariant process parameters. Chapter 7 reviews the recent research efforts in the thrust areas of functionally, compositionally graded materials and the mechanical properties of the fabricated parts. Among all powder-based AM methods, LBAM is capable of feeding different feedstock (preform) materials simultaneously to produce various alloys and functionally graded parts. Each feed rate can be adjusted to produce materials with different chemical compositions and microstructural properties as well. Accordingly, a large number of alloys have been developed using LBAM techniques over the past decade. At the same time, this also raises new challenges in process modeling, optimization, and control. Chapter 8 overviews the applications of LBAM in various industrial sectors, such as healthcare, aerospace, automotive, and others. Examples are presented, those expressing key advantages of LBAM over traditional manufacturing methods. Applications that improve biocompatibility, efficiency, safety, ergonomics, and performance are discussed. Chapter 9 discusses the economic impacts of AM in terms of three different aspects: The first involves measuring the size of AM. This includes measuring the value of the goods produced using this technology in the context of the total economy. The second aspect involves measuring the costs and benefits of using this technology. It includes understanding when AM is more cost effective than traditional manufacturing, and why it is more cost effective. It also involves understanding other advantages such as new products that might not be possible with traditional manufacturing. The last aspect of AM economics is the adoption and diffusion of this technology.

About the Editors

Dr. Linkan Bian is an assistant professor in the Industrial and Systems Engineering Department at Mississippi State University. He received his PhD degree in industrial and systems engineering from Georgia Institute of Technology in 2013, a dual master degree in statistics from Michigan State University, and a BS degree in applied mathematics from Beijing University in 2005. Dr. Bian's research interests focus on the combination of data mining and optimization methods for modeling and control of additive manufacturing processes. Other applications of his research include cybersecurity and supply chains. He has received external funding from Department of Defense, Department of Energy, National Science Foundation, and industrial companies. Dr. Bian's publications have appeared in journals such as *IIE Transactions* (*Institute of Industrial Engineering*), *Additive Manufacturing, Rapid Prototyping, IEEE Transactions* (*Institute of Electrical and Electronics Engineers*), and others. His work has received multiple awards/nominations from the Quality Control & Reliability Engineering (QCRE) division of the Institute of Industrial and Systems Engineers (IISE). He is serving as a council member of Data Mining Quality, Statistics, Reliability (DM-QSR) section within the Institute of Operations Research and Management Science (INFORMS) professional society.

Dr. Nima Shamsaei is currently an associate professor in the Mechanical Engineering Department at Auburn University. He has several years of industry experience, most notably at Chrysler Group LLC as a senior engineer and a technical leader specializing in durability test development for all Chrysler products. He received his PhD degree in mechanical engineering from the University of Toledo in 2010. His BS and MS degrees in mechanical engineering were obtained from Isfahan University of Technology and Sharif University of Technology, respectively. Dr. Shamsaei has extensive research experience in additive manufacturing, mechanical behavior of materials, design, and fatigue and failure analysis. He has published over 100 scientific papers in refereed journals and conference proceedings in the areas of fatigue and additive manufacturing. He is the founder and codirector of the

Laboratory for Fatigue and Additive Manufacturing Excellence (FAME) at Auburn University and is currently leading multiple sponsored research projects from the National Science Foundation (NSF), Department of Defense (DOD), National Aeronautics and Space Administration (NASA), and several private companies. He is a member of American Society of Mechanical Engineers (ASME), the Minerals, Metals, and Materials Society (TMS), American Institute of Aeronautics and Astronautics (AIAA), and Society of Automotive Engineers (SAE), and is actively engaged in the ASTM E08 committee on Fatigue and Fracture and the F42 committee on Additive Manufacturing. Dr. Shamsaei is the recipient of multiple awards and recognitions, including the SAE International Henry O. Fuchs Fatigue Award in 2010, the Schillig Special Teaching Award in 2015, and the American Society for Testing and Materials (ASTM) International Emerging Professional Award in 2016. He is also a consultant for multiple private companies, including Fiat Chrysler Automobiles.

Dr. John M. Usher currently serves as professor and head of the Industrial & Systems Engineering Department at Mississippi State University. He received his PhD degree in engineering sciences at Louisiana State University, where he also earned master's degrees in both chemical and industrial engineering. As well, he has a bachelor's degree in chemical engineering from the University of Florida. Prior to his academic career, John worked for Texas Instruments in the area of process development for the manufacture of multilayer printed circuit boards. John's research interests focus on production systems and systems simulation, modeling, and analysis, with applications in both manufacturing and transportation. He has published numerous papers in technical journals and conference proceedings, edited several books, and continues to serve on the editorial staff of several journals. John is a registered professional engineer in Mississippi and a senior member of IISE and SME.

Contributors

Amir M. Aboutaleb
Industrial and Systems Engineering Department
Mississippi State University
Starkville, Mississippi

Amirhesam Amerinatanzi
Mechanical Industrial and Manufacturing Engineering
University of Toledo
Toledo, Ohio

Moataz M. Attallah
School of Metallurgy and Materials
University of Birmingham
Birmingham, United Kingdom

Taniya Benny
Department of Systems Science and Industrial Engineering
Binghamton University (State University of New York)
Binghamton, New York

Linkan Bian
Industrial and Systems Engineering Department
Mississippi State University
Starkville, Mississippi

Luke N. Carter
School of Metallurgy and Materials
University of Birmingham
Birmingham, United Kingdom

Steve R. Daniewicz
Department of Mechanical Engineering
University of Alabama
Tuscaloosa, Alabama

Mohammad Elahinia
Mechanical Industrial and Manufacturing Engineering
University of Toledo
Toledo, Ohio

Ahmadreza Jahadakbar
Mechanical Industrial and Manufacturing Engineering
University of Toledo
Toledo, Ohio

Alexander Johnson
Department of Mechanical Engineering
Mississippi State University
Starkville, Mississippi

Bashir Khoda
Industrial and Manufacturing Engineering Department
North Dakota State University
Fargo, North Dakota

Narges Shayesteh Moghaddam
Mechanical Industrial and Manufacturing Engineering
University of Toledo
Toledo, Ohio

Chunlei Qiu
School of Metallurgy and Materials
University of Birmingham
Birmingham, United Kingdom

Prahalad K. Rao
Department of Mechanical and Materials Engineering
University of Nebraska–Lincoln
Lincoln, Nebraska
and
Department of Systems Science and Industrial Engineering
Binghamton University (State University of New York)
Binghamton, New York

Noriko Read
School of Metallurgy and Materials
University of Birmingham
Birmingham, United Kingdom

Michael P. Sealy
Department of Mechanical and Materials Engineering
University of Nebraska–Lincoln
Lincoln, Nebraska

Nima Shamsaei
Department of Mechanical Engineering
Auburn University
Auburn, Alabama

Samuel Tammas-Williams
Department of Materials Science and Engineering
The University of Sheffield
Sheffield, United Kingdom

Douglas S. Thomas
Applied Economics Office
National Institute of Standards and Technology
Gaithersburg, Maryland

Scott M. Thompson
Department of Mechanical Engineering
Auburn University
Auburn, Alabama

Iain Todd
Department of Materials Science and Engineering
The University of Sheffield
Sheffield, United Kingdom

Wei Wang
School of Metallurgy and Materials
University of Birmingham
Birmingham, United Kingdom

Chi Zhou
Department of Industrial and Systems Engineering
University at Buffalo
Buffalo, New York

Chapter 1

Recent Advances in Laser-Based Additive Manufacturing

Narges Shayesteh Moghaddam, Ahmadreza Jahadakbar, Amirhesam Amerinatanzi, and Mohammad Elahinia
University of Toledo

Contents

CHAPTER OUTLINE

In this chapter, we

- Introduce additive manufacturing (AM) techniques
- Describe support technologies such as software systems
- Discuss postmanufacturing processing

1.1 Introduction

Learning Objectives

- Understand the benefits of AM
- Learn the overview of processes involved in AM
- Understand the opportunities and challenges related to AM

In recent years AM techniques have been introduced as a promising method to produce metallic parts because they can overcome the common problems associated with conventional techniques. Using these methods, we can produce complex parts (e.g., porous scaffolds, free-form surfaces, and deep slots). In general, AM deals with adding material instead of its removal. First, a computer-aided design (CAD) file of the desired part needs to be created or designed. This CAD file can be completely designed by a designer, or it can be created by scanning an object. Subsequently, the CAD file will be converted into layers with specific thickness, usually ranging from 20 to 100 μm (Ravari et al., 2015), using a supporting software such as Autofab (Materialise, Louvain, Belgium). Finally, an AM machine will fabricate the part layer by layer (Gibson et al., 2010). All AM machines need a source of energy to fabricate the cross section of each layer. There are a variety of energy sources, including those based on laser, electron beam, and ultrasonic (Mici et al., 2015). We mainly focus on the laser-based additive manufacturing (LBAM) techniques.

This chapter reviews the existing AM techniques, including powder-bed and flow-based methods. Support technologies such as computers and software are then discussed. Finally, postmanufacturing processes are covered. This includes support removal techniques, surface improvement methodologies, aesthetic improvement methods, and property enhancement techniques.

1.2 AM Techniques

Learning Objectives

- Understand powder-bed-based techniques
- Learn flow-based techniques
- Understand sheet lamination

The goal of this section is to provide a review of a large number of LBAM processes. LBAM techniques can be classified into three major types: powder-bed-based or powder-bed fusion (PBF), flow-based or directed energy deposition, and sheet lamination. Figure 1.1 shows a detailed classification of LBAM processes that are used for metals.

Different types of materials are used in the AM techniques, including powder, wire, and sheet. Although powder-bed-based methods are only capable of using powder (Baufeld et al., 2011), the flow-based methods can utilize either powder or wire (Andani et al., 2014; Moghaddam et al., 2016a). Sheet is the only type of material used in sheet lamination techniques.

The use of powder is more favorable than the use of wire or sheet, and this is due to the possibility of fabricating parts with small size and high geometrical accuracy. The powder diameters are small, enabling the use of thinner layers during fabrication (Xiao and Zhang, 2007). Additionally, it is possible to produce parts with functionally graded materials (Gibson et al., 2010). From another point of view, the use of wire is a cleaner and more environmentally friendly process, because it is not associated with spreading the hazardous powder. Moreover, wire preparation is usually easier and less expensive than powder preparation. Finally, the use of wire has a higher material usage efficiency with almost zero percent waste during fabrication (Ding et al., 2015).

1.2.1 Powder-Bed Based

Powder-bed-based techniques normally consist of a laser, an automatic powder layering apparatus (e.g., blade, knife, or roller), a computer system for process control, and some accessorial mechanisms (e.g., inert gas protection system and powder-bed preheating system) (Jahadakbar, 2016; Jahadakbar et al., 2016; Amerinatanzi et al., 2016). In these methods, the powder layers are first deposited on the substrate using a blade, a knife, or a roller. Subsequently, a high laser power heats and fuses the deposited powder selectively. (Note: The duration of the laser beam radiation depends on beam size and scan speed; Kumar, 2003.) The powders that surround the solid region remain loose, and they serve as support for the next layer deposition. Then, the substrate will go down to provide space for depositing the second layer. All the steps continue until the piece is fabricated (Gibson et al., 2010) (Figure 1.2).

Table 1.1 discusses the different types of powder-bed-based techniques.

1.2.1.1 Selective Laser Sintering (SLS)

SLS, i.e., indirect laser sintering process, was developed by Dr. Joe Beaman at the University of Texas at Austin in the mid-1980s (Bandyopadhyay et al., 2015). The physical process can be either liquid-phase sintering or partial melting.

The first physical process uses two-component metal powders that are a combination of a sacrificial binder (i.e., polymer phase) and the powder particles. During laser irradiation, the powder only fuses with the binder material (Khaing et al., 2001). The liquid formed by the melted binder binds the surrounding powder such that they solidify the structure after cooling down (Kolossov et al., 2004). In this stage, powder particles remain solid and largely unaffected by the heat of

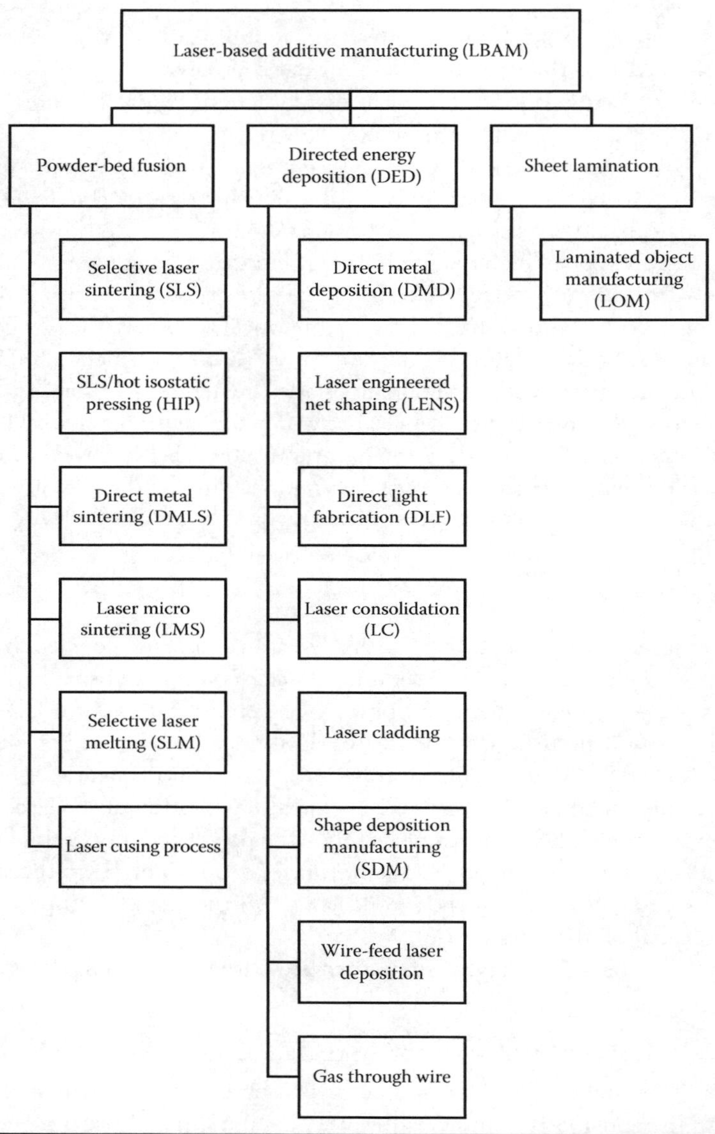

Figure 1.1 Classification of laser-based additive manufacturing (LBAM) techniques for metals. (From Kruth, J.-P. et al., *J. Mater. Process. Technol.*, 149(1), 616–622, 2007; Hu, J.Z. et al., *Mechanics of sheet metal forming*. Butterworth-Heinemann, Oxford, 2002; Campbell, G.S., Norman, J.M., *An introduction to environmental biophysics*, Springer Science & Business Media, New York, 2012; Baufeld, B. et al., *J. Mater. Process. Technol.*, 211(6), 1146–1158, 2011.)

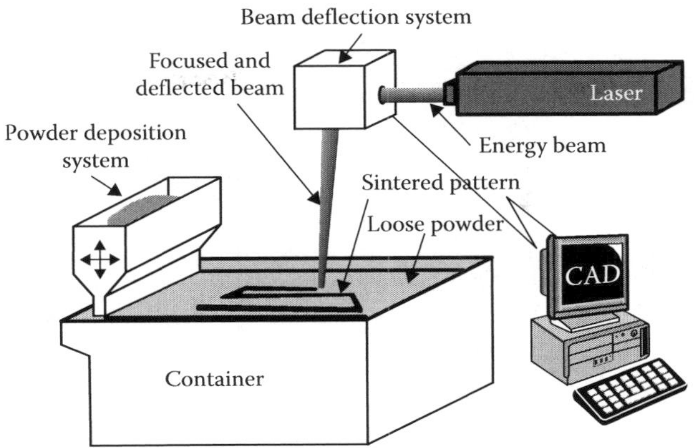

Figure 1.2 Powder-bed-based processes. (From Gibson, I. et al., Additive Manufacturing Technologies, Springer, New York, 2010.)

the laser (Bidabadi et al., 2012). The parts at this stage contain about 50% of porosity (in some cases even more) and are called green parts. After the part is fabricated, the sacrificial binder can be removed by applying a heat treatment in a thermal furnace, at temperatures in excess of 900°C. This process is called debinding (Agarwala et al., 1995; Kruth et al., 2003; Goban et al., 2013). Then, the furnace temperature is again raised till a small level of necking or sintering occurs between the powder particles. If a denser part is needed, the remaining porosity can be either filled by infiltration of a lower melting point metal or by further sintering (i.e., densifying). It is worth saying that infiltration is easier to control dimensionally, as the overall shrinkage during this process is much less than consolidation. However, infiltrated materials are always composites, while consolidated parts are made of a single material type (Jones et al., 1998) (Figure 1.3).

The latter physical process associates with the partial melting of single-component powder particles. The laser will melt the surface of powder particles, but the solid cores remain unchanged. The liquid that is formed by the melting of the surfaces will be solidified after cooling down, and as a result, the particles will be joined together (Xiao and Zhang, 2007). As a result of such partial melting, the interior section has a density of approximately 65% and can reach 80% depending on the level of laser sintering. On the other hand, the skin is approximately fully dense with the density exceeding 92%. At this level of density, the structure is impermeable completely (Jones et al., 1998) (Figure 1.4).

It is worth noting that the material selection is very important because the material's laser absorptivity depends on the laser wavelength (Gao et al., 2015).

1.2.1.2 Selective Laser Sintering/Hot Isostatic Pressing (SLS/HIP)

SLS/HIP technique is an enhanced version of the SLS method, and it was developed at the University of Texas (Knight et al., 1996). This technique is a hybrid net shape manufacturing approach that combines the freeform shaping capability of SLS with the full densification capability of HIP to produce high-value metal components at cheaper prices in shorter time (Das et al., 1999).

Previously, SLS parts could demonstrate a maximum density of 80%, while SLS/HIP parts can produce fully dense metal parts. HIP is a heat treatment technique that uses inert gas (e.g., argon) to enhance the mechanical properties of the material. A final machining step may also be applied if necessary.

Table 1.1 Different Types of Powder-Bed-Based Techniques

Powder Bed	*Layer Thickness (μm)*	*Common Materials*	*Resolution (μm)*	*Laser Type*
SLS	100–300 (Santos et al., 2006)	Almost all metallic materials (Santos et al., 2006)	>100 (Kumar, 2003)	CO_2 (Zhu et al., 2003), Nd:YAG (Fischer et al., 2005), fiber lasers (Schleifenbaum et al., 2010), and disc lasers (Kaiser and Albrecht, 2007)
SLS/HIP	100–300 (Santos et al., 2006)	InconelW 625 superalloy, Ti–6Al–4V (Knight et al., 1996; Das et al., 1997)	>100 (Kumar, 2003)	CO_2 (Zhu et al., 2003), Nd:YAG (Fischer et al., 2005), fiber lasers (Schleifenbaum et al., 2010), and disc lasers (Kaiser and Albrecht, 2007)
DMLS	20–50 (Shellabear and Nyrhilä, 2004)	Steel alloys, stainless steel, tool steel, aluminum, bronze, cobalt–chrome, (Fe, Ni)–TiC composites and titanium (Xiao and Zhang, 2007)	20 (Hänninen, 2001)	CO_2 (Zhu et al., 2003), Fiber (Schleifenbaum et al., 2010)
LMS	1–10 (Regenfuss et al., 2007)	Steel, silver, and copper (Regenfuss et al., 2007)	<30 (Regenfuss et al., 2007)	Q switched Nd:YAG in TEM00 mode (Regenfuss et al., 2007)
SLM	20–100 (Ravari et al., 2015)	Stainless steels, aluminum, copper, iron, cobalt–chrome, titanium, nickel-based alloy, and a mixture of different types of particles (Fe, Ni, Cu, and Fe_3P) (Abe et al., 2001; Kruth et al., 2004; Childs et al., 2005; Moghaddam et al., 2016b; Shayesteh Moghaddam, 2017)	20 (Ravari et al. 2015)	Nd:YAG (Delgado et al., 2012)

1.2.1.3 Direct Metal Sintering (DMLS)

DMLS was developed by EOS GmbH and Rapid Product Innovations (RPI) in 1994 (Shellabear and Nyrhilä, 2004). This process can be done by using two different powders (i.e., binary systems) or a single powder with two different grain sizes. Typical binary phase systems include Ni–Cu,

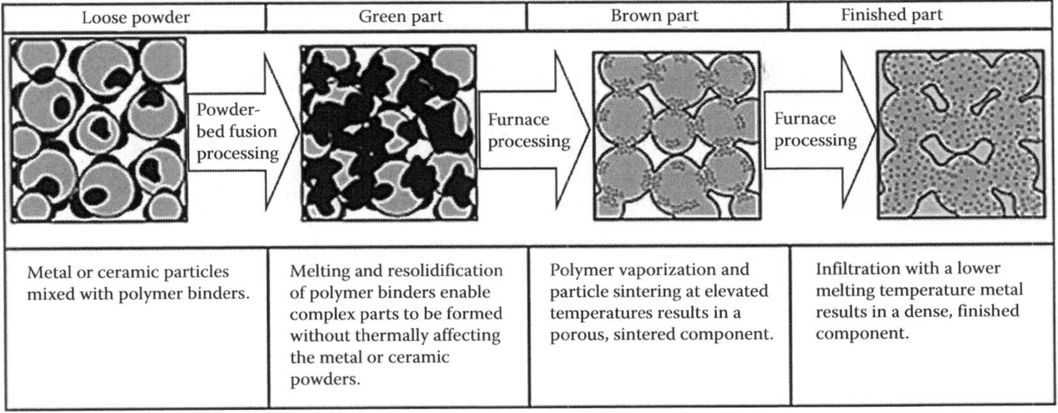

Loose powder
Green part
Brown part
Finished part
Powder-bed fusion processing
Furnace processing
Furnace processing
Metal or ceramic particles mixed with polymer binders.
Melting and resolidification of polymer binders enable complex parts to be formed without thermally affecting the metal or ceramic powders.
Polymer vaporization and particle sintering at elevated temperatures results in a porous, sintered component.
Infiltration with a lower melting temperature metal results in a dense, finished component.

Figure 1.3 The schematic of liquid-phase sintering processes. (From Jones, D.R. et al., *J. Global Optim.*, 13(4), 455–492, 1998.)

Fe–Cu, and Cu–Pb/Sn (Goban et al., 2013). Based on the level of laser power, the low melting point components are melted and employed as a matrix, and the higher melting point component sits in it (Xiao and Zhang, 2007). This process is capable of fabricating fully densified metal parts via increasing the laser power (Gibson et al., 2010).

The major advantage of DMLS is that it produces objects free from internal defects and residual stresses inside the metal. Additionally, this process overcomes the balling phenomena (it will be explained completely in the SLM section later in this chapter). The reason is that the particles inside the metal are melted together without fusing completely because the metal is not heated enough to fuse the particles completely (Kruth et al., 2004; Xiao and Zhang, 2007). The disadvantage of this method, on the other hand, is that the fabricated parts exhibit the mechanical properties of their weakest composite phase, and the functionality that is required for heavy-duty tasks is lost, and the parts are not completely dense due to the weak phase (Shellabear and Nyrhila, 2004).

Different machines have taken advantage of the DMLS technique so far, including EOSINT M 250, 270, 280, and 400 (Khaing et al., 2001). The models 250 and 270 utilize CO_2 laser, and the

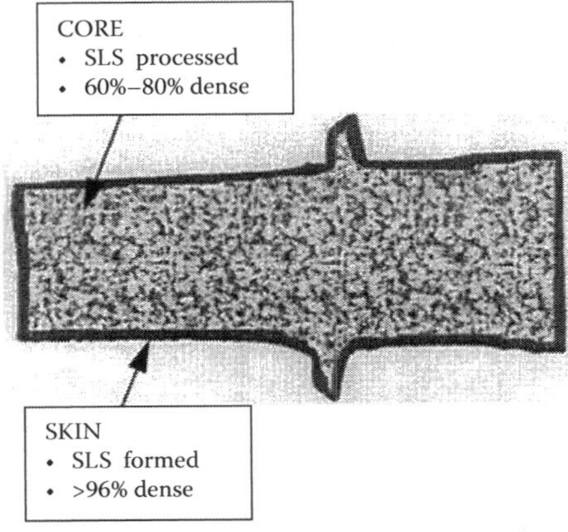

Figure 1.4 The effect of partially melting process on the resultant powders. (From Jones, D.R. et al., J. *Global Optim.*, 13(4), 455–492, 1998.)

other models often use fiber lasers. Model 270 operates in the nitrogen atmosphere, and it allows the fabrication of a wide range of materials from light alloys via steels to superalloys. Model 280 allows the fabrication of more complex geometries such as freeform surfaces and deep slots (Santos et al., 2006). This system operates in both protective nitrogen and argon atmospheres that allow the system to process a wider range of materials. Model 400 offers the manufacturing of the large metal parts on an industrial scale. This system operates in both protective nitrogen and argon atmospheres that allow the system to process a wider range of materials (Udroiu, 2012).

1.2.1.4 Laser Microsintering (LMS)

In early 2003, Laser Institute Mittelsachsen e.V. launched an innovative process called "Laser Microsintering." This technique and the equipment are marketed by 3D-Micro-mac AG, Chemnitz, Germany under the brand name microsintering. This new setup produces parts with high resolution (<30 μm) and minimal roughness (Ra = 1.5 μm), while the SLS machine could fabricate parts with a resolution of not less than 100 μm (Regenfuss et al., 2007). Another advantage of this technique is that the final parts are tension free due to the use of q-switched Nd:YAG—laser pulse (Regenfuss, 2004).

1.2.1.5 Selective Laser Melting (SLM)

SLM is another type of powder-bed-based technology that enables the fabrication of complicated parts with a density close to 100% that assures series-identical properties and eliminates the need for posttreatments (Gu et al., 2012). There are two major differences between SLM and SLS methods. First, SLM uses the integral powder without the need of adding elements with low melting points. Second, higher power laser is required for the SLM process to cause the powder to be fused completely (Mumtaz et al., 2008).

One of the problems associated with SLM is the residual stress that results from high thermal gradients in the material during the melting process. These residual stresses cause distortion of the part. However, as mentioned earlier, preheating the material and maintaining the temperature inside the chamber at a constant high level minimizes the residual stress (Mercelis and Kruth, 2006).

The main machines in the market that use SLM technique are Trumaform LM 250, MCP Realizer, and LUMEX 25C. Trumaform LM 250 is used for making complex parts such as customized implants, and the final parts can be finished by conventional processes such as machining, eroding, or grinding (Santos et al., 2006). The MCP Realizer was launched by the University of Liverpool. This machine is used for ultralight mesh structures, dental caps, crowns, and bridges in stainless steel and cobalt–chrome (Bennett and Sutcliffe, 2004). The LUMEX 25C was developed by MATSUURA. The system uses a pulsed CO_2 laser. The machine has a computer numerical code (CNC) milling system to process parts after 5–10 layers (Sutcliffe and Mardon, 1975; Santos et al., 2006).

1.2.1.6 Laser Cusing Process

The term "cusing" is derived from letter "C" and the word "Fusing". Concept Laser is a company that produces M1 cusing, M2 cusing, M2 cusing multilaser, and M3 linear machines. A special characteristic of this method is the stochastic exposure strategy. The sections of each layer, which are called island, are fabricated stochastically until the layer fabrication is finished (Figure 1.5). As a result, the inert stress will be reduced inside the part in comparison with the regular SLM method (Aliakbari and Baseri, 2012). Typical layer thicknesses for this process are 20–50 μm (Udroiu, 2012).

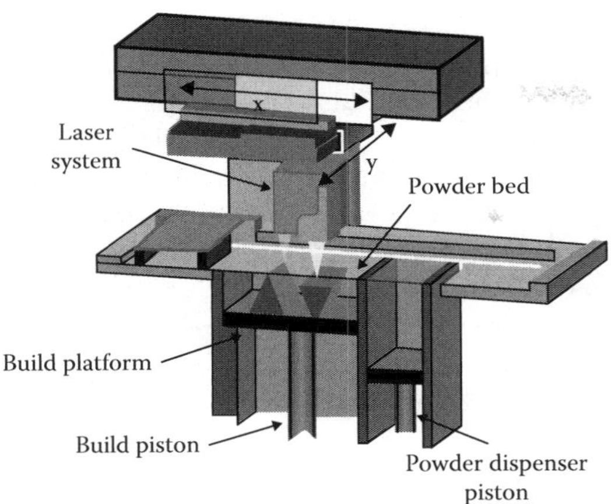

Figure 1.5 Laser cusing method. (From Aliakbari, E., Baseri, H., *Int. J. Adv. Manuf. Technol.*, 62(9), 1041–1053, 2012.)

M1 cusing is suitable for fabrication of small- to medium-sized parts. M2 cusing is in line with Atmosphères Explosibles (ATEX) guidelines, and therefore, it is possible to process aluminum and titanium alloys safely. The machine is suitable for three-shift operation. M2 cusing multilaser has a double-fiber laser to expedite the manufacturing. The M3 linear machine is easier to work with, and it can be used to process a wide variety of metals, including stainless steel and other chromium alloys. It can also be used for customer-specific materials.

These machines are capable of processing stainless and tool steel, such as high-grade steels, hotwork steels, and stainless hotwork steels. M2 is also capable of processing nonferrous alloys, such as titanium (Ti6AlV4), aluminum (AlSi12 and AlSi10Mg), cobalt–chrome alloy, and nickel-based alloys (Petrovic et al., 2011; Udroiu, 2012).

1.2.2 Flow-Based Techniques

Flow-based methods use a nozzle to deliver the material (i.e., powder or wire) directly into the laser focus zone, instead of prespreading on the substrate (Elahinia, 2015, 2016; Andani, 2014).

This series of AM processes use a powder feeder that delivers powder into a gas delivery system via nozzles. Then, the high-energy laser is delivered in the center of the nozzle and will be focused by a lens in proximity to the workpiece. Moving the system of lens and powder nozzles in the *z*-direction controls the height of the focuses of both laser and powder. The workpiece is also moved in the *x*–*y* plane under the functional area by a controller (Danforth et al., 1998).

1.2.2.1 Direct Metal Deposition (DMD)

DMD technology was developed at the University of Michigan (Maumder, 1997; Dinda et al., 2009). All DMD machines are equipped with a closed-loop controller. POM Company has introduced several machines, including DMD 5000 and DMD 105D. The DMD 5000 machine uses a 5 kW CO_2 laser, a flying optic, a gantry robot, and an *XY* table, as well as several special powder feeders. DMD 105D is equipped with a 1 kW diode/disc laser, which has a closed-loop control with five axes deposition. It is possible to produce products with dimensions of 300 mm × 300 mm × 300 mm (Kumar and Pityana, 2011).

This method is often used for making molds and dies and for repairing parts that are expensive or impossible to be repaired (Toyserkani et al., 2004; Wong and Hernandez, 2012). One of the problems associated with this method is the high residual stress that is caused by an uneven heating and cooling processes (Wong and Hernandez, 2012). Additionally, the objects created by this process may require machining for better finish. The parts fabricated with this technique have full density due to complete melting of particles (Kumar and Pityana, 2011). DMD method provides a better coating compared to other techniques such as arc welding and thermal spray, and it offers minimal distortion and good surface quality (Toyserkani et al., 2004; Dinda et al., 2009).

There are three types of material transformation to the substrate, namely, powder injection, pre-placed powder, and/or wire feeding. Powder injection is the most effective way to feed the substrate. In this method, the laser beam melts both powder particles and a thin layer of the substrate. Therefore, a bulk layer will be created on the substrate. The thickness of the layer is in the range between 0.05 and 2 mm. The angle between the powder feeder and the laser beam often ranges from 0° to 45°. However, it is shown that the most efficient way is when the powder arrives perpendicular to the substrate because it can provide much better coating (Vilar, 1999). As shown in Figure 1.6, the perpendicular feeding can be achieved by using two coaxial feeders (Gibson et al., 2010).

DMD method is known by different names, based on its application. The University of Missouri at Rolla uses the same name for this technique (Choi and Chang, 2005). In terms of coating application, it is called "laser coating," "laser powder deposition," and "laser surfacing" (Verwimp et al., 2011). There are some other typical names for this method as outlined in Table 1.2 (Toyserkani et al., 2004).

The difference between these methods is related to each process in conjunction with the process configuration and process parameters. As a result, each of these processes is used for a particular purpose based on the required functionality, morphology, and surface quality. It should be noted that the majority of these processes produce near-net shape part, which are associated with

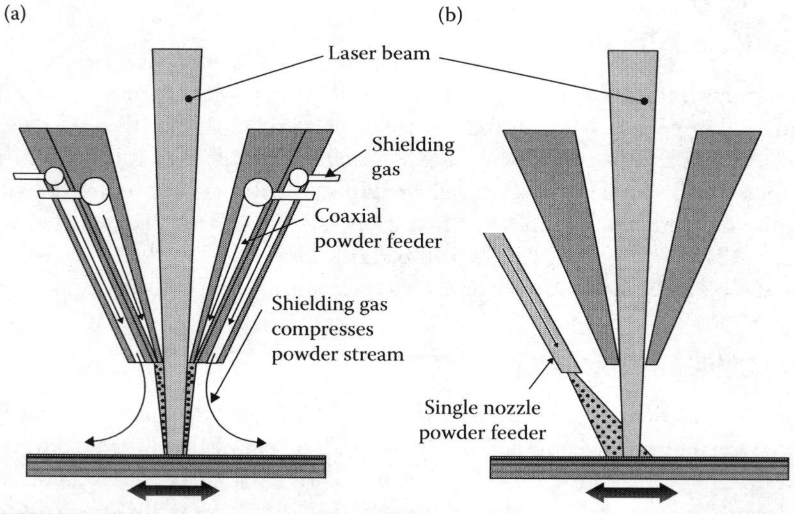

Figure 1.6 The configuration of powder nozzle to feed the substrate. (a) Coaxial feeding and (b) single feeding. (From Gibson, I. et al., *Additive Manufacturing Technologies*, Springer, New York, 2010.)

Table 1.2 Similar Methods to DMD Technique

Different Names Associated with DMD Technique	*Research Group*
Laser engineered net shaping (LENS)	Sandia National Laboratory (Atwood et al., 1998; Hofmeister et al., 1999)
Direct light fabrication (DLF)	Los Alamos National Laboratory (Lewis et al., 1994, 1997; Milewski et al., 1998)
Laser consolidation (LC)	National Research Council of Canada (Xue and Islam, 2000; Xue et al., 2000a,b, 2011; Lawrence, 2010)
Laser cladding	Rockwell International Corporation (Toyserkani et al., 2004)
Laser powder fusion (LPF)	Some industries that repair turbine blades (Bohrer et al., 2002)
Laser metal forming (LMF)	University of Missouri at Rolla (Laeng et al., 2000; Munjuluri et al., 2001) Swiss Federal Institute of Technology (Gäumann et al., 2001)
Selective laser powder remelting (SLRP)	Fraunhofer Institute (Germany) (Meiners et al., 1999)
Laser direct casting (LDC)	The University of Liverpool (Hand et al., 2000)
Laser powder deposition (LPD)	Several research groups in China (Toyserkani et al., 2004; Xie et al., 2010)
Automated laser powder deposition (ALPD)	University of Waterloo (Toyserkani and Khajepour, 2006)
Shape deposition manufacturing (SDM)	Stanford University (Fessler et al., 1996; Ramaswami et al., 1997) Carnegie Mellon University (Merz, 1997)
Laser additive manufacturing (LAM)	Minnesota University (Arcella et al., 2000)
Laser rapid forming (LRF)	Shanghai Jiaotong University (Shaohua, 2010)

additional machining. Here, we reviewed the details of four common methods: Laser Engineered Net Shaping (LENS), Direct Light Fabrication (DLF), Laser Consolidation (LC), and Laser Cladding.

1.2.2.1.1 LENS Technique

Optomec is a company that uses LENS 750 and LENS 850-MR7. LENS machines are equipped with feedback control that helps maintain quality and desired mechanical properties of the product (Kumar and Pityana, 2011). LENS technology is often used for repair, rapid prototyping and manufacturing, and manufacturing for aerospace, defense, and medical devices. Microscopy studies show the LENS parts to be fully dense with no compositional degradation. Mechanical testing reveals outstanding as-fabricated mechanical properties.

LENS 750 utilizes a power 500 W Nd:YAG laser, a three-axis computer-controlled positioning system, a controlled atmosphere box, and a coaxial powder feeder (Kumar and Pityana, 2011). The building chamber has the dimensions of 300 mm×300 mm×300 mm. The technique is often working at an oxygen level of 3–5 parts per million by using argon as back filler. The metal powders are delivered to the substrate using gravity or pressurized gas. The powder feeder utilizes multiple tubes that allow mixing of elemental powders (Liu et al., 2014; Kwiatkowska et al., 2015). Titanium alloys, nickel alloys, steels, cobalt alloys, and aluminum alloys are the main materials used for this process (Atwood et al., 1998; Santos et al., 2006; Wong and Hernandez, 2012). The powder efficiency is reported to be 80% for the laser power of 2500–3000 W and 3–4 mm spot size. However, this value was reported to be 20% for the laser power of 400–500 W and 1 mm spot size (Mudge and Wald, 2007). It should be mentioned that LENS 750 may use Fiber 1 or 2 kW. Metallic powders have higher absorptivity for Nd:YAG laser in comparison with CO_2 laser (fiber and disk have near wavelength to Nd:YAG laser) (Rahmani et al., 2012; Moghaddam et al., 2011, 2012). LENS 850-MR7 uses Fiber 1 kW, 2 kW, or other optional lasers. The building chamber is 900 mm×1500 mm×900 mm. Therefore, it produces larger parts. This model of the machine is equipped with up to seven-axis computer-controlled positioning system (Kumar and Pityana, 2011).

1.2.2.1.2 DLF Technique

DLF is a near-net shaping technique that is capable of producing complex components, with considerable cost and material savings. There are some disadvantages in this method, including the bonding defects, porosity, and the creation of heterogeneous microstructure (Parimi et al., 2014).

This technique uses a 5 kW ROFIN TR050 CO_2 laser, a four-axis CNC system, a powder feeder, a powder nozzle, and a powder recycler. The CNC unit controls the start and stop of the laser, powder feeder, and powder recycler, and the movements of the powder nozzle. The powder recycler recycles the unmelted powder for future use. The coaxial powder nozzle often uses four tubes to mix powders (Peng et al., 2005). The powder efficacy of this method is shown to be 86% (Mudge and Wald, 2007). In this technique, the undercut is done by machining, instead of tilting the part.

1.2.2.1.3 LC Technique

Accufusion is a company that uses LC 105 machine. This machine is equipped with a Nd:YAG 5 kW laser. The parts with the size of 450 mm×450 mm×450 mm can be fabricated with this machine. This machine is equipped with a control and monitoring system, and it has five-axes motion (Kumar and Pityana, 2011). This process is being investigated as a method of rapid fabrication for another project by MD Robotics, called Advanced Robotic Mechatronics System (ARMS). The goal of ARMS is to combine new technologies for the design and manufacturing of next-generation space robotic arms with the goal of achieving low mass, reduced cost, improved structural performance, and faster manufacturing time. The fabricated parts are metallurgically sound and free of porosity or cracks.

The machine that is used by the developer is equipped with a Nd:YAG laser coupled to a fiber-optic processing head. The laser has an average power of 20–300 W, and the powder feeder has a feed rate ranging from 1 to 30 g/min. The process is done at a constant room temperature, and the oxygen content is maintained below 50 ppm during the process. The system is equipped with a numerically controlled motion system with 3–5 axes. A thin-walled structure can be created using this technique. The four typical materials which have the capability of fabrication via LC are Ni-base IN-625 and

IN-738 alloys, and Co-base Stellite, and Fe-base CPM-9V tool steel (Mclean et al., 1997; Keicher and Miller, 1998; Milewski et al., 1998; Xue and Islam, 2000; Xue et al., 2000a,b).

1.2.2.1.4 Laser Cladding

One piece of equipment used for cladding is a Trumpf 3 kW CW CO_2 laser system that utilizes the blown powder cladding technique (Yellup, 1995). The power of the laser beam is set at 1.65 kW, and the beam diameter at 4 mm. In the cladding process, the laser beam was set to scan the surface of the substrate to be clad, while the hard-facing powder was fed to the spot of cladding by a pneumatic powder feeding system using argon as the carrier gas.

1.2.2.1.5 Shape Deposition Manufacturing (SDM)

This technique is often used to produce fully functional heterogeneous prototypes, multimaterial metal tooling, and electromechanical structures. The undercuts in this method occur by tilting the part. The most recent advance in the SDM technique is to mount a cladding head to FADAL VMC 6030 machine and to add Tsudacoma TTNC-301 to allow the table to have five-axes motion (Merz, 1997).

1.2.2.2 Wire-Feed Laser Deposition

This technology was introduced by the University of Manchester. This technique utilizes the use of wire to 3D objects. The process is illustrated in Figure 1.7. The advantage of this method over powder-based methods is that it offers stronger parts. It is not an ideal method to fabricate large parts (e.g., engine or other large mechanical systems) because of the slow speed of the manufacturing. This method is more friendly compared to other powder-based methods because there is no waste of material and is also more environmentally friendly (Ding et al., 2015).

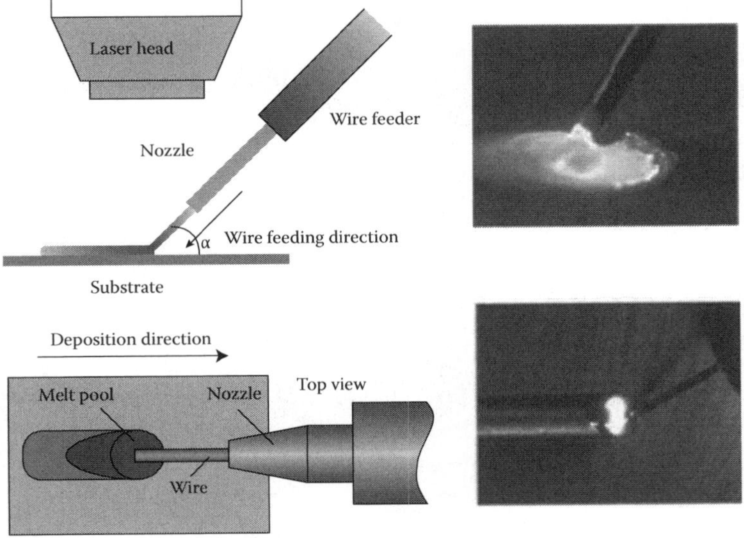

Figure 1.7 The illustration of wire-feed laser deposition technique. (From Ding, D. et al., *Robotics Computer-Integrated Manuf.* 31, 101–110, 2015.)

1.2.2.3 Gas through Wire

A new technique was recently patented by the National Aeronautics and Space Administration (NASA). The advantage of this method over other AM techniques is that it is capable of fabricating fully dense parts with different chemical compositions throughout the structure. In other words, the chemistry and alloying concentration can be locally changed by introducing a particular gas and by turning on and off rapidly. However, the other AM methods produce a uniform part, and the optimization can be done by changing the geometry. The feedstock includes a hollow wire that lets the reactive gas to be flowed through the wire as the wire is melting into the forming pool.

1.2.3 Sheet Lamination

1.2.3.1 Laminated Object Manufacturing (LOM)

This technique was first developed in 1991 by Helisys Company (Wohlers et al., 2013). LOM is a rapid prototyping technique that combines additive and subtractive techniques to build a part layer by layer subsequently from a rolling sheet (Wang et al., 1999).

As shown in Figure 1.8, a sheet of paper is deposited with a small roller. A CO_2 laser cuts the shape of each layer based on the 3D model from the CAD and STereoLithography (STL) file. The layers are bonded together by applying pressure and heat and using a thermal adhesive coating (Park et al., 2000).

There is some advantage associated with this technique. Using this technique, a full-scale, complex prototype can be created directly from a CAD image in less than a day (Klosterman, 1996; Wong and Hernandez, 2012). The large parts can be produced without distortion, shrinking, and deformation. The parts have high durability, low brittleness, and fragility. It is easy to handle and dispose of the nonreactive and nontoxic materials, which results in lower cost (Mueller and Kochan, 1999).

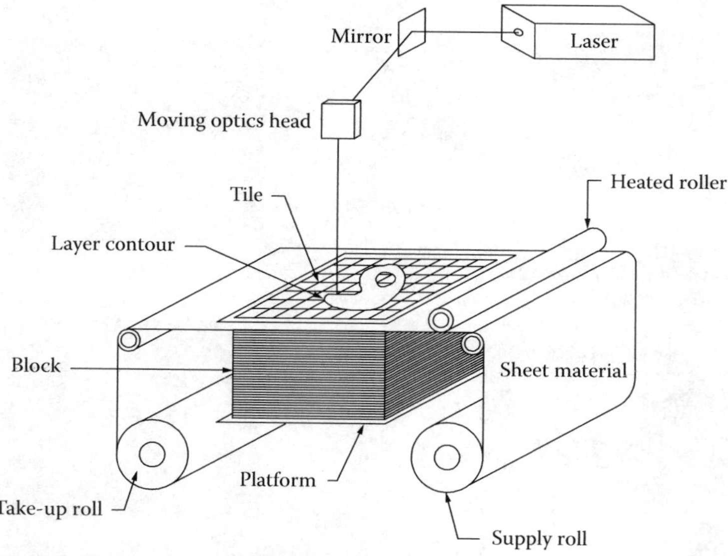

Figure 1.8 The illustration of laminated object manufacturing process. (From Park, J. et al., *Rapid Prototyp. J.* 6(1), 36–50, 2000.)

One of the major disadvantages of this technique is low surface quality, due to the stacking layers, which cause parts to be insufficient to meet general industrial purposes (Ippolito et al., 1995; Reeves and Cobb, 1997; Mahesh, 2004; Ahn et al., 2009). However, mathematical (Reeves and Cobb, 1997) and theoretical models of surface roughness have been suggested to fix the surface problem (Perez et al., 2001). Another disadvantage is the high waste of material (Murr, 2014).

1.3 Support Technologies Such as Software Systems

Learning Objectives

- Learn how computers are involved in the AM process
- Learn how software is involved in the AM process

1.3.1 Computers

AM technology is a multidisciplinary task that takes advantage of a variety of technology sectors, such as computer technology. Recent advances in computer technology (low cost of mass storage and processors, high computing power) have resulted in a faster and more reliable way for processing large amounts of data (Gibson et al., 2010). Nowadays, computers are being extensively used to assist in the development of generative process planning systems. One such area is the development of CAD in AM (Joshi, 1987).

CAD is an assistive tool in the creation, modification, analysis, and optimization of a design (Sarcar et al., 2008). CAD owes its success to Computer-Aided Manufacture (CAM), which is responsible for producing codes for numerically controlled (NC) machinery, essentially combining commands with coordinate data to select and actuate the cutting tools (Kerbrat et al., 2011).

In addition to advanced computer technology, the CAD systems have also been improved significantly. These days, most CAD systems utilize Non-Uniform Rational Basis-Splines (NURBS) (Piegl and Tiller, 2012). NURBS can define the curves and surfaces that correspond to the outer shell of a CAD model, and they can represent complex shapes without making the files too large.

1.3.2 Software

Most of the CAD/CAM software(s) are designed to cover manufacturing needs for highly integrated design environments that allow designers to work as teams and to share designs across different platforms.

An AM software should be able to cover the gap between designers viewpoint and salespeople and marketers demand. In addition, it should be able to bring all the branches of an organization together in order to produce a product out of the design. These organizations can be even in different regions of the world; therefore, high-speed Internet with protection of intellectual property and accurate transmission must be integrated (Wong and Hernandez, 2012).

In AM techniques, most CAD software slice the CAD model directly (i.e., direct slicing) and transfer the resultant contours to the Rapid Prototyping (RP) process (Jamieson and Hacker, 1995). Later, a generic format called STL was offered by 3D systems, USA, after their STL technology (Dolenc, 1994). The public domain feature of the STL file format made it applicable for all CAD vendors; and therefore, they integrated this format into their systems. STL is now a standard output for nearly all solid modeling CAD systems and has also been adopted by AM system vendors (Roscoe, 1988).

In the near future, the CAD technology is expected to move toward more functionality and capability with respect to advances in AM. To date, the focus has been on the external geometry. In future, we need to be aware of the rules associated with the optimization of the output by fully knowing the AM systems function.

1.4 Postmanufacturing Processing

Learning Objectives

- Learn how computers are involved in the AM process
- Learn how software is involved in the AM process

After fabrication by AM techniques, one or more postprocessing tasks are usually required to produce desired functionality and form. In this section, we introduce and categorize postprocessing techniques that are being used to enhance AM fabricated parts or to overcome AM limitations. These include support removal, surface texture improvements, aesthetic improvements, and property enhancements using thermal techniques.

Postprocessing plays an important role in price marketing, and therefore, competitive companies may neglect postprocessing to reduce the costs.

1.4.1 Support Removal

In AM techniques, support removal is a crucial step. In general, there are two kinds of supports for AM parts, namely, natural supports (i.e., materials that surround the part during fabrication process) and synthetic supports (i.e., rigid extra structures that connect the main part to the substrate to support and restrain the main part).

1.4.1.1 Natural Supports Removal

In some AM processes, the fabricated part is completely encapsulated in the built material and must be removed from the surrounding material right after the fabrication. PBF AM techniques require removal of the part from the surrounding powder. Sheet metal lamination processes also require removal of the surrounding sheet material.

In PBF techniques, it is necessary to let the fabricated part to cool down before removing the powders; otherwise, the nonuniform cooling causes distortion within the structure. The time required to cool down depends on the built material and size of the part. After reaching cooling down stage, there are several methods to remove parts from the surrounding powder, including using brushes, compressed air, light bead blasting, and dental cleaning tools. It should be noted that some automated loose powder removal processes have recently been developed. Several metal PBF machine manufacturers have started to include this automated process into their machines (Page et al., 2011).

1.4.1.2 Synthetic Supports Removal

Synthetic supports are not required for all AM processes, but almost all PBF AM methods need these kinds of supports. These supports can be made either out of the built material or from a secondary material. Those fabricated via built material are often processed by weaker processing parameters. However, they are still too strong to be removed by hand, and so the use of milling, band saws, cut-off blades, wire-EDM, and other metal cutting techniques is widely employed. For the supports made out of secondary materials, they are designed to be weaker, soluble in a liquid solution, or meltable at a lower temperature when compared with the built material. These techniques have been applied in some SLS processes.

Building orientation is another important factor that significantly affects support generation and removal. For instance, a horizontally fabricated thin part requires high amount of support material, even more than the amount of built material.

Support removal often follows up with witness marks on the bottom of the part, and therefore, the orientation and location of the supports are other important factors that affect the surface finish of AM part (Gibson et al., 2010; Hadi et al., 2016).

1.4.2 Surface Texture Improvements

AM parts may have undesirable surface texture features that need to be removed. These include stair-steps, powder adhesion, fill patterns from built material, and witness marks from support removal. To flatten stair-step issue, one can consider an extra thin layer thickness on the part to eliminate this error at the expense of building time. Powder adhesion is also another problem associated with the surface. Changing part orientation, powder morphology, and thermal control techniques (such as modifying the scan pattern) can control the amount of powder adhesion.

The type of postprocessing required for surface texture highly depends upon the desired surface finish outcome. For having a matte surface finish, a simple bead blasting of the surface can be useful to remove sharp corners, while for having a smooth or polished finish surface, wet or dry sanding and hand polishing may be applicable. In many cases, it is also good to paint the surface (e.g., with cyanoacrylate or a sealant) prior to sanding or polishing. Painting the surface has the benefits of sealing porosity, smoothing the stair-step effect, and making sanding and polishing easier and more effective.

For surface texture enhancements, several automated techniques have been developed and employed to smooth surface features and increase the accuracy. These techniques include tumbling for external features and abrasive-flow machining for internal features.

1.4.3 Aesthetic Improvements

In the case of using AM parts for aesthetic, artistic, or marketing purposes, the aesthetics of the final part can be considered as an important factor to be fully considered.

In most cases, surface texture improvements are enough to provide aesthetics of the final product. Some cases may require dipping the part into appropriate color containers. Due to the inherent porosity in PBF AM techniques, this technique works because of the associated high absorption. The part may need to be sealed prior to painting. Another technique can be plating the fabricated AM part with other metals such as chrome and nickel, which also strengthens the part and enhances wear resistance (Gibson et al., 2010).

1.4.4 Property Enhancements Using Thermal Techniques and Infilling

Thermal processes enhance the properties of the final AM parts (Saedi et al., 2017; Moghaddam et al., 2016). Traditional heat treatment is a common thermal technique to relieve residual stresses and form the desired microstructure. Recent special heat treatment techniques have also been applied to retain the fine-grained microstructure in the AM part, to relieve stress, and to enhance ductility (Kim, 2015).

Before the advent of AM techniques capable of direct fabrication of metals, many techniques were developed for creating metal green parts using AM. So furnace postprocessing was required to achieve dense, usable metal parts. Control of shrinkage and dimensional accuracy during furnace postprocessing is complicated because of a number of process parameters that must be optimized and the multiple steps involved.

In addition to traditional heat treatments, a number of other procedures have been developed over the years to combine AM with furnace processing to produce metal parts. One of these procedures is called SLS/HIP. As mentioned earlier, in this approach, the laser scans only the outside contours of a part during SLS process; therefore, a metal "can" filled with loose powder is made in the end. These parts are then postprocessed to full density using HIP. The SLS/HIP approach was successfully used to produce complex 3D parts in Inconel 625 and Ti–6Al–4V for aerospace applications (Das, 1999).

References

Abe, F., K. Osakada, M. Shiomi, K. Uematsu and M. Matsumoto (2001). The manufacturing of hard tools from metallic powders by selective laser melting. *Journal of Materials Processing Technology* 111(1): 210–213.

Agarwala, M., D. Bourell, J. Beaman, H. Marcus and J. Barlow (1995). Direct selective laser sintering of metals. *Rapid Prototyping Journal* 1(1): 26–36.

Ahn, B. Y., E. B. Duoss, M. J. Motala, X. Guo, S. I. Park, Y. Xiong, J. Yoon, R. G. Nuzzo, J. A. Rogers and J. A. Lewis (2009). Omnidirectional printing of flexible, stretchable, and spanning silver microelectrodes. *Science* 323(5921): 1590–1593.

Aliakbari, E. and H. Baseri (2012). Optimization of machining parameters in rotary EDM process by using the Taguchi method. *The International Journal of Advanced Manufacturing Technology* 62(9): 1041–1053.

Amerinatanzi, A., N. S. Moghaddam, A. Jahadakbar, D. Dean and M. Elahinia (2016). On the effect of screw preload on the stress distribution of mandibles during segmental defect treatment using an additively manufactured hardware. In ASME 2016 11th international manufacturing science and engineering conference, American Society of Mechanical Engineers, V002T03A015-V002T03A015.

Andani, M. T., N. S. Moghaddam, C. Haberland, D. Dean, M. J. Miller and M. Elahinia (2014). Metals for bone implants. Part 1. Powder metallurgy and implant rendering. *Acta biomaterialia* 10(10): 4058–4070.

Andersen, P. J. and S. K. Hodson (1998). Articles of manufacture and methods for manufacturing laminate structures including inorganically filled sheets, US Patent 5830548.

Arcella, F., D. Abbott, M. House and E. Prairie (2000). Titanium alloy structures for airframe application by the laser forming process. *41st Structures, Structural Dynamics, and Materials Conference and Exhibit*, Atlanta, GA.

Atwood, C., M. Griffith, M. Schlienger, L. Harwell, M. Ensz, D. Keicher, M. Schlienger, J. Romero and J. Smugeresky (1998). Laser engineered net shaping (LENS): A tool for direct fabrication of metal parts. *Proceedings of ICALEO*, Orlando, FL.

Bandyopadhyay, A. and S. Bose (eds) (2015). Additive manufacturing. CRC Press, Boca Raton, FL.

Baufeld, B., E. Brandl and O. Van der Biest (2011). Wire based additive layer manufacturing: Comparison of microstructure and mechanical properties of Ti–6Al–4V components fabricated by laser-beam deposition and shaped metal deposition. *Journal of Materials Processing Technology*, 211(6): 1146–1158.

Bennett, R.C. and C. Sutcliffe (2004). Selective laser melting—applications and developments using MCP RealizerSLM. *Proceedings of the Fourth Laser Assisted Net Shape Engineering* 1: 545.

Bidabadi, M., A. H. A. Natanzi and S. A. Mostafavi (2012). Thermophoresis effect on volatile particle concentration in micro-organic dust flame. *Powder Technology* 217: 69–76.

Bohrer, M., H. Basalka, W. Birner, K. Emiljanow, M. Goede and S. Czerner (2002). Turbine blade repair with laser powder fusion welding and shape recognition. *International Conference of Metal Powder Deposition for Rapid Manufacturing*, San Antonio, TX.

Campbell, G. S. and J. M. Norman (2012). An introduction to environmental biophysics. Springer, New York.

Childs, T., C. Hauser and M. Badrossamay (2005). Selective laser sintering (melting) of stainless and tool steel powders: Experiments and modelling. *Proceedings of the Institution of Mechanical Engineers, Part B: Journal of Engineering Manufacture* 219(4): 339–357.

Choi, J. and Y. Chang (2005). Characteristics of laser aided direct metal/material deposition process for tool steel. *International Journal of Machine Tools and Manufacture* 45(4): 597–607.

Danforth, S. C., M. Agarwala, A. Bandyopadghyay, N. Langrana, V. R. Jamalabad, A. Safari and R. Van Weeren, Rutgers and The State University (1998). Solid freeform fabrication methods. U.S. Patent 5,738,817.

Das, S., M. Wohlert, J. Beaman and D. Bourell (1997). Direct selective laser sintering and containerless hot isostatic pressing for high performance metal components. *Proceedings to the Solid Freeform Fabrication Symposium*, University of Texas at Austin, Austin, TX.

Das, S., M. Wohlert, J. J. Beaman and D. L. Bourell (1999). Processing of titanium net shapes by SLS/HIP. *Materials and Design* 20(2): 115-121.

Delgado, J., J. Ciurana and C. A. Rodríguez (2012). Influence of process parameters on part quality and mechanical properties for DMLS and SLM with iron-based materials. *The International Journal of Advanced Manufacturing Technology* 60(5–8): 601–610.

Dinda, G. P., A. K. Dasgupta and J. Mazumder (2009). Laser aided direct metal deposition of Inconel 625 superalloy: Microstructural evolution and thermal stability. *Materials Science and Engineering: A* 509(1): 98–104.

Ding, D., Z. Pan, D. Cuiuri and H. Li (2015). A multi-bead overlapping model for robotic wire and arc additive manufacturing (WAAM). *Robotics and Computer-Integrated Manufacturing* 31: 101–110.

Dolenc, A. and I. Mäkelä (1994). Slicing procedures for layered manufacturing techniques. *Computer-Aided Design* 26(2): 119–126.

Elahinia, M. (2015). Shape memory alloy actuators: Design, fabrication and experimental evaluation. John Wiley & Sons, New Jersey.

Elahinia, M., N. S. Moghaddam, M. T. Andani, A. Amerinatanzi, B. A. Bimber and R. F. Hamilton (2016). Fabrication of NiTi through additive manufacturing: A review. *Progress in Materials Science* 83: 630–663.

Fessler, J., R. Merz, A. Nickel, F. Prinz and L. Weiss (1996). Laser deposition of metals for shape deposition manufacturing. *Proceedings of the Solid Freeform Fabrication Symposium,* University of Texas at Austin, Austin, TX.

Fischer, P., V. Romano, A. Blatter and H. Weber (2005). Highly precise pulsed selective laser sintering of metallic powders. *Laser Physics Letters* 2(1): 48–55.

Gao, W., Y. Zhang, D. Ramanujan, K. Ramani, Y. Chen, C. B. Williams, C. C. Wang, Y. C. Shin, S. Zhang and P. D. Zavattieri (2015). The status, challenges, and future of additive manufacturing in engineering. *Computer-Aided Design* 69: 65–89.

Gäumann, M., C. Bezencon, P. Canalis and W. Kurz (2001). Single-crystal laser deposition of superalloys: Processing–microstructure maps. *Acta Materialia* 49(6): 1051–1062.

Gibson, I., D. W. Rosen and B. Stucker (2010). *Additive Manufacturing Technologies*, Springer, New York.

Goban, J., M. Semancík and T. Lazoríková (2013). Manufacturing of forms for injection moulding technology by rapid prototyping method. *Annals of the Faculty of Engineering Hunedoara* 11(1): 195.

Gu, D., Y. C. Hagedorn, W. Meiners, G. Meng, R. J. S. Batista, K. Wissenbach and R. Poprawe (2012). Densification behavior, microstructure evolution, and wear performance of selective laser melting processed commercially pure titanium. *Acta Materialia* 60(9): 3849–3860.

Hadi, A., K. Alipour, S. Kazeminasab, A. Amerinatanzi and M. Elahinia (2016). Design and prototyping of a wearable assistive tool for hand rehabilitation using shape memory alloys. ASME 2016 Conference on Smart Materials, Adaptive Structures and Intelligent Systems. American Society of Mechanical Engineers, V001T04A009–V001T04A009.

Hand, D., M. Fox, F. Haran, C. Peters, S. Morgan, M. McLean, W. Steen and J. Jones (2000). Optical focus control system for laser welding and direct casting. *Optics and Lasers in Engineering* 34(4): 415–427.

Hänninen, J. (2001). DMLS moves from rapid tooling to rapid manufacturing. *Metal Powder Report* 56(9): 24–29.

Hofmeister, W., M. Wert, J. Smugeresky, J. A. Philliber, M. Griffith and M. Ensz (1999). Investigating solidification with the laser-engineered net shaping (LENSTM) process. *Journal of the Minerals, Metals and Materials Society* 51(7): 1–6.

Hu, J., Z. Marciniak and J. Duncan (eds) (2002). Mechanics of sheet metal forming. Butterworth-Heinemann, Oxford, UK.

Ippolito, R., L. Iuliano and A. Gatto (1995). Benchmarking of rapid prototyping techniques in terms of dimensional accuracy and surface finish. *CIRP Annals-Manufacturing Technology* 44(1): 157–160.

Jahadakbar, A. (2016). The additively manufactured porous NiTi and Ti-6Al-4V in mandibular reconstruction: Introducing the stiffness-matched and the variable stiffness options for the reconstruction plates (Doctoral dissertation), University of Toledo, Toledo, OH.

Jahadakbar, A., N. Shayesteh Moghaddam, A. Amerinatanzi, D. Dean, H. E. Karaca and M. Elahinia (2016). Finite Element simulation and additive manufacturing of stiffness-matched NiTi fixation hardware for mandibular reconstruction surgery. *Bioengineering* 3(4): 36.

Jamieson, R. and H. Hacker (1995). Direct slicing of CAD models for rapid prototyping. *Rapid Prototyping Journal* 1(2): 4–12.

Jones, D. R., M. Schonlau and W. J. Welch (1998). Efficient global optimization of expensive black-box functions. *Journal of Global optimization* 13(4): 455–492.

Joshi, S. B. (1987). CAD interface for automated process planning. Purdue Univ., Lafayette, IN.

Kaiser, T. and G. J. Albrecht (2007). Industrial disk lasers for micro material processing–compact reliable systems conquer the market. *Laser Technik Journal* 4(3): 54–57.

Keicher, D. M. and W. D. Miller (1998). LENSTM moves beyond RP to direct fabrication. *Metal Powder Report* 12(53): 26–28.

Kerbrat-Orecchioni, C. (2011). L'impolitesse en interaction: Aperçus théoriques et étude de cas. Studii şi cercetari filologice. Seria limbi romanice (09): 142–178.

Khaing, M.W., J. Y. H. Fuh and L. Lu (2001). Direct metal laser sintering for rapid tooling: Processing and characterisation of EOS parts. *Journal of Materials Processing Technology* 113(1): 269–272.

Kim, D. B., P. Witherell, R. Lipman and S. C. Feng (2015). Streamlining the additive manufacturing digital spectrum: A systems approach. *Additive Manufacturing* 5: 20–30.

Klosterman, D. (1996). Affordable, rapid composite tooling via laminated object manufacturing. *Materials and Process Challenges: Aging Systems, Affordability, Alternative Applications* 41: 220–229.

Knight, R., J. Wright, J. Beaman and D. Freitag (1996). Metal processing using selective laser sintering and hot isostatic pressing (SLS/HIP). *Proceedings of the Solid Freeform Fabrication Symposium,* University of Texas at Austin, Austin, TX.

Kolossov, S., E. Boillat, R. Glardon, P. Fischer and M. Locher (2004). 3D FE simulation for temperature evolution in the selective laser sintering process. *International Journal of Machine Tools and Manufacture* 44(2): 117–123.

Kruth, J. P., X. Wang, T. Laoui and L. Froyen (2003). Lasers and materials in selective laser sintering. *Assembly Automation* 23(4): 357–371.

Kruth, J.-P., L. Froyen, J. Van Vaerenbergh, P. Mercelis, M. Rombouts and B. Lauwers (2004). Selective laser melting of iron-based powder. *Journal of Materials Processing Technology* 149(1): 616–622.

Kumar, S. (2003). Selective laser sintering: A qualitative and objective approach. *Journal of the Minerals, Metals and Materials Society* 55(10): 43–47.

Kumar, S. and S. Pityana (2011). Laser-based additive manufacturing of metals. In Advanced Materials Research (Vol. 227, pp. 92–95). Trans Tech Publications, Zurich, Switzerland.

Kwiatkowska, M., D. Zasada, J. Bystrzycki and M. Polański (2015). Synthesis of Fe-Al-Ti based intermetallics with the use of laser engineered net shaping (LENS). *Materials* 8(5): 2311–2331.

Laeng, J., J. Stewart and F. W. Liou (2000). Laser metal forming processes for rapid prototyping: A review. *International Journal of Production Research* 38(16): 3973–3996.

Lawrence, J. R. (2010). *Advances in Laser Materials Processing: Technology, Research and Application*, Woodhead Publishing, Cambridge, UK.

Lewis, G., J. Milewski, D. Thoma and R. Nemec (1997). Properties of near-net shape metallic components made by the directed light fabrication process. *Proceedings to the Solid Freeform Fabrication Symposium*, Univesity of Texas at Austin, Citeseer, Austin, TX.

Lewis, G. K., R. Nemec, J. Milewski, D. J. Thoma, D. Cremers and M. Barbe (1994). *Directed Light Fabrication*, Los Alamos National Laboratory, Los Alamos, NM.

Liu, Z. H., D. Q. Zhang, S. L. Sing, C. K. Chua and L. E. Loh (2014). Interfacial characterization of SLM parts in multi-material processing: Metallurgical diffusion between 316L stainless steel and C18400 copper alloy. *Materials Characterization* 94: 116–125.

Mahesh, K., D. Constantinescu and P. Moin (2004). A numerical method for large-eddy simulation in complex geometries. *Journal of Computational Physics* 197(1): 215–240.

Mazumder, J., J. Choi, K. Nagarathnam, J. Koch and D. Hetzner (1997). The direct metal deposition of H13 tool steel for 3-D components. *JOM* 49(5): 55–60.

McLean, M.A., G. J. Shannon and W. M. Steen (1997). Laser generating metallic components. *XI International Symposium on Gas Flow and Chemical Lasers and High Power Laser Conference.* International Society for Optics and Photonics, Edinburgh, UK, 753–756.

Meiners, W., C. Over, K. Wissenbach and R. Poprawe (1999). Direct generation of metal parts and tools by selective laser powder remelting (SLPR). *Proceedings of SFF*, Austin, TX.

Mercelis, P. and J. P. Kruth (2006). Residual stresses in selective laser sintering and selective laser melting. *Rapid Prototyping Journal* 12(5): 254–265.

Merz, R. (1997). Shape deposition manufacturing of heterogeneous structures. *Journal of Manufacturing Systems* 16(4): 239.

Mici, J., B. Rothenberg, E. Brisson, S. Wicks and D. M. Stubbs (2015). Optomechanical performance of 3D-printed mirrors with embedded cooling channels and substructures. SPIE Optical Engineering+ Applications, International Society for Optics and Photonics, 957306–957306.

Milewski, J., G. Lewis, D. Thoma, G. Keel, R. Nemec and R. Reinert (1998). Directed light fabrication of a solid metal hemisphere using 5-axis powder deposition. *Journal of Materials Processing Technology* 75(1): 165–172.

Moghaddam, N. S., M. T. Andani, A. Amerinatanzi, C. Haberland, S. Huff, M. Miller, M. Elahinia and D. Dean (2016a). Metals for bone implants: Safety, design, and efficacy. *Biomanufacturing Reviews*, 1(1): 1.

Moghaddam, N.S., A. Jahadakbar, A. Amerinatanzi, M. Elahinia, M. Miller and D. Dean (2016b). Metallic fixation of mandibular segmental defects: Graft immobilization and orofacial functional maintenance. *Plastic and Reconstructive Surgery Global Open*, 4(9): e858.

Moghaddam, N. S., M. T. Ahmadi, M. Rahmani, N. A. Amin, H. S. Moghaddam and R. Ismail (2011). Monolayer graphene nanoribbon pn junction. Micro and Nanoelectronics (RSM), 2011 IEEE Regional Symposium on, IEEE, 253–255.

Moghaddam, N. S., M. T. Ahmadi, J. F. Webb, M. Rahmani, H. Sadegi, M. Musavi and R. Ismail (2012). Modeling of graphene nano-ribbon Schottky diodes in the parabolic band structure limit. In AIP Conference Proceedings, ed. N. Barsoum, D. Faiman and P. Vasant, AIP, Vol. 1499, No. 1, 268–271.

Mudge, R. P. and N. R. Wald (2007). Laser engineered net shaping advances additive manufacturing and repair. Welding Journal-New York-, 86(1): 44.

Mueller, B. and D. Kochan (1999). Laminated object manufacturing for rapid tooling and patternmaking in foundry industry. *Computers in Industry* 39(1): 47–53.

Mumtaz, K.A., P. Erasenthiran and N. Hopkinson (2008). High density selective laser melting of Waspaloy®. *Journal of Materials Processing Technology*, 195(1): 77–87.

Munjuluri, B. N. R., S. Agarwal and F. Liou (2001). *Process Modeling, Monitoring and Control of Laser Metal Forming*, University of Missouri, Rolla, MO.

Murr, L. E. (2014). Additive manufacturing: Changing the rules of manufacturing.

Page, N. W. and D. Raybould (1989). Dynamic powder compaction of some rapidly solidified crystalline and amorphous powders: Compaction characteristics. *Materials Science and Engineering: A* 118: 179–195.

Parimi, L.L., G. A. Ravi, D. Clark and M. M. Attallah (2014). Microstructural and texture development in direct laser fabricated IN718. *Materials Characterization* 89: 102–111.

Park, J., M. J. Tari and H. T. Hahn (2000). Characterization of the laminated object manufacturing (LOM) process. *Rapid Prototyping Journal* 6(1): 36–50.

Peng, L., Y. Taiping, L. Sheng, L. Dongsheng, H. Qianwu, X. Weihao and Z. Xiaoyan (2005). Direct laser fabrication of nickel alloy samples. *International Journal of Machine Tools and Manufacture* 45(11): 1288–1294.

Petrovic, V., J. Vicente Haro Gonzalez, O. Jorda Ferrando, J. Delgado Gordillo, J. Ramon Blasco Puchades and L. Portoles Grinan (2011). Additive layered manufacturing: Sectors of industrial application shown through case studies. International Journal of Production Research 49(4): 1061–1079.

Piegl, L. and W. Tiller (2012). The NURBS book. Springer, Berlin, Germany.

Rahmani, M., M. T. Ahmadi, J. F. Webb, N. Shayesteh, S. M. Mousavi, H. Sadeghi and R. Ismail (2012). Trilayer graphene nanoribbon carrier statistics in degenerate and non degenerate limits. AIP Conference Proceedings, ed. N. Barsoum, D. Faiman and P. Vasant AIP, Vol. 1499, No. 1, 272–275.

Reeves, P. E. and R. C. Cobb (1997). Reducing the surface deviation of stereolithography using in-process techniques. *Rapid Prototyping Journal* 3(1): 20–31.

Ramaswami, K., Y. Yamaguchi and F. B. Prinz (1997). Spatial partitioning of solids for solid freeform fabrication. *Proceedings of the Fourth ACM Symposium on Solid Modeling and Applications*, ACM, Atlanta, GA.

Ravari, M. K., M. Kadkhodaei and A. Ghaei (2015). A microplane constitutive model for shape memory alloys considering tension–compression asymmetry. *Smart Materials and Structures* 24(7): 075016.

Regenfuss, P., L. Hartwig, S. Klotzer, R. Ebert and H. Exner (2004). Microparts by a novel modification of selective laser sintering. Technical Papers-Society of Manufacturing Engineers-All Series-.

Regenfuss, P., R. Ebert and H. Exner (2007). Laser micro sintering—A versatile instrument for the generation of microparts. *Laser Technik Journal* 4(1): 26–31.

Roscoe, S. N. (1993). Visual orientation: Facts and hypotheses. *The International Journal of Aviation Psychology* 3(3): 221–229.

Santos, E. C., M. Shiomi, K. Osakada and T. Laoui (2006). Rapid manufacturing of metal components by laser forming. *International Journal of Machine Tools and Manufacture* 46(12): 1459–1468.

Sarcar, M. M. M., K. M. Rao and K. L. Narayan (2008). Computer aided design and manufacturing. PHI Learning Pvt. Ltd., New Delhi.

Schleifenbaum, H., W. Meiners, K. Wissenbach and C. Hinke (2010). Individualized production by means of high power selective laser melting. *CIRP Journal of Manufacturing Science and Technology* 2(3): 161–169.

Shaohua, H. (2010). Effect of powder character of IN718 on laser rapid forming. *Hot Working Technology* 19: 054.

Shayesteh Moghaddam, N., A. Jahadakbar, A. Amerinatanzi, R. Skoracki, M. Miller, D. Dean and M. Elahinia (2017). Fixation release and the bone Bandaid: A new bone fixation device paradigm. *Bioengineering* 4(1): 5.

Shellabear, M. and O. Nyrhilä (2004). DMLS—Development history and state of the art. *Proceedings of the 4th Laser Assisted Netshape Engineering*, Erlangen, Germany, 4: 21–24.

Sutcliffe, P. W. and P. G. Mardon (1975). Titanium powder metallurgy. AGARD Advan. Manuf. Methods and their Econ. Implications 17(SEE N 75-22749): 14–37.

Toyserkani, E. and A. Khajepour (2006). A mechatronics approach to laser powder deposition process. *Mechatronics* 16(10): 631–641.

Toyserkani, E., A. Khajepour and S. F. Corbin (2004). *Laser Cladding*, CRC Press, Boca Raton, FL.

Udroiu, R. (2012). Powder bed additive manufacturing systems and its applications. *Academic Journal of Manufacturing Engineering*, 10(4): 122–129.

Verwimp, J., M. Rombouts, E. Geerinckx and F. Motmans (2011). Applications of laser cladded WC-based wear resistant coatings. *Physics Procedia* 12: 330–337.

Vilar, R. (1999). Laser cladding. *Journal of Laser Applications* 11(2): 64–79.

Wang, G., P. G. Boulton, N. M. Chan, M. M. Palcic and D. E. Taylor (1999). Novel Helicobacter pylori α1, 2-fucosyltransferase, a key enzyme in the synthesis of Lewis antigens. *Microbiology* 145(11): 3245–3253.

Wohlers, T. and T. Caffrey (2013). Additive manufacturing: Going mainstream. *Manufacturing Engineering* 151(6): 67–73.

Wong, K. V. and A. Hernandez (2012). A review of additive manufacturing. ISRN Mechanical Engineering, 2012.

Yellup, J. M. (1995). Laser cladding using the powder blowing technique. *Surface and Coatings Technology* 71(2): 121–128.

Xiao, B. and Y. Zhang (2007). Laser sintering of metal powders on top of sintered layers under multiple-line laser scanning. *Journal of Physics D: Applied Physics* 40(21): 6725.

Xie, S., X. Xu, G. Huang, L. Cheng and Y. He (2010). Optimization of processing parameters for laser powder deposition using finite element method. *Journal of Wuhan University of Technology-Materials Science Edition* 25(5): 832–837.

Xue, L., J. Chen and M. Islam (2000a). Functional properties of laser consolidated wear resistant Stellite 6 alloy. *Powder Metallurgy Alloys and Particulate Materials for Industrial Applications*, ed. David E. Alman and Joseph W. Newkirk, TMS, Nashville, TN, 65–74.

Xue, L., J. Chen, M. Islam, J. Pritchard, D. Manente and S. Rush (2000b). Laser consolidation of IN-738 alloy for repairing cast IN-738 gas turbine blades. *Proceedings of 20th ASM Heat Treating Society Conference*, St. Louis, MO.

Xue, L. and M. Islam (2000). Free-form laser consolidation for producing metallurgically sound and functional components. *Journal of Laser Applications* 12(4): 160–165.

Xue, L., Y. Li and S. Wang (2011). Direct manufacturing of net-shape functional components/test-pieces for aerospace, automotive, and other applications. *Journal of Laser Applications* 23(4): 042004.

Zhu, H., L. Lu and J. Fuh (2003). Development and characterisation of direct laser sintering Cu-based metal powder. *Journal of Materials Processing Technology* 140(1): 314–317.

Chapter 2

Computer-Aided Design of Additive Manufacturing Components

Bashir Khoda
North Dakota State University

Contents

CHAPTER OUTLINE

In this chapter, we

- Define the process plan for additive manufacturing (AM)
- Discuss the effect of design on AM
- Use of computer-aided design (CAD) to improve AM process

2.1 Introduction

Learning Objectives

- Understand the interaction between CAD and AM
- Learn the hierarchical location of CAD in AM process plan
- Understand the opportunities and challenges related to CAD

In AM processes, commonly known as layer manufacturing or three-dimensional (3D) printing, physical models are built layer by layer. The process plan starts with a digitized model generated by either a CAD modeler or reverse engineering (RE) techniques. A traditional digitized model contains only the geometric information with no material or topological details. The initial geometric information is then represented as a mesh model or a stereolithography (STL) model (Kulkarni et al. 2000). The monochrome STL files are generated by tessellating the outside surface of the digital model with triangle approximation. The tessellated model is sliced by a set of parallel intersecting planes perpendicular to a predefined direction vector (Khoda et al. 2011) that is commonly known as the build direction. Monochrome and close contours are generated by connecting the intersecting points between the slicing plane and the model surface, which are called layers. Tool-path is generated for individual layers in order to follow the material deposition and required support generation. Finally, by stacking the individual layers along the build direction, it will create the 3D physical model. Individual steps for AM process plan, i.e., digitizing model, build direction/orientation, slicing, tool-path generation, fabrication, and postprocessing have a hierarchical relationship and can have accumulated influence on the finished product.

The AM processes can be divided into three sequential technological steps: preprocessing (virtual), processing (actual printing), and postprocessing, as shown in Figure 2.1. Considering

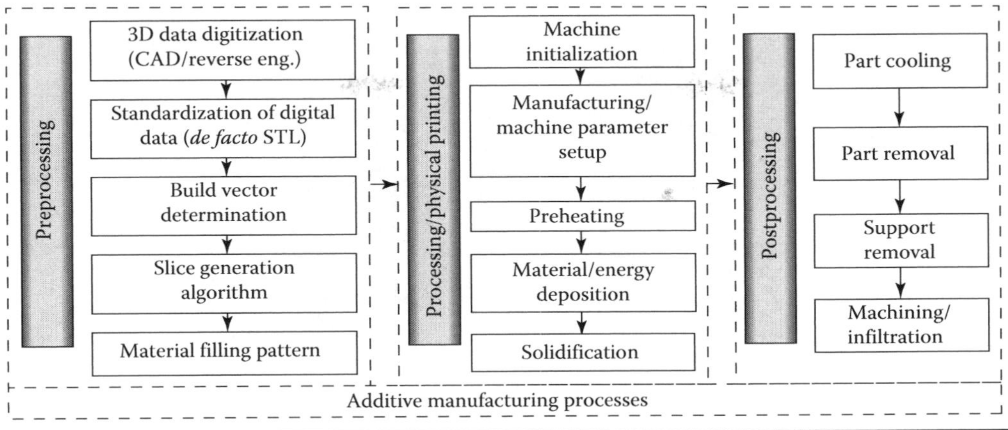

Figure 2.1 Hierarchical AM process plan. (From Ahsan, N., Khoda, B., *Additive Manuf.*, 11, 85–96, 2016.)

the desired attributes in the process and the object geometry, the process planning steps can be dramatically simplified in the layer-based manufacturing approach, which can vary the resource requirements such as build time, cost, machine/component utilization, and tool travel length. The preprocessing stage primarily uses CAD techniques to guide both the machine motion and material deposition systems. The preplanning/preprocessing stage is at the top of the AM hierarchy and can proactively alter the AM strategy into a guided desired outcome (Khoda 2014).

Each of the AM process steps is equally important and can have a significant impact on the attributes of the manufactured part. Due to their hierarchical relationship, the predecessor process steps have more influence on the finished product than their successor. In this chapter, we primarily focus on the preprocessing steps.

2.2 Data Digitization

Learning Objectives

- Understand the CAD system
- Parametric modeling
- Understand the concept of RE

The AM process starts with a valid digital model often known as a CAD model. The object or component can be represented in a digitized environment by employing forward engineering or RE techniques, as shown in Figure 2.2. In general, a 3D solid model of a product is first designed using a CAD platform and digital information is obtained, which is known as forward engineering. However, in RE, geometrical information is obtained from an existing physical object directly, and the information is converted into digital data for the downstream processes.

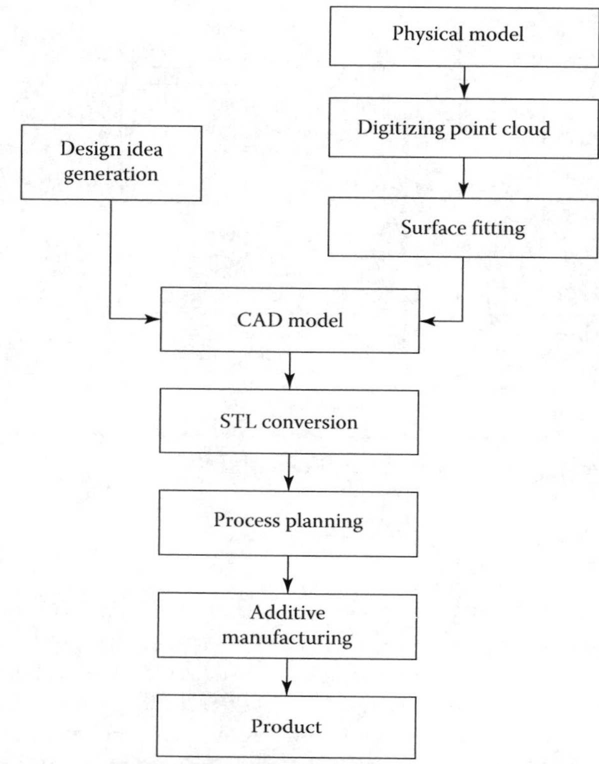

Figure 2.2 Additive manufacturing process using forward and reverse engineering of a CAD model.

2.2.1 Modeling with Primitives

The digital model can be created as a solid model (by using primitive instancing or constructive solid geometry, CSG) or a surface model (by using B-Rep), which are known as traditional design engineering or CAD. Currently, the majority of the 3D modelers are based on CSG or B-Rep (Chang et al. 2005).

Constructive Solid Geometry: A solid object is molded using volume combination occupied by overlapping 3D primitives using set Boolean operations. Primitives are solid objects defined by a very accurate mathematical definition. Primitives are generic in the sense that their shape and size are instantiated by the user-defined parameters or dimensions (Hoffmann 1992). Typical standard primitives are block, sphere, cone, cylinder, torus, wedge, etc. All these primitives are algebraic half-space, i.e., a finite closed volume in an infinite space. Half-space primitives have finite domain and are constructed by taking a sense (positive or negative) of the parameters in the defined function (Davis et al. 2007). Parts are built by combining primitives' volume in an overlapping manner and performing CSG operations, i.e., union, intersection, and difference, as shown in Figure 2.3. Objects designed with this method are represented with a binary tree, as demonstrated in Figure 2.4. Both the surface and the interior of an object are defined implicitly.

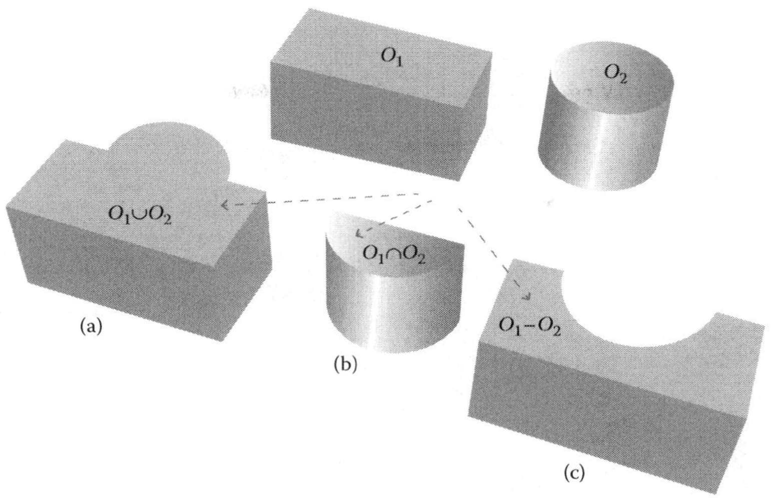

Figure 2.3 CSG operations. (a) Union, (b) intersection, and (c) difference.

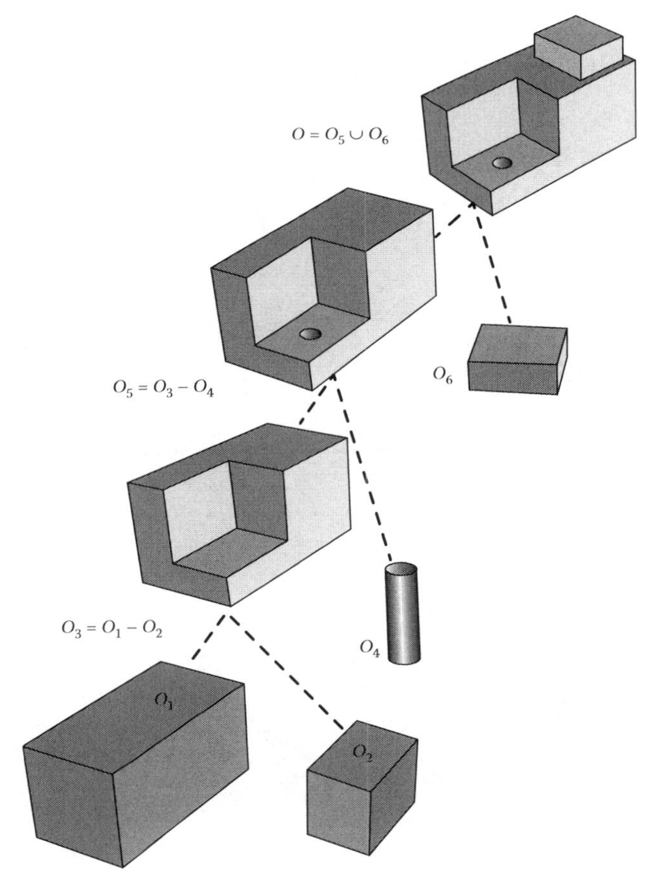

Figure 2.4 CSG tree for an object *O*.

2.2.2 Reverse Engineering (RE)

RE is a well-accepted process for product design and manufacturing of contemporary parts (Sokovic and Kopac 2006). Whereas conventional engineering transforms engineering concepts and models into real parts, RE real parts are transformed into engineering models and concepts. It recreates or clones an existing part by acquiring the surface data of an existing part using a scanner or a measurement device (Yin 2011). The RE concept can be proved useful when a physical part (natural or synthetic) exists without its design documentation. The basic process is shown in Figure 2.5 adopted from the literature (Várady et al. 1997). The phases are often overlapping and may require multiple iterations to achieve the desired digitized model.

Data acquisition is the crucial part of RE process. The shape data from an object can be captured in two ways: (1) contact or tactile method and (2) noncontact or noninvasive method. A mechanical probe is attached with the scanner head to make contact with the surface, while light, sound, or magnetic fields are used as noninvasive modalities, as shown in Figure 2.6. During the scanning process, the scanner head (contact or noncontact) moves back and forth across the unknown surface. The system then records the surface information in the form of numerical data

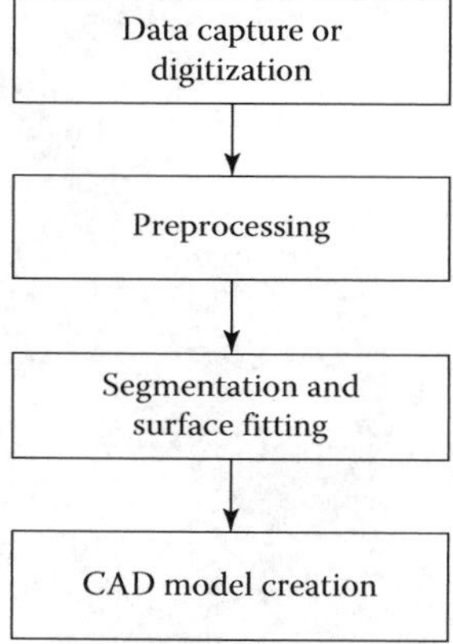

Figure 2.5 Basic phases of reverse engineering (RE). (From Várady, T. et al., *Comput. Aided Des.*, 29(4), 255–268, 1997.)

Figure 2.6 Data acquisition devices. (a) Mechanical probe, (b) noncontact laser, and (c) hybrid (contact and noncontact).

and generates a point cloud matrix (3D coordinates) (Sokovic and Kopac 2006). In the case of under the skin/surface, information need to be reconstructed but are inaccessible (i.e., porous structure) because of for many reasons; therefore, noninvasive modalities such as CT, micro-CT, MRI, and optical microscopy are often the only data acquisition options.

The data acquisition devices usually capture data in a high-resolution mode, which can introduce data redundancy. Besides, the fidelity of the acquired data may often vary based on the technology used; however, the data sets are error prone during various stages of this conversion process. The point cloud needs to be analyzed to identify and eliminate the redundant data (Chen and Ng 1997) as well as erratic data sets to increase the cloud accuracy (Lee and Woo 2000). The high-fidelity data is then segmented, and surface patches are generated using surface fitting algorithms guided by interpolation or approximation.

The surface can be constructed from the data set by using interpolation or approximation techniques. In surface interpolation, every data point is retained and the number of patches in a piecewise representation is approximately the same as the number of points (Park and Kim 1995). The success of interpolation and quality of the resulting surface depends on the configuration of input data, the selected method, parameters of interpolation, grid size, etc. Even though the surface interpolation approach is simple and straightforward, for a large set of data, the methods may soon become ineffective in both storage and computational requirements. A more compact representation of the surface is possible by filtering out small local variations from the data set. This filtration technique can be effective, especially on the bivariate or planner data subset, and piecewise linear surfaces can represent the object with reasonable accuracy, as shown in Figure 2.7. However, objects with curved regions or smooth features may require a large number of linear surface patches that may require additional point insertions using approximation technique, as shown in Figure 2.8a and b. On the other hand, higher degree surfaces can be approximated using the given data grid, which can be expressed by mathematical equations with a smaller number of points or parameters (Figure 2.8c). Continuity among the adjacent surface patches needs to be enforced to make a continuous and smooth surface, as shown in Figure 2.9. However, accurate measurement of the final digital model is a subjective process, since no prior information or design documentation of the physical part exists. The digitized CAD model is then transformed into a tessellated STL model, which is the *de facto* standard for slicing.

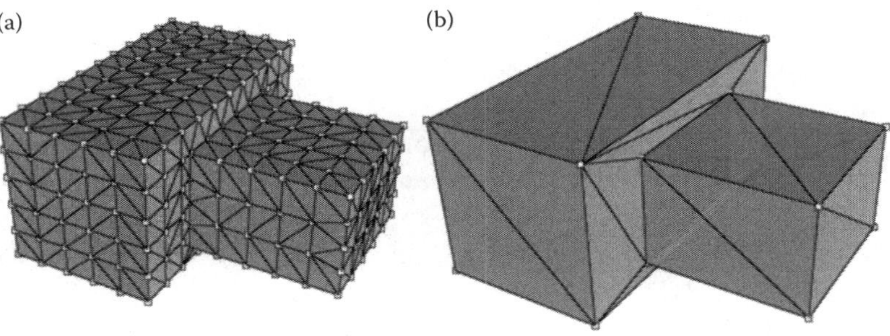

Figure 2.7 (a) Redundant planner point cloud acquired by data acquisition system and (b) reduced point set to reconstruct the surface.

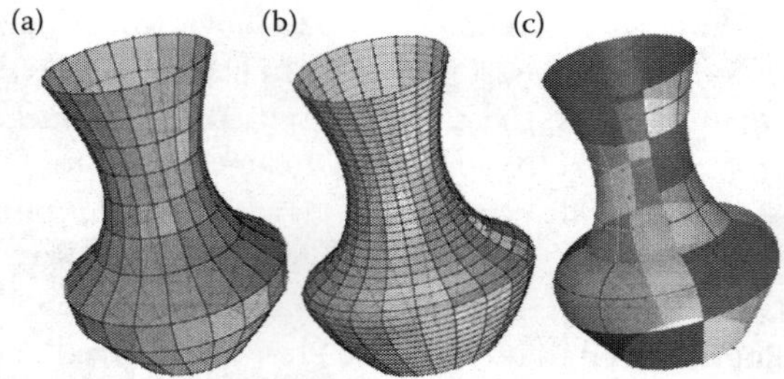

Figure 2.8 (a) Piecewise linear surface with (a) 220 points, (b) 820 points, and (c) 220 points (higher degree).

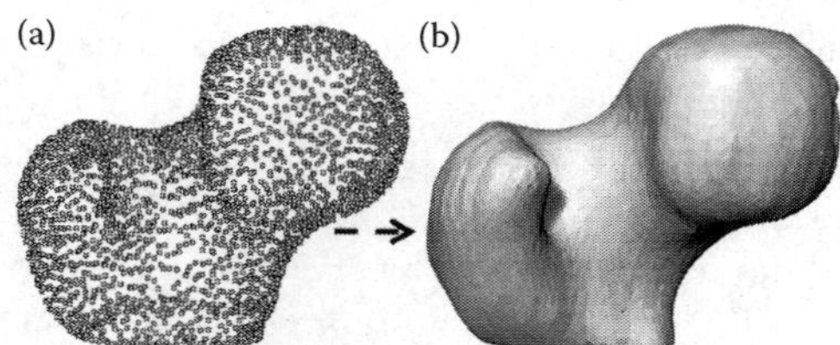

Figure 2.9 Acquired data. (a) Point cloud and (b) digital model.

2.3 From CAD to STL File

Learning Objectives

- Understand the STL transformation
- Identify the characteristics of STL model
- Learn about STL data structure
- Learn about STL model accuracy

2.3.1 Construction of STL Model

Once the desired model is constructed (by CAD or RE), the next step for AM is to convert the model into STL format. The STL format was first used by 3D systems in 1987, and almost all the current AM processes need their input in this format. The STL file is usually generated by the point set extracted from a patch work of facets representing the outside skin of the CAD model (Koc et al. 2000). The process may seem redundant especially if the RE or point clouds are used to construct the digital model. Thus, oftentimes STL models are directly generated from the acquired 3D data sets of RE process. If the RE technique use a spatially uniform data recording system, then those data points are easily stored in a matrix, as shown in Figure 2.10a. Connecting those data points in the following sequence (Equation 2.1) will create the tessellated model, as shown in Figure 2.10b.

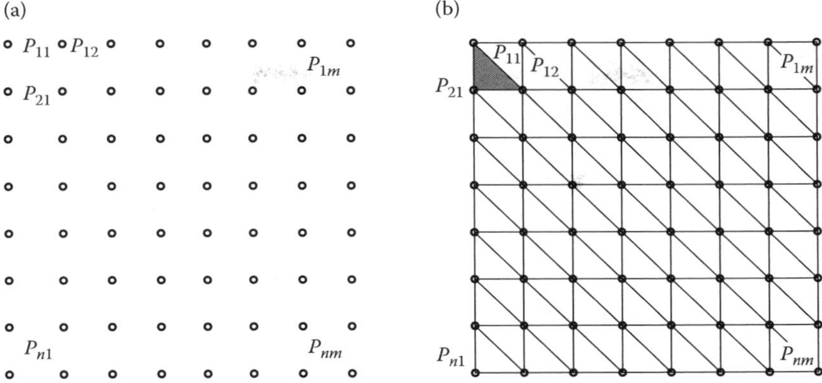

Figure 2.10 STL facet construction.

$$\left.\begin{array}{l}\text{Connect } P_{ij} \text{ and } P_{i(j+1)} \\ \text{Connect } P_{ij} \text{ and } P_{(i+1)j} \\ \text{Connect } P_{ij} \text{ and } P_{(i+1)(j+1)}\end{array}\right\} \quad \forall \ i=1,2,\ldots,n; \quad j=1,2,\ldots,m. \tag{2.1}$$

A set of triangles or facets are constructed using the extracted point sets to tessellate the boundary of the surface. The triangular surface or facets can be considered as the construction unit or patch of the triangular representation of a 3D surface geometry or STL model. Each facet is uniquely identified with a unit normal and three vertices combining 12 numerical values, as shown in Figure 2.11 and in Scheme 2.1.

Each facet is a part of the boundary between the interior and the exterior of the object. The unit normal in each facet usually defines the outward direction of the facet and hence for the object. The counterclockwise sequences of data stored for each facet also implicitly contain the outward normal direction to ensure consistency. Two vectors $\mathbf{V}_{12}$ and $\mathbf{V}_{13}$ can be calculated using the vertices of the facet using Equation 2.2.

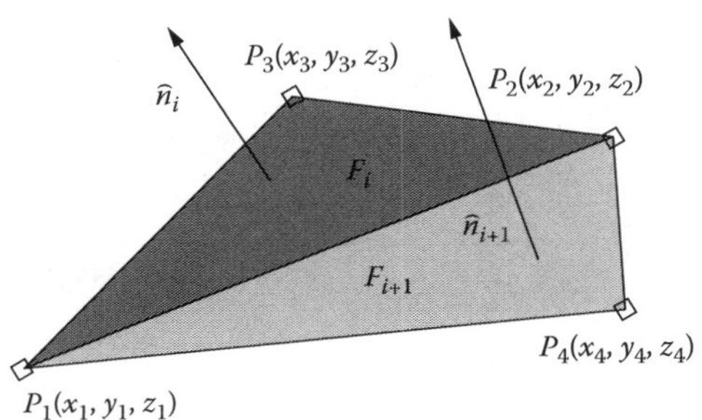

Figure 2.11 Triangular facet (F) with three points and outward normal; F_i and F_{i+1} are oriented along $\overrightarrow{P_1P_2P_3P_1}$ and $\overrightarrow{P_1P_4P_2P_1}$ as counterclockwise direction when looking from outside.

$$\mathbf{V}_{12} = \mathbf{P}_2 - \mathbf{P}_1 = (x_2 - x_1)\mathbf{i} + (y_2 - y_1)\mathbf{j} + (z_2 - z_1)\mathbf{k}$$
$$\mathbf{V}_{13} = \mathbf{P}_3 - \mathbf{P}_1 = (x_3 - x_1)\mathbf{i} + (y_3 - y_1)\mathbf{j} + (z_3 - z_1)\mathbf{k}. \quad (2.2)$$

The outward facet normal can be calculated by using the cross product of $\mathbf{V}_{12}$ and $\mathbf{V}_{13}$, as shown in Equation 2.3. Finally, the outward unit normal ($\hat{n}$) is defined by normalizing **n** using Equation 2.4.

$$\mathbf{n} = \mathbf{V}_{12} \times \mathbf{V}_{13} \quad (2.3)$$

$$\hat{n} = \frac{\mathbf{n}}{|\mathbf{n}|}. \quad (2.4)$$

2.3.2 STL Data Structure

The STL model contains three elements: vertices, edges, and faces, which are represented with geometry or 3D point coordinates. However, relationship information among the point coordinates is required to reconstruct a valid/manifold model. This spatial relationship is also defined as the topology. The amount, quality, and consistency of topological information determines the performance of STL data structure. The face-set data structure is simplest STL data structure, which is shown in Table 2.1. The data of the STL model is written by traversing all triangle or faces in sequence. This data structure contains no connectivity information. As a result, vertices and associated coordinate information are replicated bringing redundancy, which gives it a name "polygon soup."

The most commonly used ASCII STL file uses this face-set data structure. The ASCII data directory also contains an additional facet normal information. The facet normal and the coordinate of the vertices are sequenced in the same order for each consecutive triangle, as shown in Scheme 2.1.

The redundancy in face-set data structure can be improved by adding vertex index to the facet, as shown in Figure 2.13. The data retrieval or query-based performance can be further improved by constructing neighboring vertex information in an adjacency matrix, as shown in Figure 2.12.

One of the popular data structures for storing adjacency information is the edge-based data structure. Edge-based data structures store each edge pointers to both vertices and the neighboring edges table. There are several edge-based variants that differ only in the topological information they store (Bischoff et al. 2002). Among them, half-edge, data structure is most widely used today. Each edge

Table 2.1 Face-Set Data Structure

	Vertices								
Facet	P_1			P_2			P_3		
F_1	x_{11}	y_{11}	z_{11}	x_{12}	y_{12}	z_{12}	x_{13}	y_{13}	z_{13}
F_2	x_{21}	y_{21}	z_{21}	x_{22}	y_{22}	z_{22}	x_{23}	y_{23}	z_{23}
…	…	…	…	…	…	…	…	…	…
F_n	x_{n1}	y_{n1}	z_{n1}	x_{n2}	y_{n2}	z_{n2}	x_{n3}	y_{n3}	z_{n3}

```
solid < name >
    facetnormal F ) < n_x >< n_y >< n_z >
      outerloop F )
        vertex P ) < x >< y >< z >
               P ) < x >< y >< z >
        vertex P ) < x >< y >< z >
      endloop
    endfacet F )
    facetnormal F ) < n_x >< n_y >< n_z >
      outerloop F )


    endfacet F_n)
endsolid
```

Scheme 2.1 ASCII STL file format.

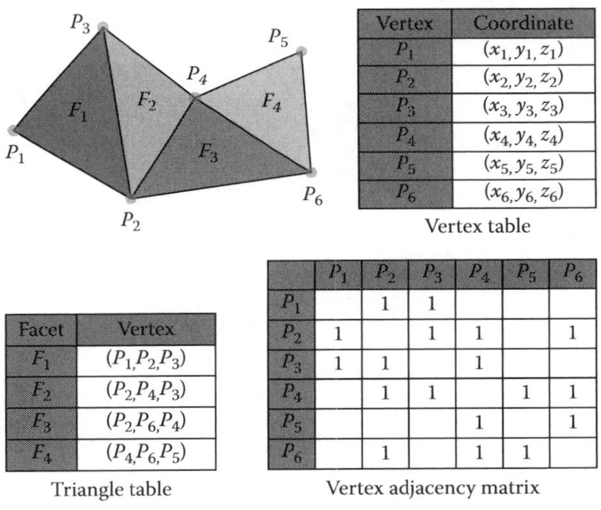

Vertex	Coordinate
P_1	(x_1, y_1, z_1)
P_2	(x_2, y_2, z_2)
P_3	(x_3, y_3, z_3)
P_4	(x_4, y_4, z_4)
P_5	(x_5, y_5, z_5)
P_6	(x_6, y_6, z_6)

Vertex table

Facet	Vertex
F_1	(P_1, P_2, P_3)
F_2	(P_2, P_4, P_3)
F_3	(P_2, P_6, P_4)
F_4	(P_4, P_6, P_5)

Triangle table

	P_1	P_2	P_3	P_4	P_5	P_6
P_1		1	1			
P_2	1		1	1		1
P_3	1	1		1		
P_4		1	1		1	1
P_5				1		1
P_6		1		1	1	

Vertex adjacency matrix

Figure 2.12 Indexed face-set data structure and adjacency matrix.

splits into two, each of which is referred to as a *half-edge,* as shown in Figure 2.13. Each facet is oriented counterclockwise and each half-edge has six pieces of incident information: incident facet, target vertex, source vertex, previous half-edge, next half-edge, and opposite vertex, as shown in Figure 2.13.

2.3.3 Error in STL Model

The digital model is converted into the STL model by transforming the polynomial surface into first-order piecewise linear approximation. This introduces a deviation from the actual surface, which is known as chordal error or chord height deviation, as shown in Figure 2.14. Furthermore, the triangular surfaces in an STL file are not very information-rich compared to a higher level representation such as a nonuniform rational B-spline (NURBS) surface. Therefore, an STL file

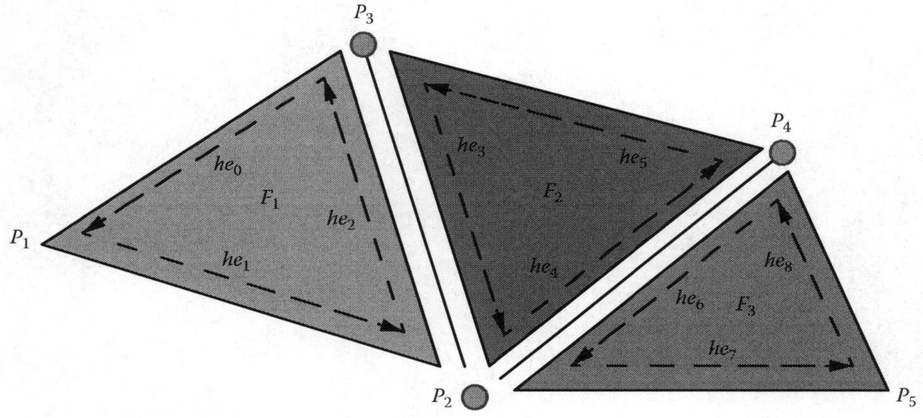

Half Edge	Incident Facet	Start Vertex	End Vertex	Previous HE	Next HE	Opposite Vertex
0	1	3	1	2	1	2
1	1	1	2	0	2	3
2	1	2	3	1	0	1
3	2	3	2	5	4	4
4	2	2	4	3	5	3
5	2	4	3	4	3	2
6	3	4	2	8	7	5
7	3	2	5	6	8	4
8	3	5	4	7	6	2

Figure 2.13 Half-edge data structure.

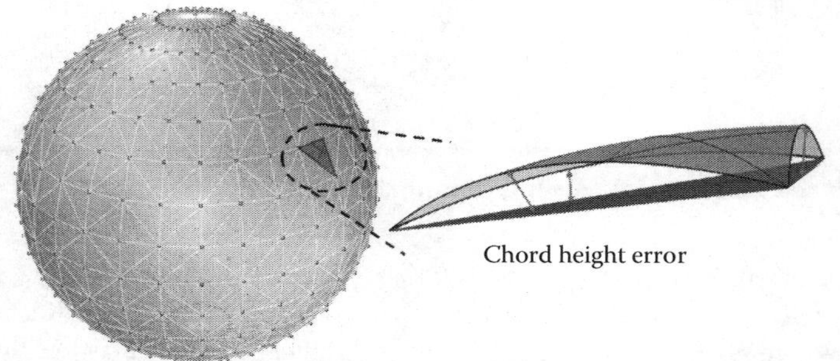

Figure 2.14 Chord height error in STL model.

requires thousands of triangles to represent a model as a "close" approximation compared to a few NURBS surfaces that represent the same model in a mathematically precise format (Koc et al. 2000). As a result, STL representation can be cumbersome in both space and computational requirements. Both space and computational issues can be addressed by reducing points from the point cloud, which will increase the chord height deviation and subsequently affecting the model accuracy, as shown in Figure 2.14. Two common criteria for point reduction are: (1) percentage data reduction and (2) the maximum bounded error.

Traditional CAD models contain both geometric information and topological information that ensure a two-step model validity. In contrast, less topological information is transferred into the simplified STL model, reducing its fidelity. This may result in truncation errors, dangling faces, or puncture gaps in the model. Besides, inconsistent or incomplete facet normal is common during the automated STL construction, which may introduce an erratic nature on the surface and cause inconsistencies in the subsequent processes, e.g., slicing.

To ensure the fidelity of closed STL model, most algorithm use manifold criterion. The face incident to a vertex forms a closed "fan." In other words, the union of all faces adjacent to a vertex can be continuously deformed to a disk, as shown in Figure 2.15. In addition, each triangular facet must share two vertices with each of its adjacent triangles, which is known as the vertex-to-vertex rule (Chen and Ng 1997). The problem vertex is shown in Figure 2.16a. The violation can be corrected by either adding triangulation (add face) or eliminating the problem vertex (edge collapse), as shown in Figure 2.16b, c. By sharing vertices P_1 and P_2 by the adjacent facets F_i+F_{i+1} ensures connectivity in the generated surface. Adjacent patches must also be explicitly stitched together using geometric continuity conditions to generate a valid surface model.

The simple but effective construction techniques make the STL model popular, and it does not require sophisticated CAD systems for visualization. However, defective or nonmanifold STL model is very common especially when the freeform-shaped STL model is constructed with spatially unstructured point sets. Euler characteristic is most commonly used to identify leak in STL files.

$$F - E + V = 2(1 - G). \tag{2.5}$$

Here, F, E, and V represent the total number of face, edge, and vertices, respectively. G is genus that corresponds to through holes. For example, a cube has no genus and a torus has one.

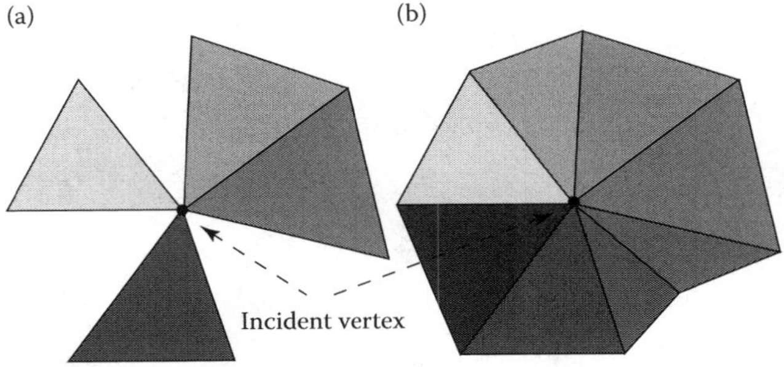

Figure 2.15 (a) Nonmanifold and (b) manifold segment.

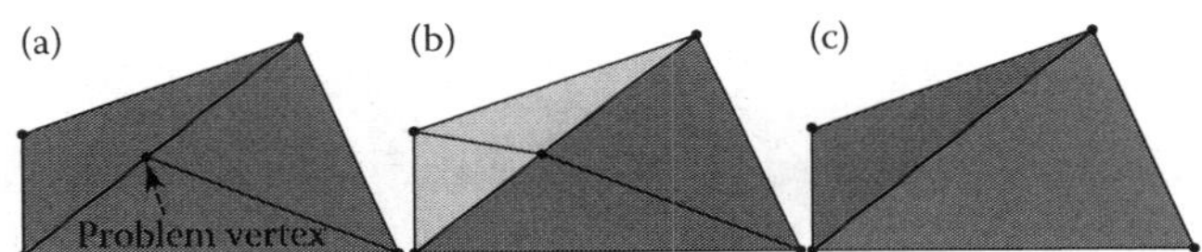

Figure 2.16 (a) Vertex-to-vertex rule violation and (b, c) corrected triangulation.

2.4 Additive Manufacturing File (AMF) Format

Learning Objectives

- Understand the AMF format
- Identify the characteristics of AMF model
- Learn about AMF model accuracy

The 3D CAD models are constructed with geometry and topological data and often possess additional information such as material, color, texture, and features. However, during CAD to STL format conversion, only the basic geometric information is stored neglecting other relevant information such as color, texture, and material. This neglected information could be used in downstream processing (selecting build direction, analyze the model for support creation, etc.) or to verify the solidity of the converted model. On the other hand, multimaterial AM strategies are being investigated for improved functionality and performance (Gibson et al. 2010). The monochrome STL file format cannot support these evolving potentials and therefore needs to be modified or replaced. There is a growing need for a standard interchange file format that can support multimaterial geometries in full color with functionally graded materials and microstructures. AMF is introduced to fill the shortcomings of STL and accommodate the multimaterial or heterogeneous part system.

2.4.1 What Is AMF Model

In 2009, the American Society for Testing and Materials tasked ASTM Committee F42 on Additive Manufacturing Technologies to address the growing gap between what 3D printers can fabricate and what STL files can describe. The goal was to create a common open standard (nonproprietary) file format platform with the ability to specify geometry with high fidelity and small file sizes, multiple materials, color, and microstructures. The committee has led to an effort by international standardization organizations to create a new AM language, leading to the ISO/ASTM 52915 AMF specification (Hod 2014). This standard describes a framework for an interchange format to address the current and future needs of AM technology. In order to be successful across the field of AM, AMF file format was designed to address the following concerns (ASTM 2013).

2.4.1.1 Technology Independence

The file format shall describe an object in a general way such that any machine can build it to the best of its ability. It is resolution and layer thickness independent and does not contain information specific to any one manufacturing process or technique. This does not negate the inclusion of properties that only certain advanced machines support (e.g. color, multiple materials, and so forth), but these are defined in such a way as to avoid exclusivity.

2.4.1.2 Simplicity

The AMF file format is easy to implement and understand. The format can be read and debugged in a simple ASCII text viewer to encourage understanding and adoption. No identical information is stored in multiple places.

2.4.1.3 Scalability

The file format scales well with increase in part complexity and size and with the improving resolution and accuracy of manufacturing equipment. This includes being able to handle large arrays of identical objects, complex repeated internal features (e.g., meshes), smooth curved surfaces with fine printing resolution, and multiple components arranged in an optimal packing for printing.

2.4.1.4 Performance

The file format should enable reasonable duration (interactive time) for read-and-write operations and reasonable file sizes for a typical large object.

2.4.1.5 Backward Compatibility

Any existing STL file can be converted directly into a valid AMF file without any loss of information and without requiring any additional information. AMF files are also easily converted back to STL for use on legacy systems, although advanced features will be lost. This format maintains the triangle-mesh geometry representation to take advantage of existing optimized slicing algorithm and code infrastructure already in existence.

2.4.1.6 Future Compatibility

To remain useful in a rapidly changing industry, this file format is easily extensible while remaining compatible with earlier versions and technologies. This allows new features to be added as advances in technology warrant, while still working flawlessly for simple homogeneous geometries on the oldest hardware.

2.4.2 Archiving AMF Model

The formatting of any AMF file follows the general Extensible Markup Language (XML) syntax, where the code is divided into markup and content. XML is an ASCII text file comprising a list of elements and attributes, which is used for creating, viewing, manipulating, parsing, and storing AMF files. There are five top-level elements in the AMF file format that define five different characteristics of the model. They can be considered as library of characteristic, where each of them are associated with an identification (ID) number.

<object>—The object element defines a volume or material. At least one object element shall be present in the file. Additional objects are optional.

<material>—The optional material element defines one or more materials for printing with an associated material ID. If no material element is included, a single default material is assumed.

<texture>—The optional texture element defines one or more images or textures for color or texture mapping each with an associated texture ID.

<constellation>—The optional constellation element hierarchically combines objects and other constellations into a relative pattern for printing. If no constellation elements are specified, each object element will be imported with no relative position data.

<metadata>—The optional metadata element specifies additional information about the object(s) and elements contained in the file.

Only a single "object" element is required for a fully functional AMF file.

2.4.2.1 Geometry Specification of AMF

The top-level *<object>* element specifies a unique *id* and contains two child elements: *<vertices>* and *<volume>*. The *<vertices>* element contains the 3D point coordinates of all vertices of the object with an implicitly assigned sequence. The *<volume>* element contains the topological information under its child element *<triangle>* that tessellates the surface of the volume. Up to this point, this basic AMF file is compatible with the STL file.

However, other physical characteristics such as vertex normal, edge normal, color, material, and texture can be stored in the AMF file as optional. Each piece of optional information is inserted under the corresponding feature element in the AMF file, as shown in Scheme 2.2. For example, vertex normal information is inserted explicitly under *<vertices>* element and material ID information is imported from the *<material>* element library under the *<volume>* element. On the other hand, color information can be associated with vertex, triangle, or material.

2.4.2.2 Restrictions on Geometry

All geometry shall comply with the following restrictions.

1. Every triangle shall have exactly three different vertices.
2. Triangles may not intersect or overlap, except at their common edges or common vertices.
3. Volumes shall enclose a closed space with nonzero volume.
4. Volumes may not overlap.
5. Every vertex shall be referenced by at least three triangles.
6. Every pair of vertices shall be referenced by zero or two triangles per volume.
7. No two vertices can have identical coordinates.
8. The outward direction of triangles that share an edge in the same volume must be consistent.

2.4.2.3 Smooth Geometry

All the detailed information stored in the digital document will provide flexibility to the end user as well as to the designer. The optional information module can be activated based on the user profile, functional requirement, or AM capabilities. For example, with the vertex unit normal information under the *<vertices>* element, it will allow the user to construct nonplanner triangular patch with nonlinear parametric Hermite curve interpolation, as shown in Figure 2.17. Hermite curve construction requires four parameters: two endpoints P_1 and P_2 and two tangent vectors P_1^t and P_2^t at two endpoints. The tangent vector at vertex P_1 can be computed such that it is perpendicular to the given vertex normal $\widehat{N}_1$ and resides in the plane defined by that normal and the vector $\overline{P_1P_2}$ using the following Equation 2.6.

$$P_1^t = \left|\overline{P_1P_2}\right| \frac{(\widehat{N}_1 \times \overline{P_1P_2}) \times \widehat{N}_1}{\left\|(\widehat{N}_1 \times \overline{P_1P_2}) \times \widehat{N}_1\right\|}. \tag{2.6}$$

```
<?xml version="1.0"?>
<amf unit="millimeter">

  <material id="1">
    <metadata type="Name">StiffMaterial</metadata>
      <color>
        <r>0</r>
        <g>z</g>
        <b>1-z</b>
      </color>
  </material>

  <texture id="1" width="10" height="26" type="grayscale">
    TWFuIGlzIGRpc3RpbmdlaXNoZWQsIG5vdCB
    vbmx5IGJ5IGhpcyByZWFzb24sIGJ1dCBieS
    B0aGlzIHNpbmdlbGFyIHBhc3Npb24gZnJvb
    SBvdGhlciBhbmltYWxzLCB3aGljaCBpcyBh
    ...
  </texture>


  <object id="0">
    <mesh>
      <vertices>
        ...
      </vertices>
      <volume materialid="1">
        <color>
          <r>0.9</r>
          <g>0.9</g>
          <b>0.2</b>
          <a>0.8</a>
        </color>
        ...
        <triangle>
          <v1>0</v1>
          <v2>1</v2>
          <v3>3</v3>
          <texmap rtexid="1" gtexid="2" btexid="3">
            <utex1>0.1</utex1>
            <utex2>0.21</utex2>
            <utex3>0.15</utex3>
            <vtex1>0.65</vtex1>
            <vtex2>0.72</vtex2>
            <vtex3>0.91</vtex3>
          </texmap>
        </triangle>
      </volume>
    </mesh>
  </object>
</amf>
```

Scheme 2.2 A sample AMF file format. (From ASTM, Standard specification for additive manufacturing file format (AMF). ISO/ASTM 52915:2013(E) Version 1.1, ASTM International, West Conshohocken, PA.)

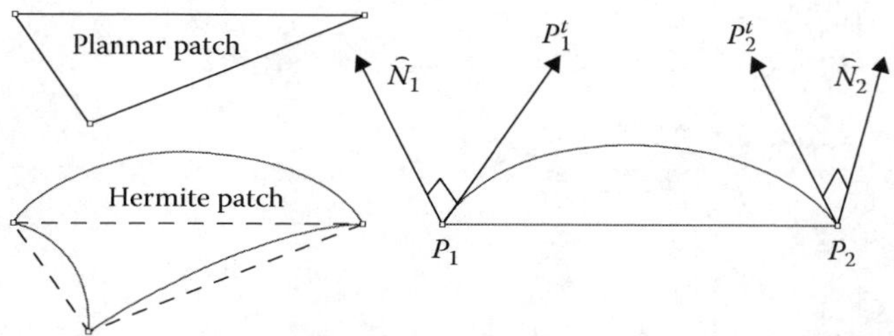

Figure 2.17 Plannar vs. Hermite patch with Hermite curve interpolation.

Degree three Hermite curve H can be mathematically expressed with the parameter u as:

$$H(u) = au^3 + bu^2 + cu + d = CU^u$$

where

$$C = [a\ b\ c\ d] = \begin{bmatrix} a_x & b_x & c_x & d_x \\ a_y & b_y & c_y & d_y \\ a_z & b_z & c_z & d_z \end{bmatrix};\ U = [u^3\ \ \mathrm{u}^2\ \ \mathrm{u}\ \ 1];\ u \in [0,1]. \tag{2.7}$$

By definition, the Hermite curve must pass through the two endpoints $H(0) = P_1$; $H(1) = P_2$; and the tangent defines the curve direction at the endpoints $H'(0) = P_1^t$; $H'(1) = P_2^t$. After simplifying, the Hermite curve equation becomes

$$H(u) = (u^3,\ u^2,\ u,\ 1)\begin{bmatrix} 2 & -2 & 1 & 1 \\ -3 & 3 & -2 & -1 \\ 0 & 0 & 1 & 0 \\ 1 & 0 & 0 & 0 \end{bmatrix}\begin{bmatrix} P_1 \\ P_2 \\ P_1^t \\ P_2^t \end{bmatrix}. \tag{2.8}$$

This will provide a better approximation of the surface with nonlinear point interpolation instead of a linear planer triangle, as shown in Figure 2.17.

2.4.2.4 Future of AMF

Future features that will be added in AMF file format are as follows:

1. Provisions for dimensional and geometric tolerances
2. Provisions for surface roughness
3. Provisions for support structure
4. Provisions for functional representations
5. Provisions for voxel representations

6. Provisions for copyright protection and watermarking
7. Provisions for surface textures and coatings
8. Provisions for more compact Vertex Coordinate and Mesh Encoding

For further interest on AMF, readers can follow the ref (AMF 2015). There are other evolving interchange file formats such as 3D Manufacturing Format or 3MF by Microsoft (http://3mf.io/), Common Layer Interface (CLI) (http://cadexchanger.com/products/cl), JT (ISO 14306) (http://www.iso.org/iso/catalogue_detail.htm?csnumber=60572), etc.

2.5 Build Direction

Learning Objectives

- Understand the effect of build direction (BD)
- Analyze the digital model with respect to the BD
- Learn about the relationship between surface quality and BD
- Constructing support volume

Depending upon the STL digitizing techniques, the model may or may not require postprocessing and is ready for slicing along the build direction or part orientation. The build direction can be defined by the perpendicular vector on the imaginary plane for material deposition. It can also be considered as the part augmentation vector between the bottom and top layers of fabrication. Build orientation defines the cross section, where material is accumulated that has a strong relationship with the structural properties of additively manufactured parts.

Therefore, build direction is a crucial factor to be considered and cannot be decoupled from the AM process plan.

2.5.1 Effect of BD in Fabrication

The most common assumption about build direction is that, it affects the build time, surface quality, and the volume of support structure required during fabrication (Xu et al. 1999; Pandey et al. 2007). But oftentimes build direction is not the sole parameter that affects those factors.

2.5.1.1 Build Time

The conventional knowledge indicates that increasing the number of layers in AM will increase the build time. However, that may not be true in every circumstance. For example, the object in Figure 2.18 with two different build directions $\overrightarrow{BD_i}$ and $\overrightarrow{BD_j}$ defines the conventional wisdom. The STL file of the object is analyzed with the commercial extrusion-based "Makerbot Desktop" software (Makerbot), and the result is shown in Table 2.2. The resulting disparity in fabrication time may come from contour plurality (Ahsan et al. 2015). Slicing a concaved surface may result in layers that will contain more than one disjoint closed contour. This phenomenon is termed contour plurality. This may also happen for objects with internal hollow features. The number of layers with contour plurality is fully dependent upon the build direction. For the same object, the overall contour plurality can vary with different build directions $\overrightarrow{BD}$, as shown in Figure 2.18.

Table 2.2 Build Direction Analysis by Makerbot Software (Makerbot)

	$\overrightarrow{BD_i}$ (%)	$\overrightarrow{BD_j}$ (%)
Number of layer (%)	100	114
Time required (%)	108	100
Contour plurality (% layers)	80	0
Total number of contour vs. layer (%)	155	100

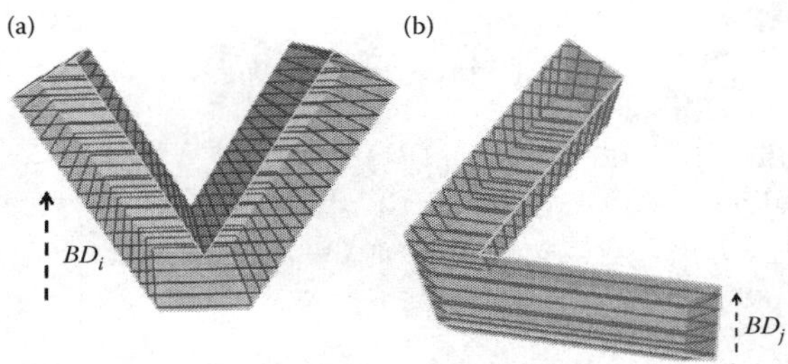

Figure 2.18 Effect of build direction. (a) Contour plurality and (b) single contour sets.

2.5.1.2 Surface Quality

Because of the discrete layer-based approach, every AM part inherits the staircase effect on the surface, as shown in Figure 2.19. The staircase effect between the consecutive layers impairs the physical surface quality of AM objects. The deviation introduced by the staircase effect is measured with the cusp height, which is defined as the maximum distance from the manufactured part's surface perpendicular to the model surface (Alexander et al. 1998). The difference between the nominal design (CAD model) and the STL surface (chord height error) is also shown Figure 2.19, which is accumulated over the cusp height error.

Surface quality of additively manufactured objects largely depends on the build direction, which yields the object orientation during fabrication. There will be almost no staircase effect for an object oriented along a build direction when the build direction is parallel or perpendicular to the surface of the object. Thus, the angle (θ) between the facet normal and build direction can be used as a measure for such cusp height deviation using the following equation (Ahsan et al. 2015), which is shown in Figure 2.20.

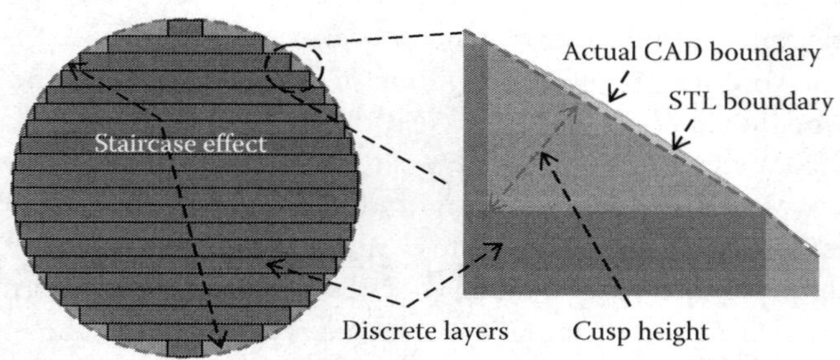

Figure 2.19 Staircase effect and cusp height.

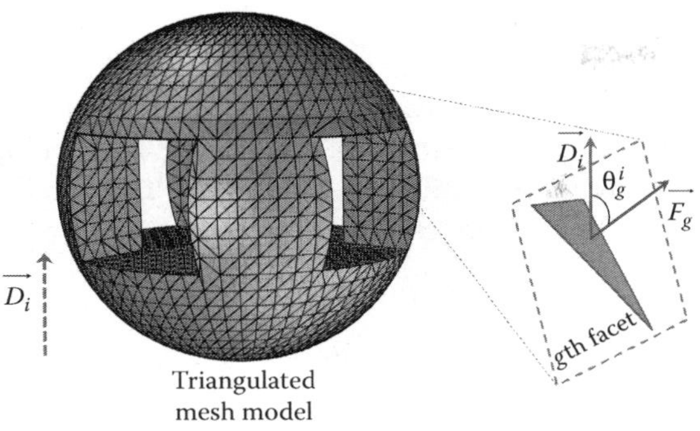

Figure 2.20 Facet normal and surface roughness. (From Ahsan, A.M.M.N. et al., *Comput. Aided Des.*, 69, 112–125, 2015.)

$$\text{Error\_Index}^i = \begin{cases} 0, & \text{when } \theta_g^i = 0°,\ 90°,\ \text{or } 180° \\ \sum_{g=0}^{G} \left|\tan\theta_g^i\right|, & \text{when } 0° < \theta_g^i \le 45°,\ \text{or } 135° < \theta_g^i < 180°, \\ \sum_{g=0}^{G} \frac{1}{\left|\tan\theta_g^i\right|}, & \text{when } 45° < \theta_g^i < 90°,\ \text{or } 90° < \theta_g^i \le 135° \end{cases} \quad \forall\, g = 0,1,2,\ldots,G. \tag{2.9}$$

Here, Error_Index^i is the cusp height error index for an *i*th build direction, θ_g^i is the angle for the *g*th facet, and *G* is the total number of facet.

2.5.1.3 Support Volume

In AM process, materials are accumulated/cured in 2.5D layers and stacked successively to generate the 3D part. The material in each layer must be connected and supported fully or in-part by the previous layer. This is necessary to ensure surface continuity and structural integrity in the fabricated object. However, some surface segment or stacks may be "floating" in mid-air especially where the facet normal is downward in direction. These regions are called overhangs and need support from below by additional sacrificial material. Support material can be broadly classified into two categories: (1) material that surrounds the part as a naturally occurring by-product of the build process (natural supports) and (2) rigid structures, which are designed and built to support, restrain, or attach the part being built to a build platform (synthetic supports) (Gibson et al. 2010).

The powder- and sheet-based AM processes primarily provide natural supports since the part being built is fully encapsulated in the build material. On the other hand, AM processes, which do not naturally support parts, require synthetic supports for overhanging features.

However, synthetic supports are also required to resist distortion or collapse in powder-based process for metals. Based on the build direction, most 3D printing software provide automated generation of synthetic support structure using its own algorithm, as shown in Figure 2.21. As seen in the figure, the physical characteristic of the support volume can vary significantly on the process parameter (i.e., build direction) as well as optimization technique.

To design and generate the support structure, first the points on the object surface that requires support are determined. For STL model, the outward facet normal is calculated using Equation 2.4. For the parametric surface $S(u,v)$, the outward normal direction at any point on the surface can be calculated by Equation 2.10 (Habib and Khoda 2017), as shown in Figure 2.22.

$$\mathbf{n} = \frac{\partial S(u,v)}{\partial u} \times \frac{\partial S(u,v)}{\partial v} \quad : u,v \in [a,b] \tag{2.10}$$

The normal vector is compared with the build direction vector $\overrightarrow{BD}$, and the angle is measured to determine its support requirement. Once the points are identified, then the process plan for

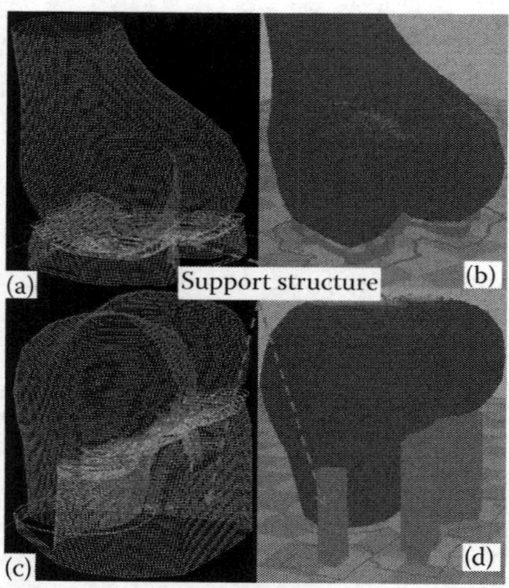

Figure 2.21 Support structure for same object with different build directions using two commercial FDM software: with Stratasys (Stratasys) dimension (a, c) and with Cura (Cura) (b, d).

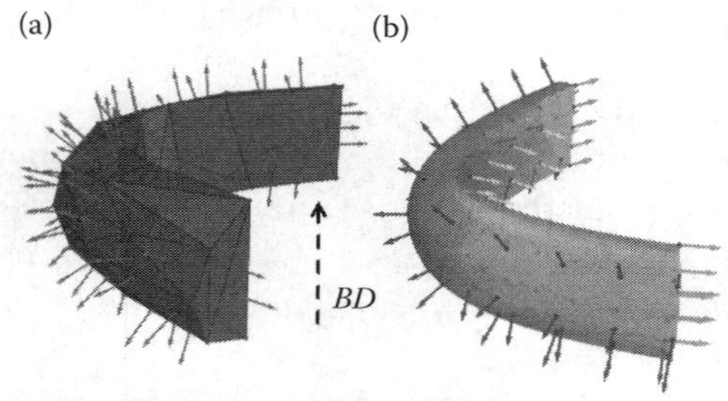

Figure 2.22 (a) Facet normal constructed at the centroid of the facet. (b) Surface normal constructed on the parametric surface.

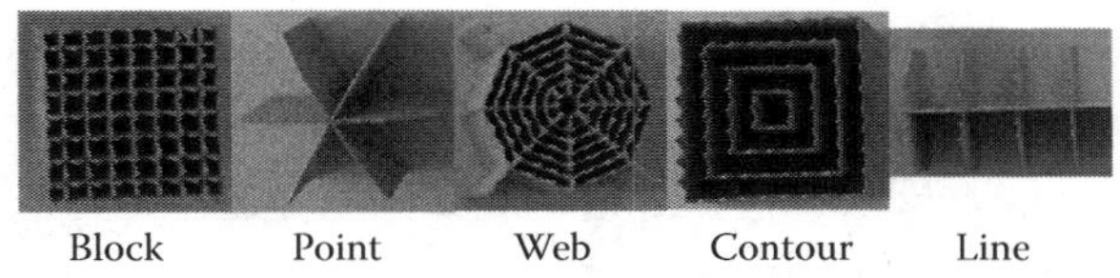

Figure 2.23 Different support geometry for metal and powder-based additive manufacturing. (From Krola, T.A. et al., *Proceedings of the Solid Freeform Fabrication Symposium,* Houston, TX, 2012.)

the support structure needs to be created. The peripheral points from the support required point set can be used to generate the support surface contour. Projecting the support surface contour on the object floor will introduce bulk support. Support structures are sacrificial, i.e., they need to be removed from the overhanging part after completing the fabrication. Supports may be located in difficult-to-reach regions within the part. Another issue with support material is that it creates marks or textures at the junctures and may not be desirable at all. Thus, it should be designed in such a way that will make it easier to remove without damaging the surface quality of the contact surface. Moreover, the material consumption for support structure should be minimized to reduce the overall fabrication time and cost as well as postprocessing of the fabricated part.

The design of support structure along a given build direction is a challenge during fabrication. Different algorithms are proposed to design and create the support structure. In fact, computation for support structure can be significantly rigorous for its multifunctional attribute requirements. A support volume reduction framework is introduced as clever support (Vanek et al. 2014). First, the build direction is optimized for minimal support. Then the support structure is progressively built while attempting to minimize the overall length of the support using a greedy heuristic. The resulting tree-like support structure reduces both time (~30%) and support volume (~40%). Other different basic forms of supports are displayed in Figure 2.23 (Krola et al. 2012)

2.6 Model Slicing

Learning Objectives

- Understand the slicing process
- Learn about uniform and nonuniform slicing algorithms
- Understand the error introduced through slicing

Once the build direction is defined, the model requires slicing. The 3D object can be sliced following the rectilinear bounding box methodology, as shown in Figure 2.24. The rectilinear bounding box for the object **O** is constructed with the six extreme axis value sets, $\{x_{min}, x_{max}, y_{min}, y_{max}, z_{min}, z_{max}\}$ in the domain R^3. The eight corner points of object **O** are aligned with the corresponding coordinate system, as shown in Figure 2.24. The planes perpendicular to the build direction vector $\vec{D}_b$, are defined as bottom and top planes. The parametric equation for the bottom plane can be represented by Equation 2.11, where t and u are the two parameters.

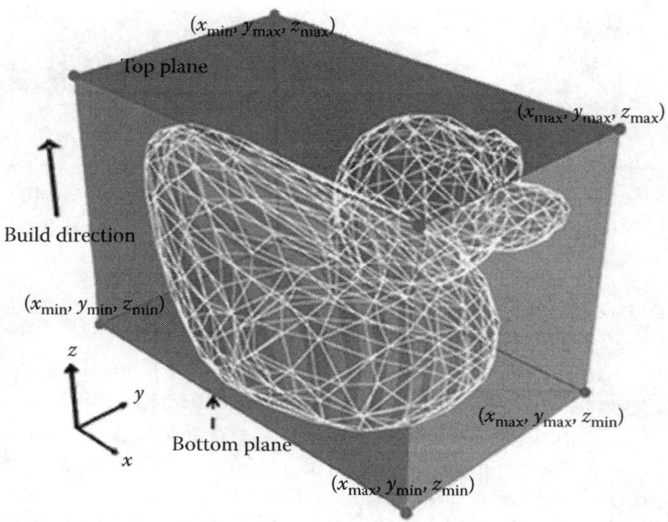

Figure 2.24 Bounding box construction for three-dimensional model.

$$\begin{cases} x = x_{\min} + t(x_{\max} - x_{\min}) \\ y = y_{\min} + u(y_{\max} - y_{\min}) \quad \forall \ t \in [0, 1]; \ \ u \in [0, 1] \\ z = z_{\min} \end{cases} \tag{2.11}$$

The 3D model is sliced by a set of parallel and intersecting planes perpendicular to the build direction $\vec{D}_b$, as shown in Figure 2.24. The bottom and the top plane, bind the number of parallel slicing planes. Each of these planes will intersect with sets of STL facet, and the intersection points can be calculated by the following Equation 2.12, as shown in Figure 2.25.

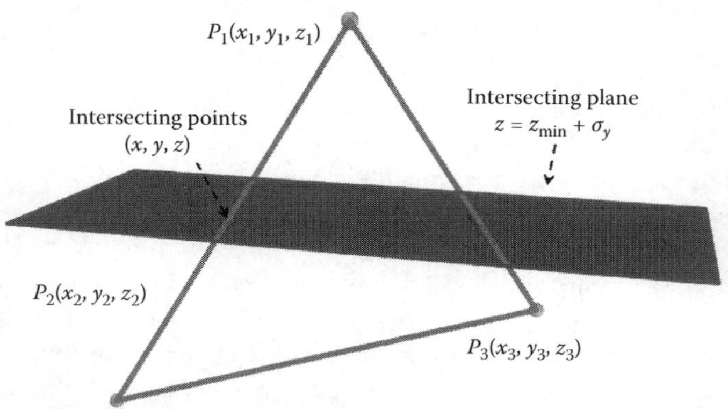

Figure 2.25 Slicing plane and triangular (STL) facet intersection.

$$\begin{cases} x = x_1 + \dfrac{(z - z_1)\times(x_2 - x_1)}{z_2 - z_1} \\ y = y_1 + \dfrac{(z - z_1)\times(y_2 - y_1)}{z_2 - z_1} \\ z = z_{\min} + \sigma_i. \end{cases} \quad \forall\ x_{\min} \le x \le x_{\max};\quad y_{\min} \le y \le y_{\max};\quad z_{\min} \le z \le z_{\max} \tag{2.12}$$

Intersecting points are inserted at a location where the boundary of the facet and the plane intersect, as shown in Figure 2.26. The distance between the parallel planes is the same in uniform slicing, which is the layer thickness. Such discretization creates surface roughness in the fabricated part, which is usually defined as staircase error.

Reducing the layer thickness may reduce the staircase error effect in the fabricated part. However, layer thickness is restricted to the depth of material curing capability in the AM equipment. Followed by the staircase effect, the AM process may also cure excess or inadequate material volume in each layer. This volume difference between the nominal (CAD) models and their physical counterparts is defined as deposition inaccuracies (Mani et al. 1999). Three different categories of deposition inaccuracies may result from uniform slicing, which are shown in Figure 2.27, and can be described by Equation 2.13 (Mani et al. 1999).

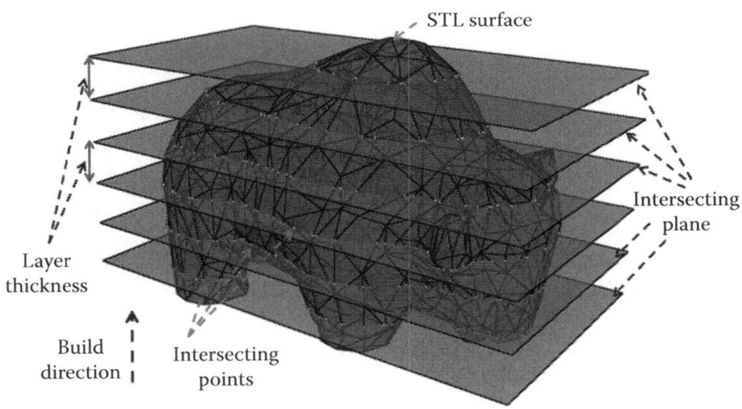

Figure 2.26 Intersecting point generation through model slicing along the build direction.

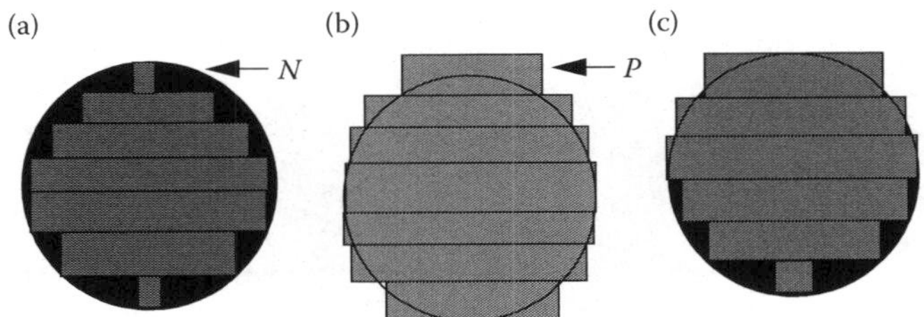

Figure 2.27 Deposition inaccuracies. (a) Case 1, (b) Case 2, and (c) Case 3. (From Mani, K. et al., *Comput. Aided Des.*, 31(5), 317–333, 1999.)

$$\begin{aligned}&\text{Case 1: } P \subseteq N \quad \text{(inadequate material volume)}\\&\text{Case 2: } N \subseteq P \quad \text{(excess material volume)}\\&\text{Case 3: } N \not\subset P \;\text{ and }\; P \not\subset N \quad \text{(partially excess and inadequate).}\end{aligned} \tag{2.13}$$

Adaptive slicing may reduce both staircase and deposition inaccuracy in AM part (Figure 2.27), but may require more computational resources and time to fabricate. Thus, a region-based adaptive slicing algorithm is proposed (Mani et al. 1999) that provides the user the flexibility to pick between the physical surface quality and resource consumption. They combine thin shell and sparse/thick internal layer in slicing algorithm for faster and accurate AM part building. However, most AM processes use a uniform slicing technique to match the physical system capability. The material delivery or curing system needs to be modified to implement such an advanced algorithm.

The intersections points for the set of slicing plane are calculated by Equation 2.12, which are used to generate the contours or layer, as shown in Figure 2.28. Closed 2D contours are generated by connecting these intersecting points consecutively in a linear fashion. As a result, the contours are a piecewise linear curve, as shown in Figure 2.29a. All contour curves are assumed to be a simple planner closed curve, i.e., a planner curve does not intersect itself other than its start and endpoints and has the same (positive) orientation.

Most AM processes use numerical control (NC) systems or their derivatives, where point sets are used as the input during the motion control. The machine movement will

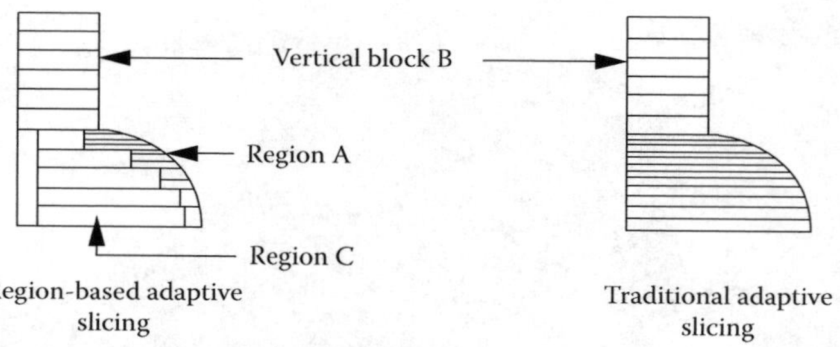

Figure 2.28 Region-based adaptive slicing vs. traditional adaptive slicing. (From Mani, K. et al., *Comput. Aided Des.*, 31(5), 317–333, 1999.)

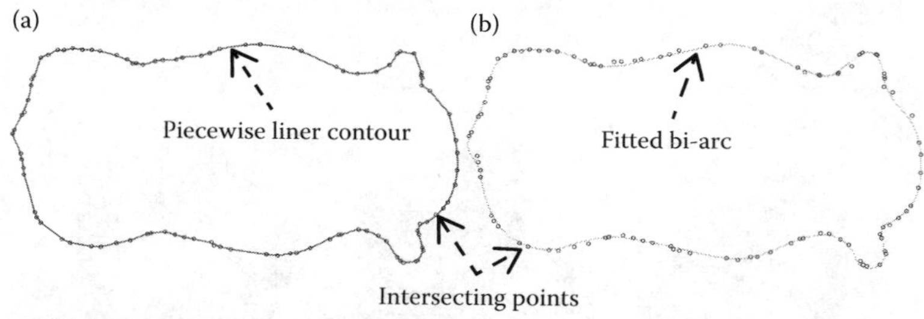

Figure 2.29 Contour generation from the plane-facet intersecting points: (a) piecewise linear curve and (b) fitted bi-arc. (From Khoda, A.B., Koc, B., *Comput. Aided Des.*, 45(11), 1276–1293, 2012.)

be generated by linear interpolation between points from the piecewise linear curve. As a result, this may create intermittent motion, while fabrication and nonuniform material curing may occur along the contour (Khoda et al. 2013b). Uniform motion can be achieved with intervening algorithm along the transition between the line segments. Alternatively, a point reduction algorithm can be used for minimizing this over deposition effect. A max-fix biarc curve-fitting technique is applied by Khoda et al. (Khoda and Koc 2013) that simultaneously reduces points and uses accurate circular interpolation suitable for NC control system, as shown in Figure 2.29b.

2.7 Tool-Path/Deposition Pattern

Learning Objectives

- Understand the tool-path generation process
- Learn about different tool-path pattern
- Learn about object hollowing

Tool-path/deposition/scan pattern is the final stage of the AM preprocessing plan. The material deposition path aka tool-path needs to be designed in CAD for an individual slice with a raster angle and hatch density to fill the internal space. The extruder or laser/light head needs to be guided through the predefined contours or tool-path in order to cure material and construct the layers. One of the major constraints of AM processes is the speed, i.e., it consumes a significant amount of time to fabricate parts (Ahsan et al. 2015). Even though new product/prototype development cycle time is considerably lower, the lead fades away as the quantity to be produced increases. As a result, the viability of AM processes is oftentimes limited toward mass-customized, low-quantity, small size (Hansen et al. 2013) products. Two common approaches for speeding up AM is to reduce the deposition volume by (1) hollowing the object inside and (2) designing the object with macroholes pattern or porosity.

Current material deposition techniques or tool-path are primarily differentiated into fine peripheral contours and course infill deposition to satisfy better aesthetic look. Peripheral deposition is critical for the surface quality. On the other hand, the most traditional assumption about the infill or internal space is that, it affects the build time and the volume of material required during fabrication with little contribution to the value of the prototype. As a result, infill deposition has not attracted the AM user community and is more or less considered a proprietary pattern for the process software.

Object hollowing or shelling technique is primarily performed by 3D inward offsetting, which is mainly adapted from NC machining tool-path. Generally, all facets from the 3D model are copied and moved to perpendicular inward direction at a distance of desired offset thickness. To generate the valid closed 3D shell model, it is necessary to identify all the intersections and gap profiles for trimming and filling. This process can be quite complex and erroneous, since STL files oftentimes may inherit inaccuracy and number of facet can be increased with the complexity of the object geometry.

A vertex-based STL file offsetting technique is described by Koc et al. (Koc and Lee 2002). In the method, the weighted average vertex normal vector is calculated from the normal vector of the vertex sharing triangles from the STL file. Each of the vertices from the STL file are then translated toward the weighted normal direction with a user-defined distance or shell thickness.

Self-intersections, loops, and irregularities are determined and eliminated by identifying their loop direction to ensure valid offset contour of the 3D model. The thickness or offset distance of the shell is considered as user-defined parameter although it can be an important parameter for self-supporting hollow object. Alexander and Dutta (2000) considered the shell thickness as the design parameter and proposed adaptive or localized wall-thickening technique. They measure wall constructability property (WCP) for any facet as a function of angle between the facets normal and build direction. They also prove that WCP exists for any facet if the angle is not perpendicular and can support itself by thickening the wall. Finally, the adaptive wall thickness is measured as a function of WCP. This methodology maximizes the accuracy of the fabricated self-supporting shell while minimizing the building time and material uses.

Recently, a spatial enumeration method through voxelization is proposed (Yoo 2009) for large number of triangle meshes or STL model, as shown in Figure 2.30. First, the model is bounded and the space is divided with 3D cells or voxel. A radial basis function is used to assign value to each voxel considering the distance from the triangle mesh. Cells are refined following a distance field-based function, and the valid cells are subdivided to ensure better resolution. Both uniform and nonuniform offsetting can be performed in this method with higher accuracy. All these hollowing techniques can successfully reduce the build time and cost by reducing the deposition volume digitally. However, physical stability or imbalance from self-weight, shape, or volume of the fabricated part is not considered in those virtually hollowed model.

Hollowing technique provides better solution for saving material and fabrication time, while could create physical and structural instability as well as shape retention problem once it is fabricated. Traditionally, machine manufacturers use a "propitiatory" filling pattern to put material in the hollowed section, which is often defined as scaffolding. Perpendicular raster angles with 0°–90° infill pattern between layers is the most commonly used in extrusion-based AM processes, as shown in Figure 2.31. However, other angle combinations have been reported optimum for certain part of geometry considering its functionality (Habib et al. 2015), as shown in Figure 2.32.

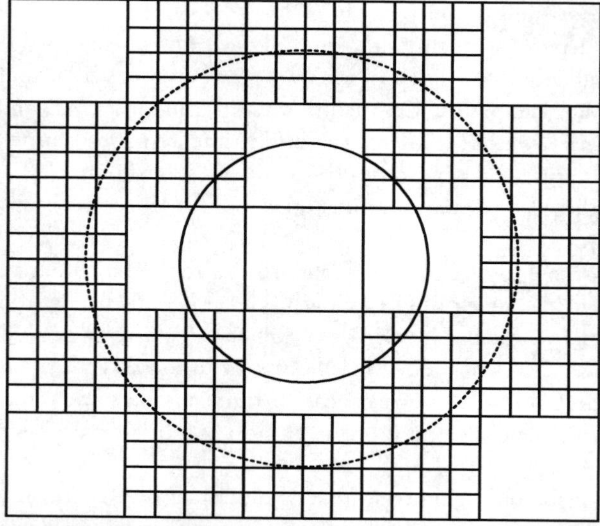

Figure 2.30 Voxelization and their refinement. (From Yoo, D.-J., *Int. J. Precision Eng. Manuf.*, 10(4), 131–142, 2009.)

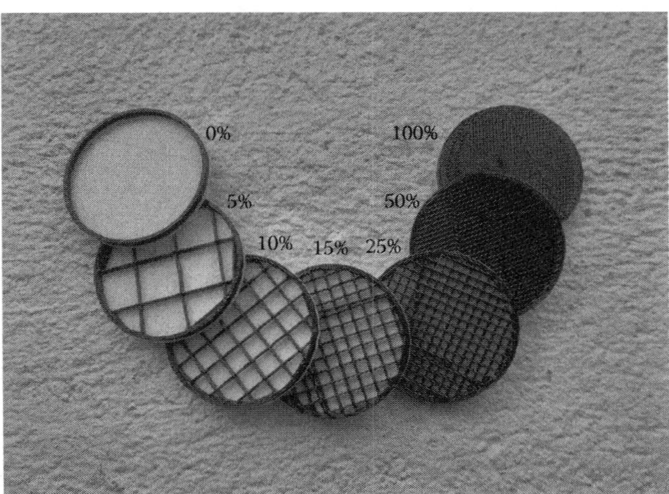

Figure 2.31 0°–90° infill pattern with varying material density by controlling the filament spacing (From http://de3dmakerij.nl/.)

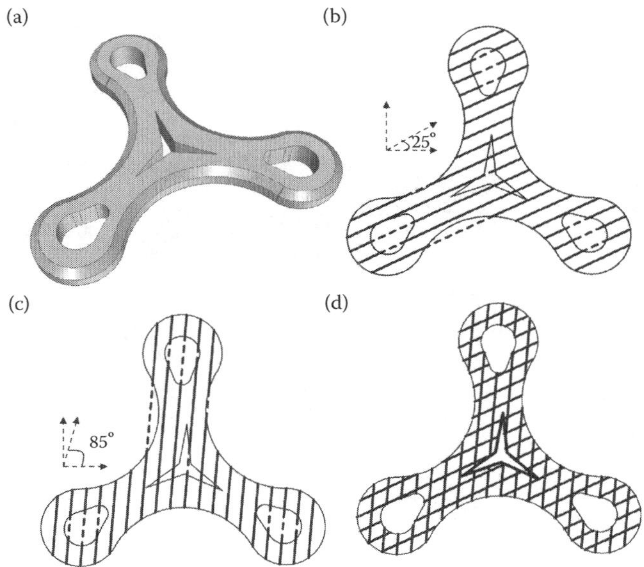

Figure 2.32 Optimum infill pattern for bilayer pattern with minimum time. (From Habib, A. et al., *Proc. Manuf.*, 1, 378–392, 2015.)

The maker community has also considered other periodic unit patterns for infill, as shown in Figure 2.33. Major manufacturers, such as Stratasys, provide few qualitative options (sparse, dense, or solid) for their user to choose. Other AM processes such as powder bed fusion, sheet lamination, vat polymerization, and binder jetting also use a similar periodic pattern during part fabrication. However, removing the uncured material from the internal space can be tedious and time consuming and may oftentimes require minor design modification to avoid trapped material volume.

Figure 2.33 Infill patterns at varying densities. Left to Right: 20%, 40%, 60%, 80%. Top to Bottom: honeycomb, concentric, line, rectilinear, Hilbert curve, Archimedean chords, Octagram spiral (Formslic3r.org.)

The heuristic selection process of the scaffolding pattern and the shell thickness from the limited options provided by the manufacturer may introduce random strength, but is far from being optimal. Moreover, such synthesized structures are disconnected from the physical stability and shape retention and may not be well adapted to high stress acted upon the fabricated part. The infill deposition during AM fabrication creates an internal mesostructure, which is often determined by considering the time, support materials, and the surface finish of the fabricated part. However, there is a strong correlation between the physical properties of the part with the deposition of the scan pattern. Both extrinsic and intrinsic properties can be controlled or influenced with the mesostructure. Thus, with the shifting of AM paradigm toward the functional object, the significance of the infill or scan pattern is gaining traction. For example, predetermined layout patterns are tested for the structural integrity (Es-Said et al. 2000) and the attributes (i.e., pore size and geometry) (Domingos et al. 2011) to measure their effect. Adaptive layout patterns (Khoda and Koc 2013) has recently been proposed by Khoda et al. to achieve the desired attributes, i.e., porosity. Besides, a multidirectional parametric deposition orientation (Khoda et al. 2013a) is proposed by considering the accessibility and porosity of the internal region, as shown in Figure 2.34.

As AM processes can fabricate virtually everything or anything, creative designer started designing unconventional object and shape with freeform geometry. While fabricating, the equipment uses natural or synthetic support (Gibson et al. 2010) for stability; those supports are removed during the postprocessing period leaving only the physical realization of the virtual design. Even though synthesized structures with shell and frame/scaffold is an integrated part of the AM process plan, the overall physical stability or imbalance of the finished part is generally not considered within its scope. As a result, illusion about AM is disrupted with unstable or toppled printed model instead of standing as intended but never planned. A skin-frame combinatorial structure (Prevost et al. 2013) is proposed for shape balancing the 3D additively manufactured part. Their "Make it Stand" optimization technique is formulated as an energy minimization, improving stability by modifying the volume and hence the center of gravity of the object, while preserving its surface details. The iterative optimizer assists users in producing properly balanced designs by interactively deforming an existing model. Both upright and suspended object balancing is performed by minimally modifying the model using interior carving and surface deforming technique.

Another skin-frame structure is proposed for minimizing the material volume and hence the printing cost for custom printing (Wang et al. 2013). The cost-effective 3D frame modeling is performed with node-based struts that significantly reduces the material volume while preserving the shape and the stability. Printability of the strut fame is ensured by defining threshold strut radius. The proposed technique optimizes both topology and geometry in sequence until the frame volume does not decrease. Total number of strut is minimized during topology optimization using sparsity optimization. Subsequently, the position of the internal nodes and strut radius are refined during the geometry optimization. As a result, self-supporting frame is modeled with minimal material volume.

A tensor-based hole pattern (Andrade et al. 2016) is proposed on the freeform surface object. The methodology utilizes a physically based optimization of bubbles known as bubble packing

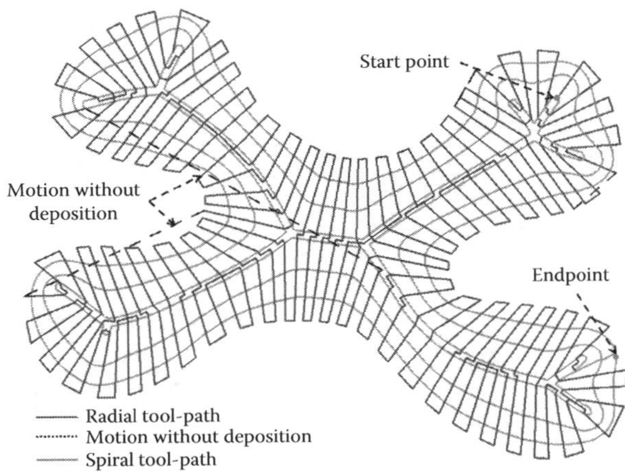

Figure 2.34 Radial and spiral tool-path proposed by Khoda et al. (2013a).

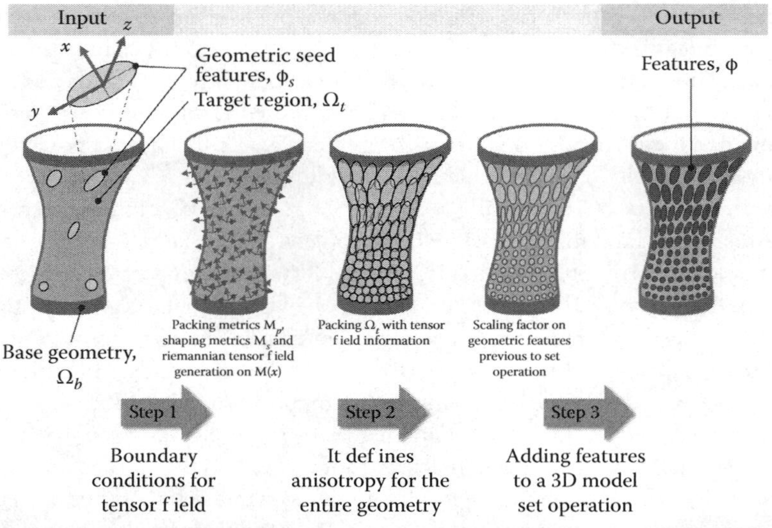

Figure 2.35 Tensor-based anisotropic bubble packing algorithm. (From Andrade, D. et al., *Proc. Manuf.*, 5, 944–957, 2016.)

technique in conjunction with anisotropic interpolation of seeds based on the Laplace–Beltrami operator. The proposed algorithm takes the target surface and "seed features," as input and generates size and directionality of the anisotropic pattern. The methodology is comparable to the Cell cycle (by Nervous System), which is an iterative physics simulations tool and bases its designs on natural occurring of cellular structures (Figure 2.35).

2.8 Summary

This chapter introduced the process planning of AM at the preprocessing stage. This stage starts with design conceptualization and ends by generating the AM process control instructions. Each of these activities is discussed in this chapter along with its technical details and the potential effects on the part fabricated by AM. The directional activities in the preprocessing stage have a hierarchical relationship, where error accumulates along the downstream activities. The outcome of this stage is information, which is fully reversible based on the user requirement. Information from this stage is directed toward the processing stage, where actual fabrication starts occurring and the process becomes irreversible.

2.9 Review Questions

- What are the design features of the 3MF file format?
- Consider an engineering part with five different build directions and explain their effect on the support volume.
- Scan a simple 3D object and extract its points. Use those points to make the STL file in the lab.
- How does curve fitting during layer contour generation affect the material curing of AM processes?
- What kind of geometry will benefit from uniform and nonuniform slicing?

- What kind of physical modifications are required in the AM machine at the laboratory to attain the benefit from nonuniform slicing?
- Consider an engineering part and identify the pros and cons of different infill patterns for that object.
- What are the factors that determine the effectiveness of hollowing inside techniques?
- What are the potential effects of feature patterning on the freeform surface? Discuss them with respect to the physical properties, AM processes, and materials.

References

Ahsan, A. M. M. N., M. A. Habib and B. Khoda (2015). Resource based process planning for additive manufacturing. *Computer-Aided Design* 69: 112–125.

Ahsan, N. and B. Khoda (2016). AM optimization framework for part and process attributes through geometric analysis. *Additive Manufacturing* 11: 85–96. doi:10.1016/j.addma.2016.05.013.

Alexander, P., S. Allen and D. Dutta (1998). Part orientation and build cost determination in layered manufacturing. *Computer-Aided Design* 30(5): 343–356.

Alexander, P. and D. Dutta (2000). Layered manufacturing of surfaces with open contours using localized wall thickening. *Computer-Aided Design* 32(3): 175–189.

AMF. Additive manufacturing file format. https://en.wikipedia.org/wiki/Additive_Manufacturing_File_Format (Accessed November 2015).

Andrade, D., W. He and K. Shimada (2016). Automated generation of repeated geometric patterns for customized additively manufactured personal products. *Procedia Manufacturing* 5: 944–957.

ASTM (2013). Standard specification for additive manufacturing file format (AMF). ISO/ASTM 52915:2013(E) Version 1.1. West Conshohocken, PA: ASTM International.

Bischoff, M., S. Steinberg, S. Bischoff and L. Kobbelt (2002). OpenMesh—A generic and efficient polygon mesh data structure. *OpenSG Symposium*. https://www.graphics.rwth-aachen.de/media/papers/openmesh1.pdf.

Chang, T.-C., R. A. Wysk and H.-P. Wang (2005). *Computer-Aided Manufacturing*. Englewood Cliffs, NJ: Prentice Hall.

Chen, Y. H. and C. T. Ng (1997). Integrated reverse engineering and rapid prototyping. *Computers & Industrial Engineering* 33(3–4): 481–484.

Command Line Interface (CLI). http://cadexchanger.com/products/cli (Accessed April 2017).

Cura. Cura 3D printing software. https://ultimaker.com/en/products/cura-software (Accessed November 2015).

Davis, J. E., M. J. Eddy, T. M. Sutton and T. J. Altomari (2007). Converting boundary representation solid models to half-space representation models for Monte Carlo analysis. *Joint International Topical Meeting on Mathematics & Computation and Supercomputing in Nuclear Applications*. Monterey, CA: American Nuclear Society, LaGrange Park.

Domingos, M., F. Chiellini, S. Cometa, E. D. Giglio, E. Grillo-Fernandes, P. Bartolo and E. Chiellini (2011). Evaluation of in vitro degradation of PCL scaffolds fabricated via BioExtrusion. Part 2: Influence of pore size and geometry. *Virtual and Physical Prototyping* 6(3): 157–165.

Es-Said, O. S., J. Foyos, R. Noorani, M. Mendelson, R. Marloth and B. A. Pregger (2000). Effect of layer orientation on mechanical properties of rapid prototyped samples. *Materials and Manufacturing Processes* 15(1): 107–122.

Gibson, I., D. W. Rosen and B. Stucker (2010). *Additive Manufacturing Technologies: Rapid Prototyping to Direct Digital Manufacturing*. New York: Springer.

Habib, A., N. Ahsan and B. Khoda (2015). Optimizing material deposition direction for functional internal architecture in additive manufacturing processes. *Procedia Manufacturing* 1: 378–392.

Habib, M.A. and B. Khoda (2017). Attribute driven process architecture for additive manufacturing. *Robotics and Computer-Integrated Manufacturing* 44: 253–265.

Hansen, C. J., R. Saksena, D. B. Kolesky, J. J. Vericella, S. J. Kranz, G. P. Muldowney, K. T. Christensen and J. A. Lewis (2013). Inkjet printing: High-throughput printing via microvascular multinozzle arrays. *Advanced Materials* 25(1): 2–2.

Hod, L. (2014). AMF tutorial: The basics (Part 1). *3D Printing and Additive Manufacturing* 1(2): 85–87.

Hoffmann, C. M. (1992). *Geometric and Solid Modeling*. San Francisco, CA: Morgan Kaufmann Publishers Inc.

JT. JT file format specification for 3D visualization. http://www.iso.org/iso/catalogue_detail.htm?csnumber=60572 (Accessed November 2015).

Khoda, A. B. and B. Koc (2012). Designing controllable porosity for multifunctional deformable tissue scaffolds. *Journal of Medical Devices* 6(3): 031003.

Khoda, A. K. M., I. T. Ozbolat and B. Koc (2011). A new functionally gradient variational porosity architecture for hollowed scaffold design. *Journal of Biofabrication* 3(3): 1–15. doi:10.1088/1758-5082/3/3/034106.

Khoda, A. K. M. B. and B. Koc (2013a). Functionally heterogeneous porous scaffold design for tissue engineering. *Computer-Aided Design* 45(11): 1276–1293.

Khoda, A. K. M., I. T. Ozbolat and B. Koc (2013b). Designing heterogeneous porous tissue scaffolds for additive manufacturing processes. *Computer-Aided Design* 45(12): 1507–1523.

Khoda, B. (2014). Process plan for multimaterial heterogeneous object in additive manufacturing. *3D Printing and Additive Manufacturing* 1(4): 210–218.

Koc, B. and Y.-S. Lee (2002). Non-uniform offsetting and hollowing objects by using biarcs fitting for rapid prototyping processes. *Computers in Industry* 47(1): 1–23.

Koc, B., Y. Ma and Y. S. Lee (2000). Smoothing STL files by Max-Fit biarc curves for rapid prototyping. *Rapid Prototyping Journal* 6(3): 186–205.

Krola, T. A., M. F. Zaehb and C. Seidela (2012). Optimization of supports in metal-based additive manufacturing by means of finite element models. *Proceedings of the Solid Freeform Fabrication Symposium*, Houston, TX.

Kulkarni, P., A. Marsan and D. Dutta (2000). A review of process planning techniques in layered manufacturing. *Rapid Prototyping Journal* 6(1): 18–35.

Lee, K. H. and H. Woo (2000). Direct integration of reverse engineering and rapid prototyping. *Computers & Industrial Engineering* 38(1): 21–38.

Makerbot. Makerbot 3D printer manufacturer. http://www.makerbot.com/ (Accessed November 2015).

Mani, K., P. Kulkarni and D. Dutta (1999). Region-based adaptive slicing. *Computer-Aided Design* 31(5): 317–333.

Microsoft. 3D manufacturing format (3MF) by Microsoft. http://3mf.io/ (Accessed November 2015).

Pandey, P. M., N. Venkata Reddy and S. G. Dhande (2007). Part deposition orientation studies in layered manufacturing. *Journal of Materials Processing Technology* 185(1–3): 125–131.

Park, H. and K. Kim (1995). An adaptive method for smooth surface approximation to scattered 3D points. *Computer-Aided Design* 27(12): 929–939.

Prevost, R., E. Whiting, S. Lefebvre and O. Sorkine-Hornung (2013). Make it stand: Balancing shapes for 3D fabrication. *ACM Transactions on Graphics* 32(4): 1–10.

Sokovic, M. and J. Kopac (2006). RE (reverse engineering) as necessary phase by rapid product development. *Journal of Materials Processing Technology* 175(1–3): 398–403.

Stratasys Inc. 3D printer manufacturer. http://www.stratasys.com/ (Accessed November 2015).

Vanek, J., J. A. G. Galicia and B. Benes (2014). Clever support: Efficient support structure generation for digital fabrication. *Computer Graphics Forum* 33(5): 117–125.

Várady, T., R. R. Martin and J. Cox (1997). Reverse engineering of geometric models—An introduction. *Computer-Aided Design* 29(4): 255–268.

Wang, W., T. Y. Wang, Z. Yang, L. Liu, X. Tong, W. Tong, J. Deng, F. Chen and X. Liu (2013). Cost-effective printing of 3D objects with skin-frame structures. *ACM Transactions on Graphics* 32(6): 1–10.

Xu, F., H. T. Loh and Y. S. Wong (1999). Considerations and selection of optimal orientation for different rapid prototyping systems. *Rapid Prototyping Journal* 5(2): 54–60.

Yin, Z. W. (2011). Direct generation of extended STL file from unorganized point data. *Computer-Aided Design* 43(6): 699–706.

Yoo, D.-J. (2009). General 3D offsetting of a triangular net using an implicit function and the distance fields. *International Journal of Precision Engineering and Manufacturing* 10(4): 131–142.

Chapter 3

Microstructural and Mechanical Properties of Metal ALM

Moataz M. Attallah, Luke N. Carter, Chunlei Qiu, Noriko Read, and Wei Wang

University of Birmingham

Contents

CHAPTER OUTLINE

In this chapter, we

- Describe the microstructural development in Ni-superalloys, Al-alloys, and Ti-alloys.
- Describe the development in texture, grain, and solidification structures.
- Summarize the mechanical properties of the additive layer manufacturing (ALM) structures.
- Highlight the influence of the process parameter and postprocessing on the microstructural and mechanical properties development.

3.1 Introduction

Learning Objectives

- Identify the key alloy classes manufactured using ALM and the key technologies employed in their manufacturing.
- Highlight the key microstructural and mechanical properties characteristics of metal ALM structures.

The majority of metal ALM research focuses on three alloy types—Ni-superalloys, Ti-alloys, and Al-alloys—due to their heavy usage in a wide spectrum of manufacturing sectors (e.g., aerospace, automotive, and medical implants), as well as their high value, compared to ferrous alloys (e.g., stainless steels). Some specific alloys (e.g., IN718 and Ti-64) are workhorse alloys in the aerospace sector; hence, a large number of studies are investigating their ALM. Powder-based technologies, such as Selective Laser Melting (SLM), Direct Laser Deposition (DLD), and Electron-Beam Selective Melting (EBSM), and wire-based technologies, such as Wire and Arc Additive Manufacturing (WAAM), have been successfully utilized to fabricate these alloys. The microstructure of ALM structures demonstrates interesting solidification structures and texture development. As such, this chapter focuses on the microstructure-property development in metallic ALM structures, focusing primarily on Ni-superalloys, Ti-alloys, and Al-alloys, which are the most commonly used alloys in metal ALM.

Previous work has shown that the microstructural development depends on the process parameters, which can be used to control the microstructure, especially the texture and grain size. Nonetheless, the microstructure always demonstrates a high degree of anisotropy due to the nature of the directional thermal fields and rapid cooling rates experienced during the process, which is reflected in the mechanical properties obtained during the process. Postprocessing (e.g., using heat treatment or hot isostatic pressing) has typically been employed, and it usually results in an improvement in the mechanical properties. However, it does not appear to reduce the anisotropy in mechanical properties in the samples built at different build orientations, nor does it improve the properties of the samples with difficult to "heal" defects, such as surface connected cracks and lack of fusion defects.

3.2 Nickel-Based Superalloys ALM

Learning Objectives

- Understand the impact of Ni-superalloys metallurgy on the ALM processability.
- Explain the microstructural development in SLM, DLD, and EBSM technologies, highlighting the differences.
- Discuss the interaction between the process parameters, microstructure, and mechanical properties development.

Nickel-base superalloys have been critical in the advancement of aerospace technologies, specifically in raising the turbine entry temperature (TET) of gas turbine engines and thus improving thermodynamic efficiency. The recent advancements in Ni-superalloys ALM have been recognized by the aerospace industry, resulting in a significant interest in producing ALMed Ni-superalloy components. In broad terms, several potential advantages of ALMed Ni components can be identified, especially the economic low-volume production for engine tests (compared to investment casting) and the potential for producing cooled components with optimized cooling passages. This section explores the current published research in regarding the AM of Ni-superalloys, their microstructure, mechanical properties, and process parameters.

3.2.1 Metallurgy of Ni-Superalloys

While the term "superalloy" can also be applied to nickel-iron-base materials and the niche group of cobalt-base superalloys, Ni-base superalloys have only one base alloying element, nickel, and can generally be divided into the following classes:

γ' *strengthened*: Hardened by the precipitation of the coherent γ' phase (Ni_3Al or Ni_3Ti) typically heat treated to form in a fine cuboidal structure and optimum creep strength.

γ'' *strengthened*: These Nb-containing alloys are hardened by the precipitation of the Body-Centered Tetragonal (BCT) γ'' phase (Ni_3Nb) as typified by IN718 where the γ'' phase is precipitated as coherent disc-shaped particles. γ' strengthening can occur in addition to that provided by γ''.

Solid-solution strengthened nickel-base superalloys: Typified by IN625 and HastelloyX where material strength is achieved through solid-solution strengthening elements (e.g., Co, Cr, Fe, Mo, W, Ta, Re).

Generally, the microstructure of Ni-based superalloys consists of three major phases [1]:

- The matrix *gamma*(γ) austenite, which has a face-centered cubic (FCC) structure.
- The precipitation hardening phases; *gamma prime* (γ') Ni_3(Al, Ti) or γ'' Ni_3Nb.
- The carbide particles (MC, $M_{23}C_6$, M_6C, and M_7C_3). Carbide-forming "M" elements include Cr, Ti, Nb, Hf, Fe, W, Mo, or Ta. Depending on the size, the carbide precipitates can contribute to the strength, help in grain size control, and provide grain boundary strengthening in high temperatures.

In addition, the presence of certain alloying elements or the extended exposure at high temperatures may lead to the formation of additional phases, such as:

- The orthorhombic *delta*(δ) Ni_3Nb intermetallic compound in Nb-containing superalloys. Depending on its size, volume fraction, and morphology, this phase can be beneficial or detrimental to the high-temperature mechanical properties.
- The hexagonal *eta*(η) ordered Ni_3Ti phase, which forms intergranularly in alloys with high Ti/Al ratio following extended exposure.
- The tetragonal borides (M_3B_2), which form in the presence of 0.03% B with "M" possibly being Mo, Ta, Ni, Ni, Fe, or V. Their impact on the microstructure and properties is similar to that of the carbides.
- The cubic nitrides (MN), which forms in Ti-, Ni-, or Zr-containing alloys.
- The tetragonal *sigma*(σ) (Fe, Cr, Mo, Ni), which forms following extended exposure between 540°C and 980°C, has detrimental effects on the mechanical properties.

The typical structure, composition, and elemental roles of Ni-base superalloys are well documented in the literature. The reader is referred to other textbooks that cover this subject [2].

3.2.2 Ni-Superalloys ALM Microstructure

There is a growing body of research characterizing the microstructure of AM Ni-base superalloys. The typical microstructures will be described here according to the AM method.

3.2.2.1 SLM Microstructure

SLM microstructure is largely governed by the small laser spot size (typically in the range 50–100 μm), rapid laser scan speeds, and relatively low overall power (current mid-size machines have a typical max. laser power of ≈400 W). The SLM-processed material is, therefore, subject to much localized rapid heating/cooling and many reheat cycles.

SLM-fabricated Ni-superalloys are prone to a highly directional columnar grain structure; this is due to the partial remelting of grains with each subsequent laser pass, combined with the large heat sink effect of the previously deposited material and build substrate. This heat sinking effect results in a highly directional thermal gradient with heat flowing along the build axis and, therefore, grain elongation similar to that found in a directionally solidified casting process. This epitaxial solidification of grains during SLM is described by Das [3]. These characteristic columnar grains can be clearly seen in SLM-fabricated Nimonic 263 [4] (Figure 3.1).

Also visible in Figure 3.1 is the "fish-scale"-like weld pattern caused by each scan pass of the laser. This weld pattern is a common feature of SLM-fabricated Ni-superalloys and is also reported

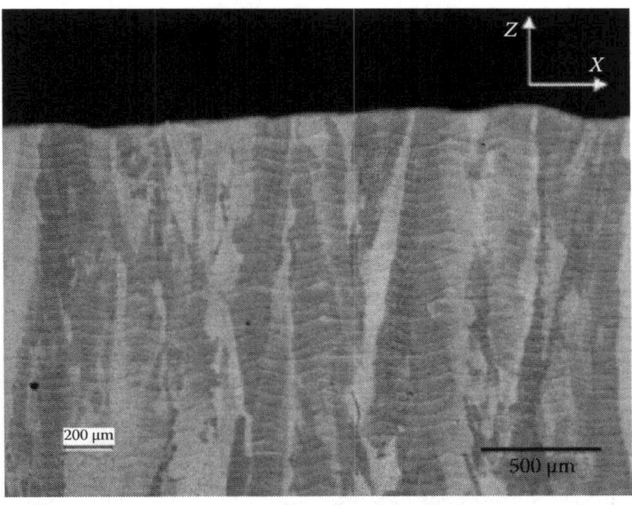

Figure 3.1 SEM micrograph showing SLM-fabricated Nimonic 263 with clearly visible columnar grains elongated in the build direction (*Z*) and "fish-scale" weld pattern. (From Vilaro, T. et al., *Mater. Sci. Eng. A*, 534, 446–451, 2012.)

in other studies (e.g., IN718 [5] and Hastelloy X [6]). In all cases, the columnar grains were shown to grow epitaxially across these weld patterns from one layer to the next. The grains vary in width from approximately 10 to 100 μm and can be >1 mm in length. The fish-scale pattern itself originates from the difference in dendrite formation within different parts of the molten pool and also from the elemental segregation across the melt-pool (as observed in IN625 by Anam et al. [7], specifically Mo and Nb segregation). The micrograph shown in Figure 3.2 illustrates that the dendritic structure changes dramatically between neighboring molten pools depending on the cooling regime of that specific region; in this case, from a cellular to a very fine dendritic structure. These variations across the melt-pool are effected by chemical etchants to varying degrees and

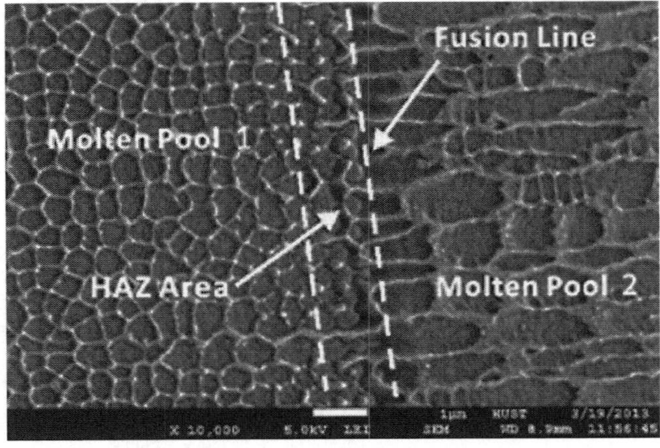

Figure 3.2 Micrograph of the interface between two molten pools presented by Li et al. [9] in SLM-fabricated IN625.

result in the characteristic fish-scale pattern. The variation of this fine (~1 μm) dendritic structure with respect to the processing parameters has also been observed in small cuboidal samples of IN718 by Qingbo and Dongdong [8]. It was observed that a faster scan speed resulted in fragmented dendrites, while increasing the laser power refined their columnar nature with respect to the build direction. The variation in dendritic formation was identified by Li et al. [9] in IN625 and also by Rickenbacher et al. [10] in IN738 (see Figure 3.3), showing the absence of SEM-distinguishable γ′, due to the rapid cooling rates. Figure 3.4 shows this variation with processing parameters in IN718 SLM.

The regular columnar structure gives an idealized view of SLM; the columnar structure obtained by Vilaro et al. [4] in Figure 3.1 was formed from a regular back-and-forth scan strategy with the same laser path direction (shown by the regular fish-scale pattern) for a small regular test coupon. In most SLM-fabricated components, the sample geometry varies hugely, thus influencing the heat flow, and accordingly influencing the grain structure. Likewise, very few commercially available SLM systems operate using a simple back-and-forth laser scan pattern, which is identical for all the layers. Most SLM manufacturers have propriety laser scanning strategies designed to be unique for their equipment, which in turn dramatically alters and complicates this oversimplified view of SLM-fabricated grain structure.

The influence of scan strategies is dramatically illustrated in the micrograph presented by Anam et al. [7], as shown in Figure 3.5. Here the rotating "stripe" strategy is used where the scan vectors are rotated by 67° with each successive build layer, resulting in a "basket weave" appearance of the melt-pool patterns on the *X–Y* plane. Within this research, Anam et al. compared the rotating "stripe" strategy to a "block" strategy where the scan direction is simply rotated by 90°

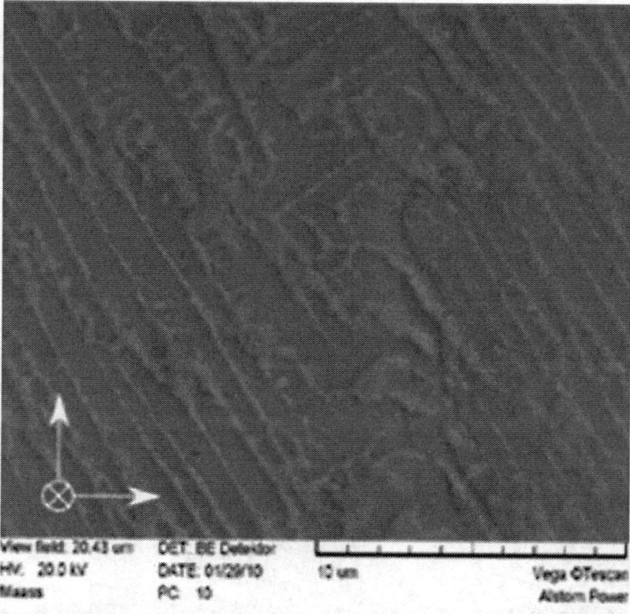

Figure 3.3 Fine dendritic structure of IN738 in the SLM-fabricated condition observed by Rickenbacher et al. [10]. Sample has been etched to reveal the γ′ phase; however, none is visible.

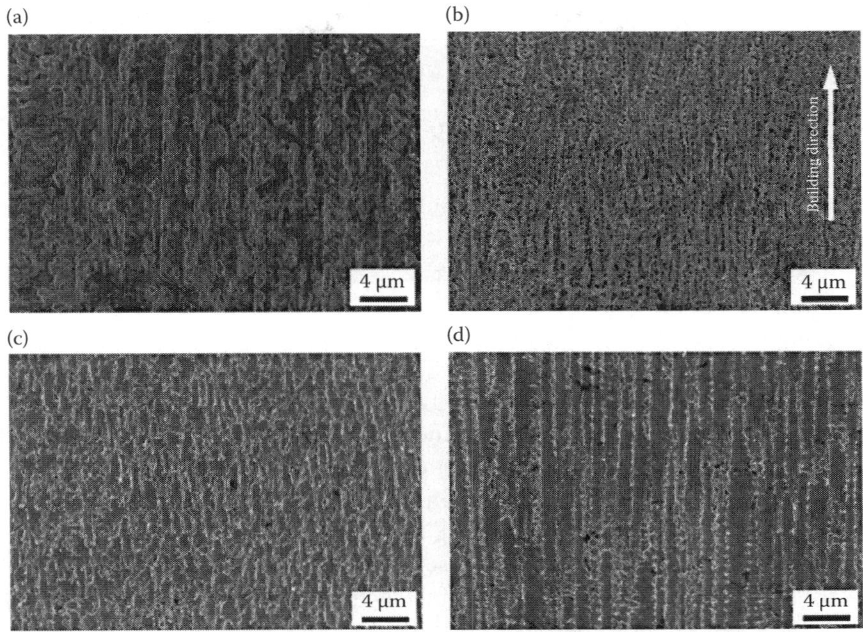

Figure 3.4 SEM images showing characteristic microstructures of SLM-processed Inconel 718 parts at different laser energy densities, laser powers and scan speeds: (a) 180 J/m, 110 W, 600 mm/s; (b) 275 J/m, 110 W, 400 mm/s; (c) 300 J/m, 120 W, 400 mm/s; and (d) 330 J/m, 130 W, 400 mm/s. (From Qingbo, J., Dongdong, G., *J. Alloys Compd.*, 585, 713–721, 2014.)

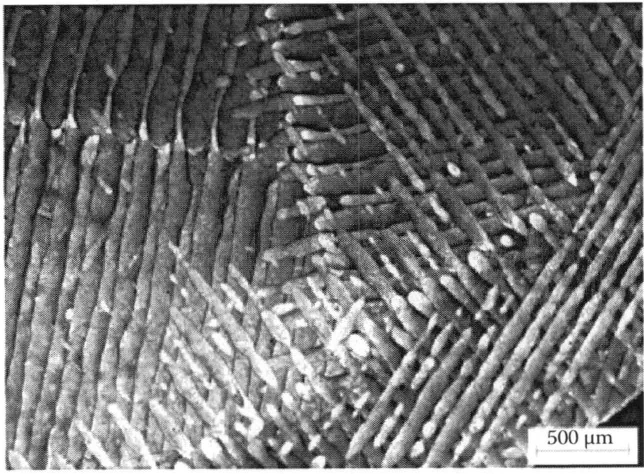

Figure 3.5 Micrograph of SLM-fabricated IN625 using the rotating stripe strategy; note the interlaced melt-pool pattern. (From Anam, M.A. et al., *Proceedings of 25th Annual International Solid Freeform Fabrication Symposium,* 2014.)

with each layer. They reported that, while both strategies resulted in elongated grains, the "stripe" strategy shows longer grains than the "block" strategy. Given that other than this variation both samples were nominally the same (i.e., produced using the same key power/speed parameters, same powder/machine, etc.), it might be unexpected that such a subtle variation as the laser scan strategy would have such a noticeable effect on the grain structure. The observation that a change in the scan direction with each layer can disrupt the growth of the very large columnar grains was also reported by Liu et al. [11] in SLM-fabricated IN718.

The influence of laser scanning strategy is also discussed by Carter et al. [12] when examining SLM of CM247LC. Within this study, the laser is scanned in a randomized "island" regime where each slice is subdivided into regular square islands, each melted using a regular back-and-forth pattern, but the overall order of which are randomized. The island pattern is shifted between layers; however, it was observed that the interference pattern generated by the borders of these islands produces a distinct bimodal grain structure, fine and chaotic, where the island borders line up and are large and columnar in the island center. It was suggested that this is due to the local heat flow within the island and the effective "quenched" nature at the island orders where heat flows rapidly to cool the surrounding material. This bimodal structure is illustrated in Figure 3.6, and both the *XY* and *XZ* planes are shown in Figures 3.7 and 3.8.

The grain structure of SLM-fabricated Ni-superalloys is strongly linked to the texture. Ni solidifies preferentially in the ⟨001⟩ direction, and when directionally solidified, or single crystal cast, the resulting material shows a marked {001} texture along the heat flow axis [13]. The directional heat flow, also responsible for forming the elongated grains associated with SLM-fabricated material, typically results in a highly textured microstructure. This texture is presented by Li et al. [9] in IN625. The Electron Backscatter Diffraction (EBSD) maps and corresponding pole figures shown in Figure 3.9 illustrate how the randomly orientated equiaxed structure of the substrate material transitions into the {001} textured material of the SLM-deposited material.

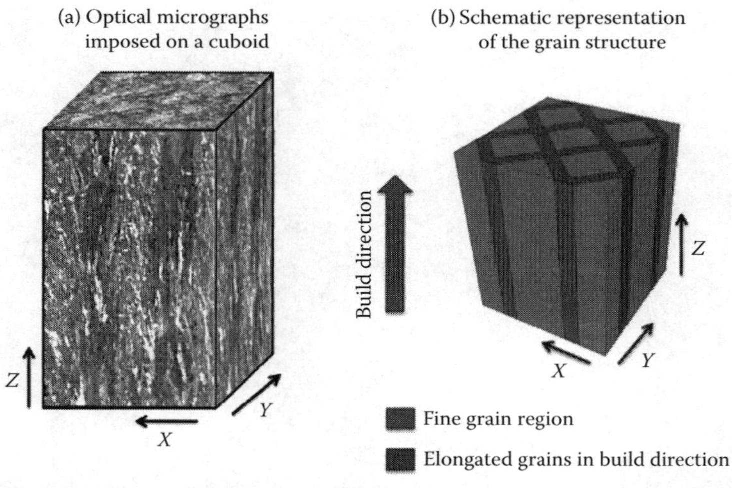

Figure 3.6 Diagram showing how the grain structure in the transverse (*X–Y*) and longitudinal (*X–Z*) planes relate to each other. (a) A 3D representation constructed from optical micrographs and (b) the relationship showing schematically. (From Carter, L.N. et al., *J. Alloys Compd.*, 615, 338–347, 2014.)

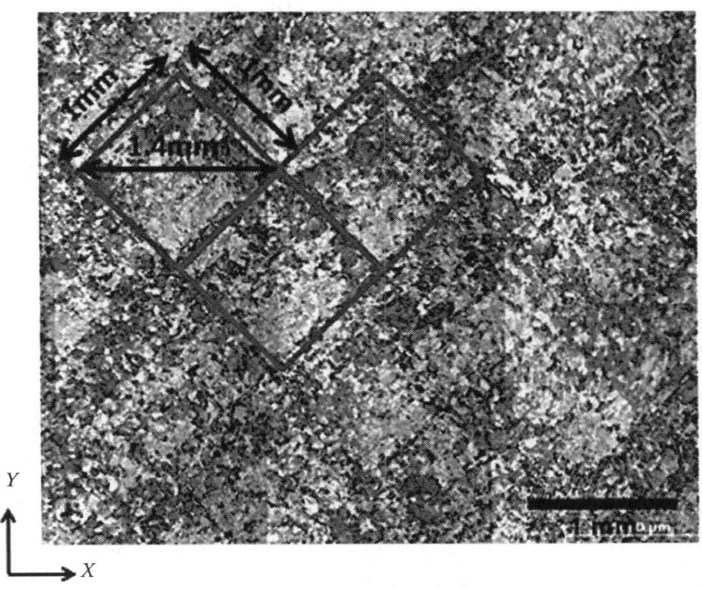

Figure 3.7 Micrograph showing the 1 mm repeating pattern in the *X*–*Y* plane. (From Carter, L.N. et al., *J. Alloys Compd.*, 615, 338–347, 2014.)

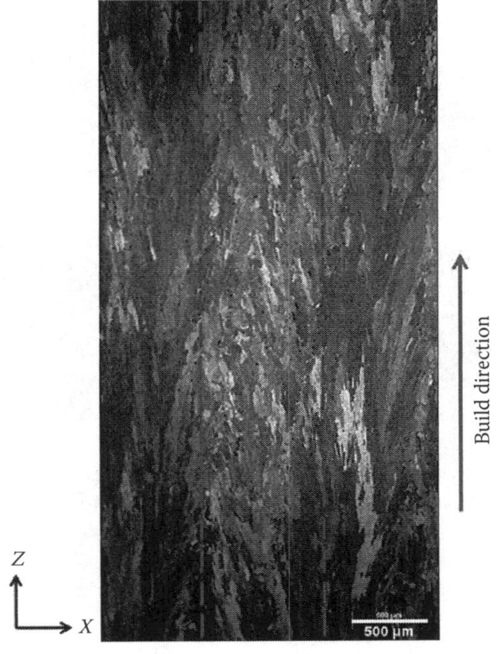

Figure 3.8 Optical micrograph in the *X*–*Z* plane showing the long columnar grains with the fine-grained region down the middle of the image in the build direction. (From Carter, L.N. et al., *J. Alloys Compd.*, 615, 338–347, 2014.)

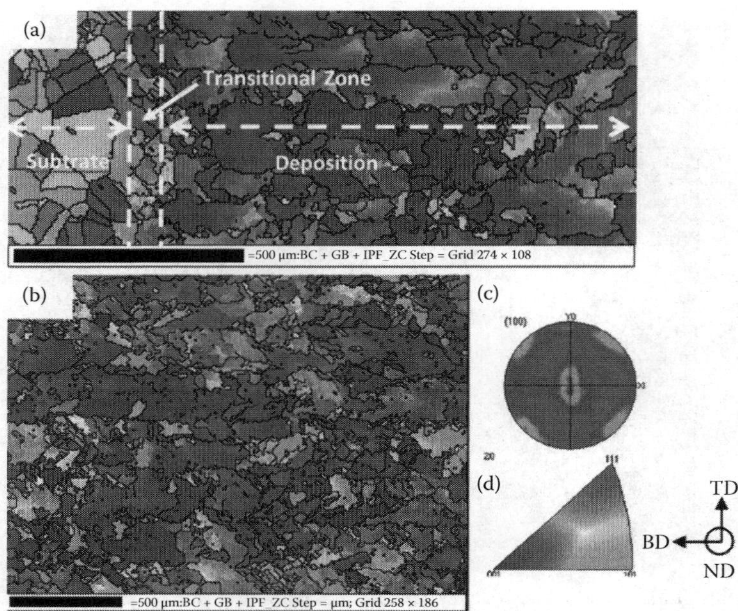

Figure 3.9 Microstructure morphology of the SLM-formed Inconel 625 superalloy: (a) inverse pole figure (IPF) colored map of the *Y–Z* section, (b) inverse pole figure (IPF) colored map of the X–Y section, (c) (100) pole figure of the IPF image of the SLM-formed Inconel 625 part, and (d) the index map of the IPF and the reference coordinate. (From Li, S. et al., *J. Mater. Sci. Technol.*, 31, 946–952, 2015.)

This link between the characteristic columnar structures of SLM-fabricated material and highly textured material was further illustrated in the work of Carter et al. [12]. As discussed previously, the SLM-processed CM247LC produced using the island scan strategy showed a bimodal structure with large columnar regions divided into a repeating square pattern bordered by fine-grained regions. The EBSD map, as shown in Figure 3.10, illustrates how the long columnar

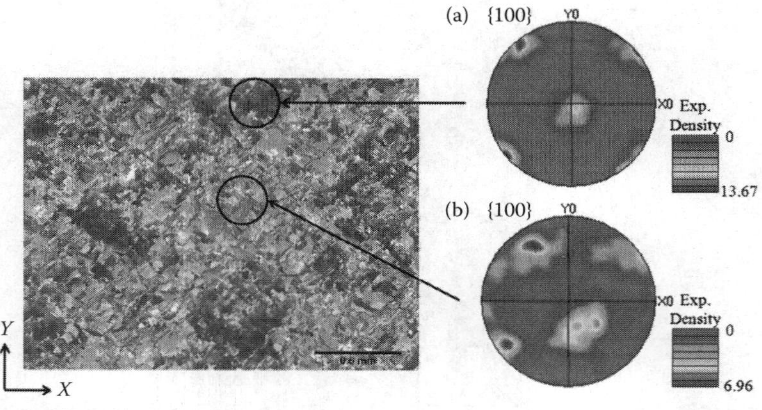

Figure 3.10 {100} pole figures corresponding to (a) area of strong texture and (b) less defined orientation with respect to the "*Z*" build direction. (From Carter, L.N. et al., *J. Alloys Compd.*, 615, 338–347, 2014.)

regions (viewed from the *XY* plane) show an affinity for the ⟨001⟩ direction, while the more chaotic fine-grained boundary regions have a much less distinct texture. This further illustrates the importance of the laser scan strategy in governing the microstructure of the processed Ni-superalloys, but also highlights the future potential to use the SLM process not only to construct complex three-dimensional components but also tailor the microstructure to meet the requirements of the final component.

In terms of the other secondary phases, carbides have a key role in controlling the microstructure and properties in Ni-base superalloys. Fine carbides form at grain boundaries, improving the resistance to grain boundary sliding. They can, however, be brittle in their undesirable forms (typically coarse, grain boundary films, or "script-like"), resulting in material failure and brittle behavior.

One of the characteristics of the SLM process is the rapid solidification of the material, which does not leave sufficient time for the precipitation of carbides, or indeed the diffusion of the carbide-forming elements through the material. The subgranular, fine dendritic structure, discussed previously, shows bright phases both at the interdendritic and the intergranular regions. This can be seen in Figures 3.3 and 3.4, and also in the study by Carter et al. [14], in relation to crack formation in CM47LC. Although not thoroughly investigated, these regions do appear to show the segregation of the carbide-forming elements, showing as a bright interdendritic phase as seen in Figure 3.11.

The carbides can precipitate out at the grain boundaries in discreet particles following any postfabrication heat treatments. This was shown by Li et al. [9] in heat-treated SLMed IN625 (Figure 3.12) and also by Carter et al. [14] in CM247LC following Hot Isostatic Pressing (HIPping). This ability to grow the carbides following postfabrication thermal treatments allows the "customization" of the microstructure in order to optimize the mechanical properties.

In addition, the coherent γ' and γ'' phases are critical for creep resistance of Ni-base superalloys. Their precipitation as discrete fine precipitates induced by heat treatment inhibits the movement of dislocations through the material, which significantly improves creep performance and high-temperature mechanical performance in general. Given the important role that these phases play in the final material, they are rarely discussed in research regarding the SLM fabrication of

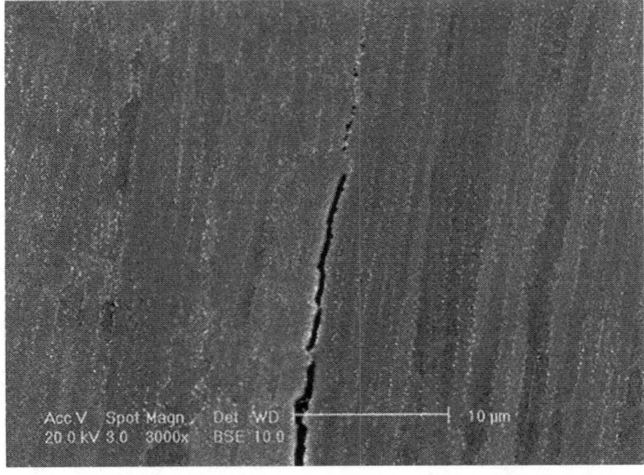

Figure 3.11 BSE SEM micrograph of SLM-fabricated CM247LC showing bright interdendritic and intergranular phases, indicating the presence of the heavier carbide-forming elements.

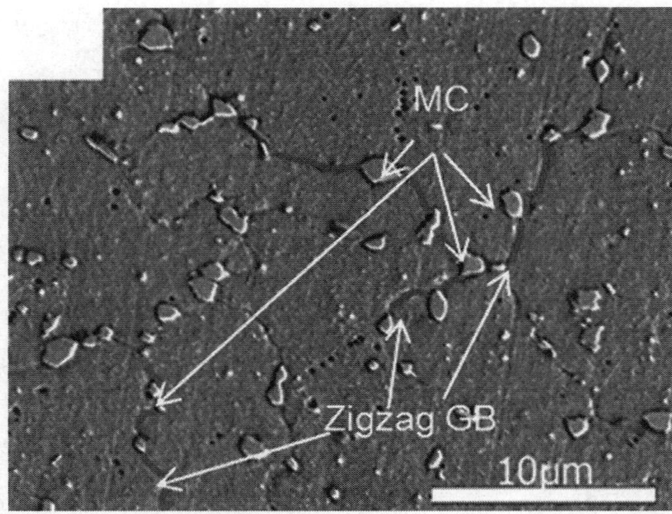

Figure 3.12 Micrograph showing GB MC carbides in SLM-fabricated IN625 following heat treatment. (From Li, S. et al., *J. Mater. Sci. Technol.*, 31, 946–952, 2015.)

these alloys. The reason for this is illustrated clearly in Figure 3.3, which shows a micrograph for as-SLMed IN738LC, where no typically cuboidal γ' phase is present as the γ' forming elements have been held in solution due to the very rapid solidification associated with SLM [10]. Typically, material samples are subject to a heat treatment prior to mechanical testing (usually the standard heat treatment used for the cast material), which forms the γ'/γ'' particles in the desired way.

Clearly, the characterization and understanding of the as-fabricated microstructure of SLM-processed Ni-superalloys is a field currently in development. What can be drawn from this study is that the microstructure requires studying on many different scales. Figure 3.13 shows a summary of the key elements involved in this microstructure.

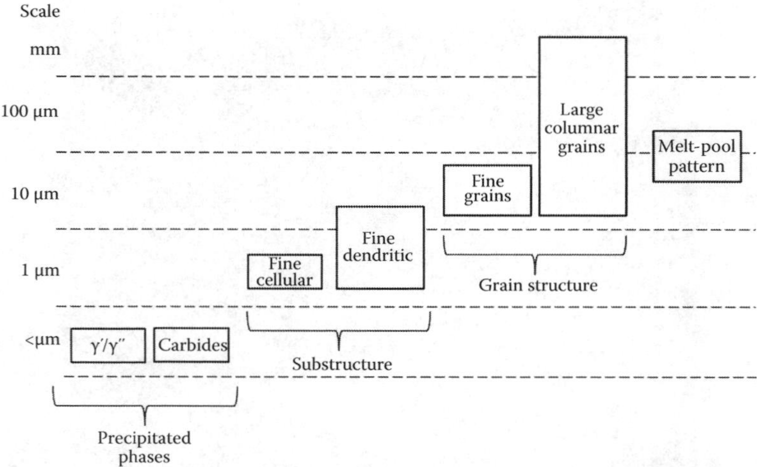

Figure 3.13 Diagram illustrating the essential elements of an SLM-fabricated microstructure of a Ni-base superalloy.

3.2.2.2 DLD Microstructure

The laser blown powder processes (also referred to as DLD, DLF, Laser Engineered Net Shaping [LENS]) share many characteristics with the SLM process. The key differences conceptually are DLD is intended for bulk addition of material and DLD structures are generally not intended to be truly "netshape" (i.e., a machining process is required to finish the part). From a microstructural perspective, the key features in contrast to SLM are the higher laser power (>400 W), large laser spot size, larger melt-pool (>1 mm ∅), relatively slower laser travel speeds (<1000 mm/s), simple back-and-forth laser paths, and the generally large overall component size, with the exception to its use in repair.

The larger melt-pool and the greater overall power input results in slower material solidification and cooling rates. As such, the grain structure of DLDed Ni-base superalloys is typically much larger than that of comparable SLM-processed material. As with SLM-processed material, the substrate and the previously deposited layers act as a heat sink governing the axis to heat flow, and due to the partial remelting and epitaxial solidification of previously solidified grains, the structure is often highly columnar and directional. The research presented by Liu et al. [11] shows this highly directional grains structure in IN718 (Figure 3.14); note the grain size of approximately 100–200 μm in width and several millimeters in length. As much of the published research (and potential application) relating to this process involves what can be simplified to the construction of basic walls and block structures, this columnar structure is much more widely reported and consistent when compared with the highly variable microstructures (due to wide variations in geometry type) seen in SLM.

As with SLM the microstructure, various aspects of the DLD microstructure are governed by the laser scan path. However, due to the relatively simple back-and-forth paths typically employed, these microstructure variations are much simpler to identify and characterize. The work presented by Dinda et al. [15] on IN625 (Figure 3.15) shows how the grain growth direction varies with the laser movement direction. The large columnar grains tilted in the direction of laser travel for each of the deposited layers due to the influence of the laser movement on the thermal gradient and dominant heat flow direction. Furthermore, the study by Liu et al. [11]

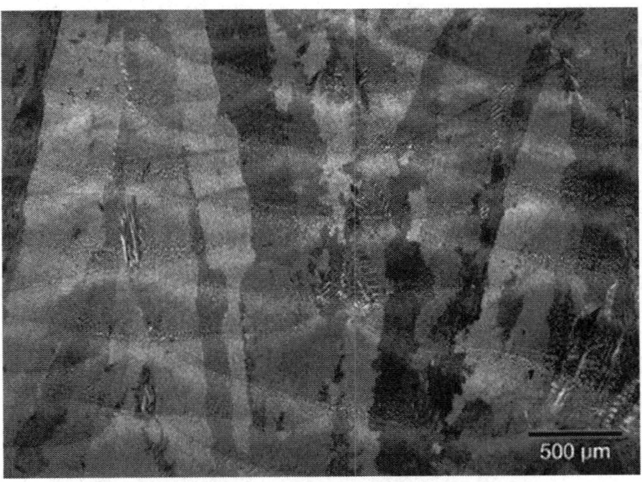

Figure 3.14 Micrograph showing DLD-fabricated IN718 using a simple scan pattern. (From Liu et al., *J. Alloys Compd.*, 509, 4505–4509, 2011.)

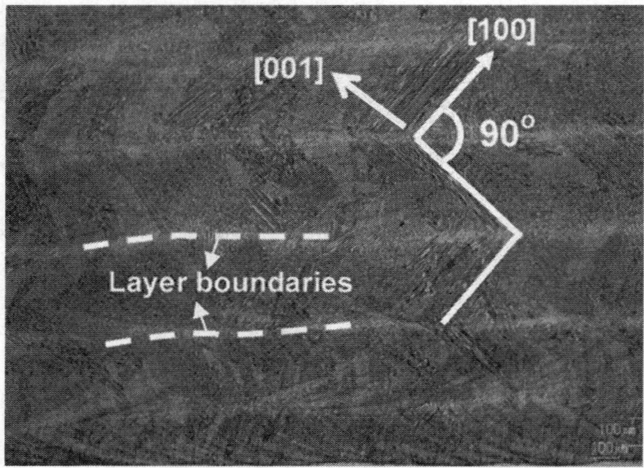

Figure 3.15 Micrograph showing DLD-deposited IN625 using a back-and-forth scan pattern. Note how the growth direction of the elongated grains is influenced by the laser movement direction. (From Dinda, G.P. et al., *Mater. Sci. Eng. A*, 509, 98–104, 2009.)

shows that the columnar structure is disrupted (but still present) when alternating scan directions with each layer are used.

Figure 3.16 shows a different effect caused by the laser path [16]; the section shows DLD-fabricated IN718 produced using a simple back-and-forth laser scan path sectioned perpendicular to the scan path direction. The material shows long columnar gains, where the center of the tracks align and a fine-grained region at the edge of the tracks, where the solidifying melt-pool is cooled rapidly due to the surrounding cooler material.

Still, aspects of the grain structure can be governed by the process parameters. The research presented by Parimi et al. [17] (IN718) in Figure 3.17 shows three EBSD maps using (1) 390 W with cooling between each layer (unidirectional deposition), (2) 390 W, but no interlayer cooling (bidirectional deposition), and (3) 910 W with no interlayer cooling. Note how the increase in laser power (and reduction in interlayer) cooling results in larger grains, more columnar in nature, with strong epitaxial growth along the build direction.

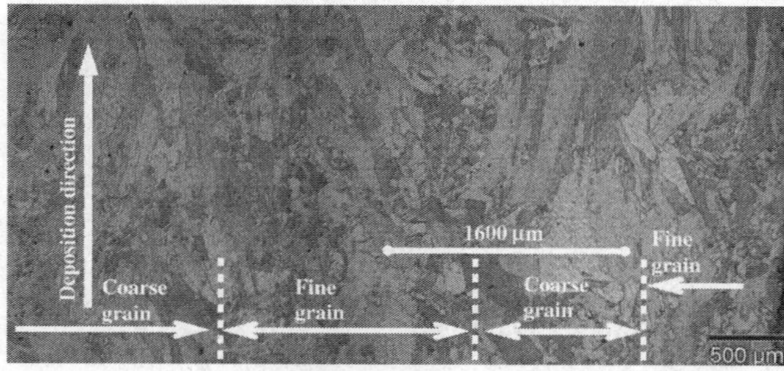

Figure 3.16 Micrograph showing regions of columnar and fine grains in DLF-fabricated IN718. (From Liu, F. et al., *Optics Laser Technol.*, 43, 208–213, 2011.)

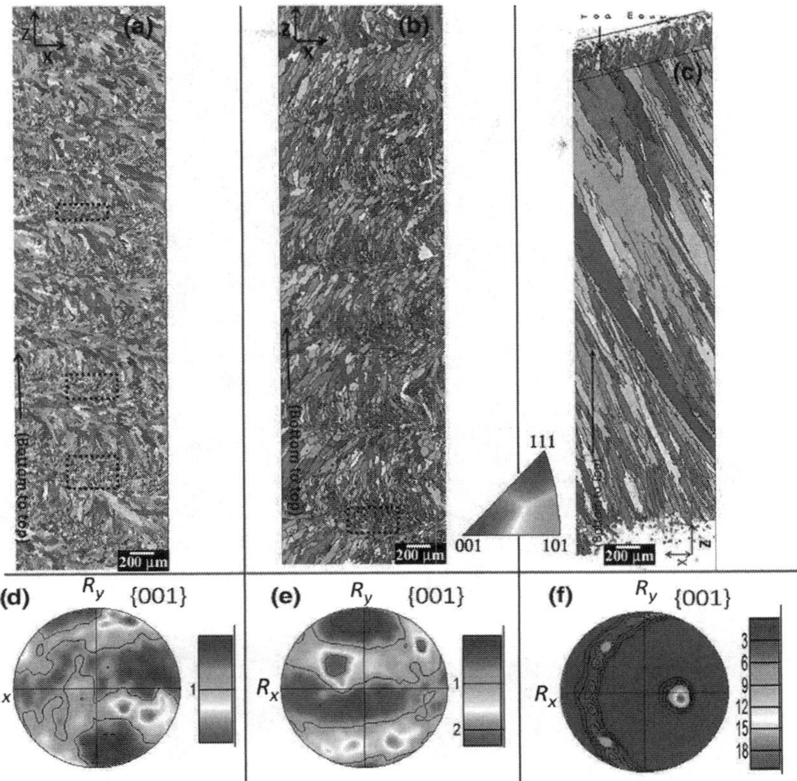

Figure 3.17 EBSD maps of DLD-fabricated IN718 showing (a) interlayer cooling 390 W, (b) no interlayer cooling 390 W, and (c) no interlayer cooling 910 W. (From Parimi, L.L. et al., *Mater. Charact.*, 89, 102–111, 2014.)

As with SLM-fabricated Ni-base superalloys discussed previously, there exists a solidification structure within the typical elongated grains on DLD-fabricated material. The research by Chen and Xue [18], Zhao et al. [19], and Amato et al. [5] all note this solidification structure. In contrast to the rather chaotic dendritic/cellular substructure noted in SLM-fabricated material, the substructure within DLDed Ni-base superalloys is typically composed of regular elongated dendrites. This differentiation is clearly illustrated in Figure 3.18, which shows long columnar grains in DLDed IN738 crossing many build layers (marked by the weld pattern of light and dark bands) [18]. Within these grains, the elongated dendritic structure is visible.

As with the SLM-fabricated material, the occurrence of columnar grains also promotes the strong {001} texture. This was seen in the pole figure of the highly columnar sample (c) presented by Parimi et al. [17] (Figure 3.17). In addition, it is generally observed that the γ'/γ'' phases are not easily identifiable in the as-fabricated structure due to the rapid cooling nature of the process. Chen and Xue [18] suggest that DLDed IN738 is primarily a supersaturated solution of the γ' forming elements. The research presented by Yuan et al. [20] goes further and uses transmission electron microscopy (TEM) techniques to identify ultra fine γ'' in DLD-fabricated IN718. Figure 3.19 shows the γ'' identified in the interdendritic regions by diffraction pattern and the corresponding dark-field TEM micrograph is shown in Figure 3.20 with the oval γ'' particles visible in the submicron scale.

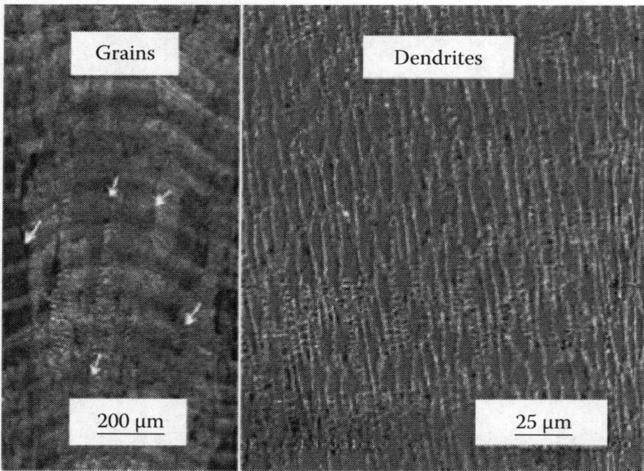

Figure 3.18 Micrograph showing columnar grains with "fish-scale" weld pattern (left) and the fine dendritic structure within the grains under higher magnification (right) within IN738 processed by DLD. (From Chen, J., Xue, L., *Mater. Sci. Eng. A*, 527, 7318–7328, 2010.)

Figure 3.19 TEM diffraction pattern identifying the γ″ phase in DLDed IN718. (From Yuan, D. et al., *Metall. Mater. Trans. A*, 45, 4470–4483, 2014.)

γ″ alloys are prone to the formation of the Nb-rich Laves and needle-like δ phase when incorrectly processed. Excessive amounts of this phase will impair room-temperature tensile ductility and creep properties, as they can act as weak points within the material, but also tie-up Nb, which is critical in forming the strengthening γ″ phase. Parimi et al. [17] identified both these phases in the as-fabricated structure of DLDed IN718 (Figure 3.21), as did the research by Qi et al. [21]. Odabasl et al. [22] reported that coarser laves precipitates are typically associated with slower cooling rates and higher energy inputs during laser processing. Although detrimental, laves can be essentially eliminated in the final material with the use of a postfabrication heat treatment. However, this does add another stage in the processing route of the material. Figure 3.22 summarizes the typical microstructures of as-deposited DLD IN718.

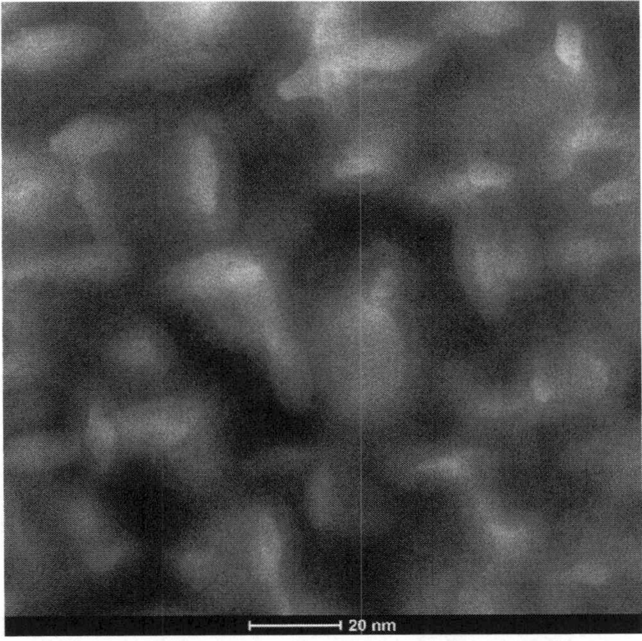

Figure 3.20 Darkfield TEM micrograph showing the ultrafine oval γ″ precipitate in DLDed IN718. (From Yuan, D. et al., *Metall. Mater. Trans. A*, 45, 4470–4483, 2014.)

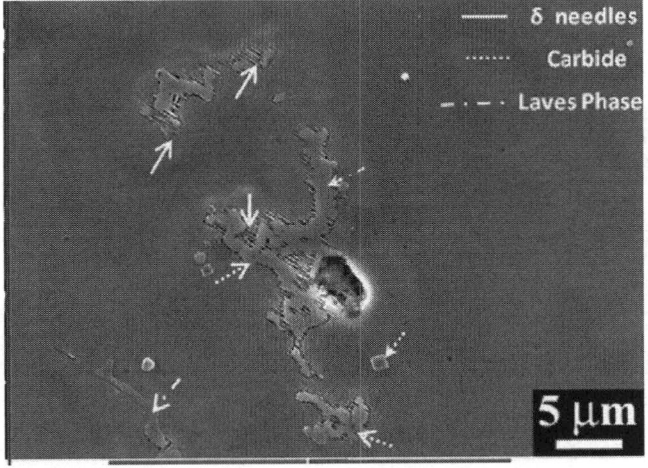

Figure 3.21 Micrograph showing the detrimental laves and δ phases in DLDed IN718. (From Parimi, L.L. et al., *Mater. Charact.*, 89, 102–111, 2014.)

3.2.2.3 EBSM Microstructure

The majority of literature relating to the EBSM relates to the processing of titanium alloys; however, some interesting research has been carried out, specifically by Murr et al. [23] relating to the EBSM of René 142. The microstructure is very similar to that of other AM techniques

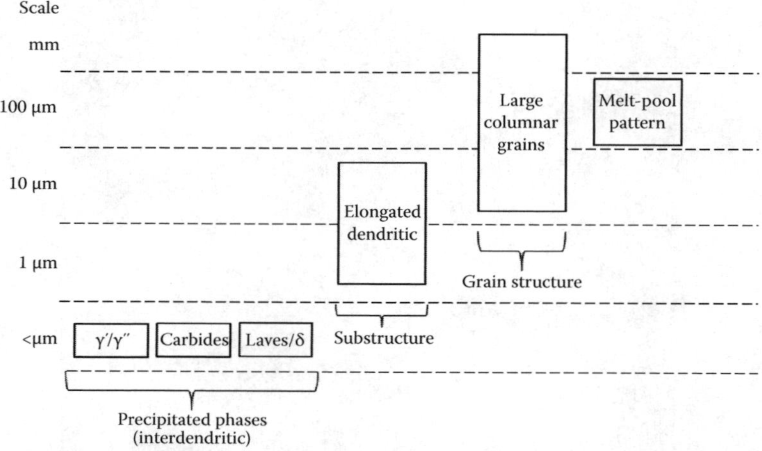

Figure 3.22 Diagram showing typical microstructural features and associated scale of DLDed Ni-base superalloys.

already discussed. Figure 3.23 shows the characteristic columnar grain structure. In contrast to the other ALM techniques discussed, the entire build chamber is kept at an elevated temperature during the EBSM process. This has the effect of *in situ* aging the microstructure, and can mitigate the formation of residual stresses during ALM. Figure 3.24 shows the cuboidal γ' structure in the as-fabricated state, showing a stark contrast to the SLM and DLD microstructures, which display a rapidly cooled microstructure with superfine γ'/γ'' precipitates.

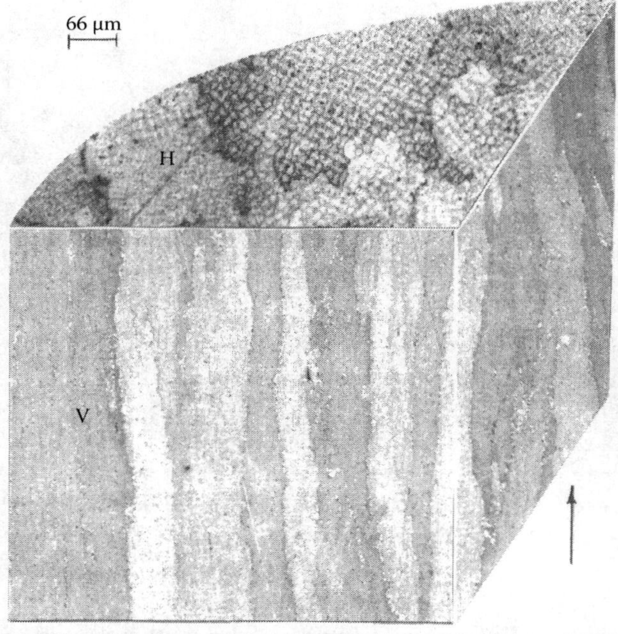

Figure 3.23 Composite 3D optical micrograph of René 142 formed by EBSM-showing the large columnar grain structure. (From Murr, L.E. et al., *Acta Mater.*, 61 (2013):4289–4296, 2013.)

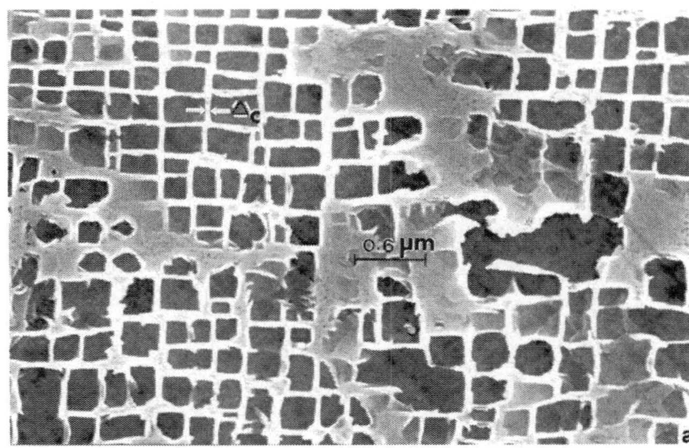

Figure 3.24 Micrograph showing the cuboidal γ′ structure in EBSM-fabricated René 142. (From Murr, L.E. et al., *Acta Mater.*, 61 (2013):4289–4296, 2013.)

3.2.3 Mechanical Properties

The mechanical properties of ALM processed nickel-based superalloys are currently difficult to objectively assess. The variations in microstructure due to the processing technique, process parameters, and part geometry all contribute to an uncertainty regarding the final mechanical behavior of the material. Nickel alloys are also typically subject to standard heat treatments following forming (e.g., casting), currently most studies in the ALM of these materials use postfabrication heat treatments based on those recipes used for the cast alloys, which may not be appropriate for the initial microstructure. To add to the difficulties, there is currently no standard test procedure for the formation of the mechanical test samples; for example, size of initial piece, machined or as-fabricated surface, postfabrication treatments, build orientation, etc.

Because of the previously discussed directionality within the microstructure, samples for mechanical testing are fabricated in both parallel and perpendicular (often referred to as "vertical" and "horizontal") orientations with respect to the build axis (*Z*-direction). Often, the mechanical properties, similar to the microstructure, show an anisotropy with respect to the build axis.

Several studies investigating the ALM of IN718 (both via SLM and DLD) provide room temperature tensile data and act as a good case study for ALM-fabricated Ni-superalloys. These results have been collated in Figures 3.25 and 3.26. Within this collection of data, it is impractical to do precise comparisons as the heat treatments and testing methods vary hugely between the studies; general observations, however, can be made. The tensile properties generally compare favorably against the nominal wrought values for IN718 either outperforming or being within 80% of those standard values. In the three studies compared, the SLM material performed the least favorably in terms of tensile strength; however, the HIPping and heat treatment applied produced excellent elongation above 25%. The as-DLDed material showed very low tensile strength, likely due to poor precipitation morphology. This was dramatically improved by aging treatment designed to refine the precipitation size and morphology. Furthermore, the study by Azer and Ritter [21] showed that solution treatments and homogenization treatments in addition to aging improve the ductility of the processed material.

One of the more complete studies relating to the mechanical properties of a nickel superalloy produced by SLM is presented by Rickenbacher et al. [10] relating to IN738LC. This is a

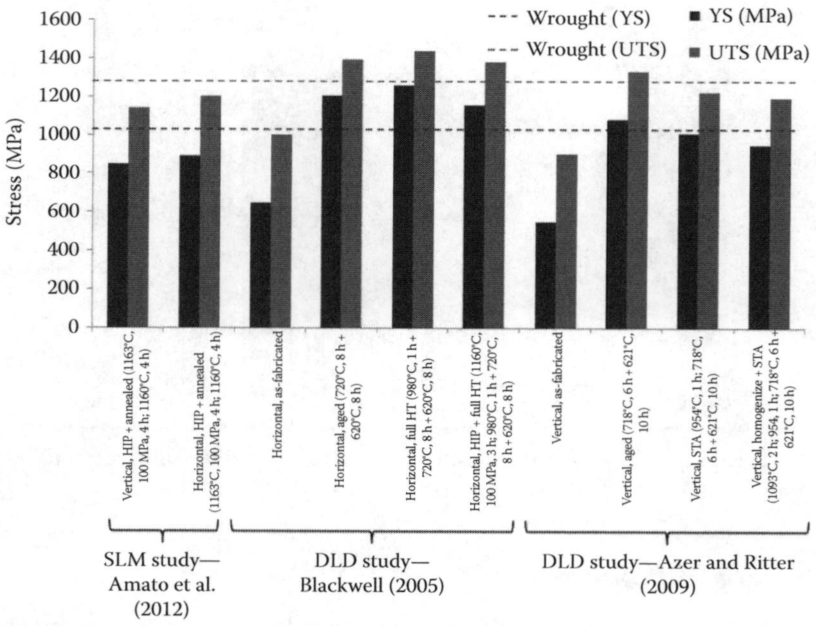

Figure 3.25 Collated tensile (Yield and Ultimate) data for IN718 formed via SLM and DLD from various studies [5,21,26]. Dashed lines indicate typical values for wrought IN718 (AMS5662).

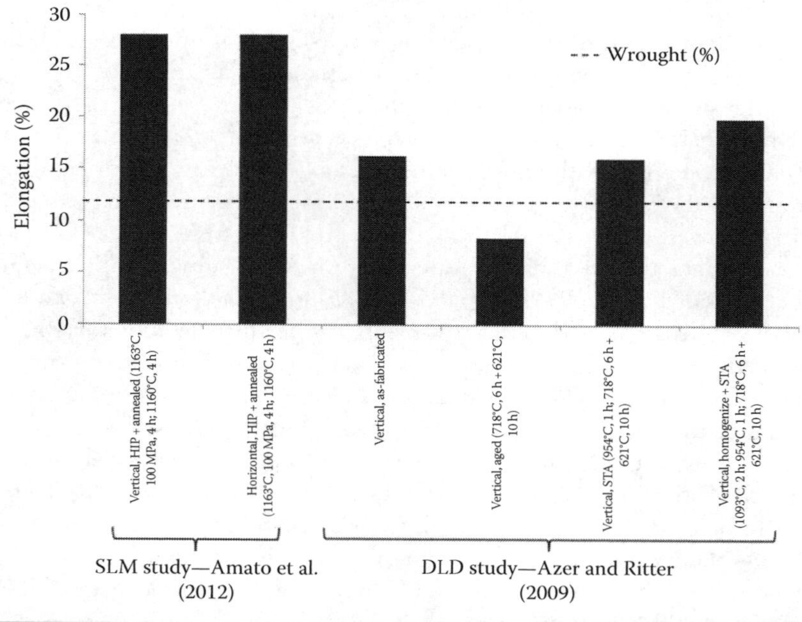

Figure 3.26 Collated tensile (% elongation) data for IN718 formed via SLM and DLD from various studies [5,21]. Dashed line indicates typical values for wrought IN718 (AMS5662).

particularly interesting study as IN738LC is precipitating hardened by the γ′ (Ni_3Al) phase and prone to weld-induced cracking. The tensile samples were produced in both horizontal and vertical orientations, and tensile tested under both room temperature and elevated (850°C) conditions. The material was HIPped (*conditions not specified*) and fully heat treated (solution treatment 1220°C, 2 h; aging 850°C, 24 h) prior to testing. Figure 3.25 shows the collated tensile data from this study. Under both testing conditions, the horizontal samples marginally outperform their vertical counterparts in terms of yield and ultimate tensile strength (UTS); however, the vertically built specimens show a much greater ductility than those fabricated in the horizontal orientation due to the grain elongation in that direction. The yield strength (YS) and UTS of the SLM-fabricated material compare well with the cast material, although the elongation for the horizontal samples is significantly poorer. Rickenbacher et al. present a direct comparison of the SLM-processed material (in both orientations) against the cast material for elongation to 1% plastic strain (see Figure 3.27); again, the vertical specimens outperform the horizontally built ones meeting similar properties to the cast material.

IN625 is a commonly laser processed material as it avoids typical problems related to grain-boundary cracking being hardened primary from solid solution additions rather than by phase-precipitation. The study by Ganesh et al. [24] presents some room-temperature tensile properties (horizontal) of DLD-deposited IN625. The deposited material shows YS of 540 MPa and UTS of 690 MPa, which are comparable to the typical values presented for cast material, 350 and 710 MPa, respectively. The deposited material does show marginally poorer ductility (36%) when compared to the cast material (48%). The study conducted by Zhao et al. [25] regarding René88DT also showed favorable tensile properties when compared to typical cast material when a pre heat treatment HIPping treatment was performed on the material in order to close and residual porosity and defects.

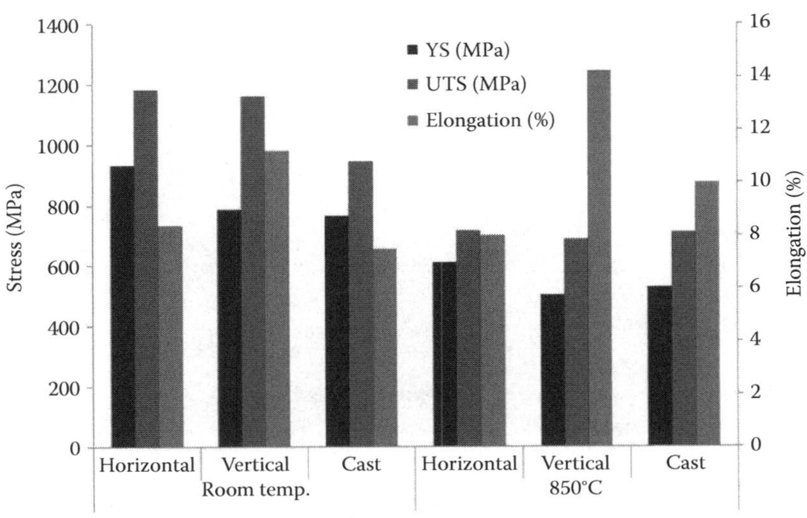

Figure 3.27 Tensile properties of IN738LC formed by SLM (HIPped and Heat Treated) presented by Rickenbacher et al. [10].

3.2.4 Conclusions on ALM of Ni-Superalloys

Although these studies cover only a small handful of alloys and do not fully explore the mechanical properties of ALM-formed nickel-base superalloys, they do present some interesting general conclusions. The ALM-formed material (via either DLD or SLM) can be processed to produce tensile properties comparable to traditionally formed material when the correct heat treatments and HIPping (where necessary) are performed. It is also obvious that, while the applications of traditional heat treatment "recipes" are an obvious starting point, these must be adapted for the condition of the as-fabricated material and the requirements of the final application. Finally, it is apparent that the build orientation does influence the mechanical properties of the material, specifically with respect to ductility.

3.3 Al-Alloys ALM

Learning Objectives

- Understand the factors affecting the ALM processability of Al-alloys.
- Describe the microstructural development due to SLM.
- Explain the mechanisms for defect formation in Al-alloys SLM.
- Discuss the interaction between the process parameters, microstructure, and mechanical properties development.

Most studies of Al-alloys AM focused on the use of SLM, with very limited and recent work on WAAM [27]. The studies have typically investigated SLM of cast alloys (e.g., Al–Si and Al–Si–Mg). Few studies investigated other alloy classes (e.g., Al–Cu and Al–Zn), or Al-based Metal Matrix Composites (MMCs), such as AlSi10Mg–Fe_2O_3 composites [28–30].

3.3.1 Metallurgy of Al-Alloys

Al-alloys are used in the transportation sectors (aerospace, automotive, marine, etc.) due to the abundance of its ores, to the high strength-to-weight ratio of its alloys and good corrosion resistance. However, Al-alloys processing using ALM are known to be difficult due to the following factors:

- *Susceptibility to hydrogen-induced porosity formation*: Al is known to have a high diffusivity for hydrogen in the molten state [31,32], which results in the formation of hydrogen-induced porosity during solidification [33].
- *Oxidation*: Al has a high susceptibility for oxidation. Al-oxide has a much higher melting point than pure Al.
- *Thermal conductivity and reflectivity*: Al has a high thermal conductivity, which makes it rather difficult to retain the heat in the melt-pool. Al-alloys have high reflectivity so they absorb a low-energy fraction from the incident laser energy [34].
- *Cracking in some specific alloy*: Some Al-alloys are highly susceptible for hot cracking. The resistance to hot cracking is high at both high and low percentage of alloying elements [35].

Depending on the alloy chemistry, Al-alloys can contain different secondary phases, which can be classified into the following types [36,37]:

1. *Precipitation/hardening phases*: Fine precipitates which form following a solution treatment and aging process. Typical phases include Mg_2Si, Al_2Cu, $MgZn_2$, Al_2CuMg, and others. They form as coherent or semi-coherent structures, depending on the aging temperature and time.
2. *Constituent particles (primary precipitates)*: 1–10 μm sized inclusions that form during solidification from the trace impurity elements (e.g., Fe, Mn, Si, etc.). They are generally insoluble, as their melting temperatures are higher than the α-Al.
3. *Divorced eutectic phases*: Phases that form within the interdendritic regions during solidification, such as Si, Al_3Mg_2, and Mg_2Si. They can be dissolved by homogenization at 500°C–600°C, forcing the alloying elements back into solution.
4. *Dispersoids (secondary precipitates)*: Submicron incoherent precipitates that form during post solidification thermal treatments. They contribute to grain size control due to their size, and possibly to strength via Orowan strengthening.

3.3.2 ALM Microstructural Development

The microstructure and properties of SLMed Al-alloys components strongly depend on the alloy type and on the SLM processing parameters. Unlike other materials, the influence of the processing conditions and their influence on the mechanisms of consolidation are not fully understood. The following section summarizes the current state of knowledge on the influence of process and material parameters on the microstructure in Al-alloys processed using SLM.

3.3.2.1 Microstructure of SLM Al-Alloys

During SLM, melting occurs in the layer exposed to the laser beam, followed by shrinkage and solidification. To improve the bonding between the layers, remelting typically occurs across a number of layers in all orientations (Figure 3.28) [38]. The melt-pool undergoes solidification by nucleation and epitaxial growth from the previously solidified grains in the remelted surface of the previous layer, resulting in melt-pool depth (observed in the top layer) that is ~5–10 times the powder layer thickness (Figure 3.28c). On the *X–Y* plane (top surface), the melt-pool width can also be observed, which is also wider than the incident laser beam spot size (typically ~50 μm). In most cases, the thermal footprint of the laser is much larger than the optical footprint. In addition, similar to SLMed Ni-superalloy, a fish-scale pattern can also be observed along the build direction.

The rapid and directional solidification during SLM leads to the formation of a cellular dendritic microstructure, with cellular dendrites of size <1 μm (Figure 3.29a). The cell size is not uniform throughout the whole build, not even within the melt-pool in itself, especially at the boundary of the melt-pool where the cells are slightly coarser (Figure 3.29b). The melt-pool boundary is also associated with chemical segregation that makes the border of the melt-pool distinguishable. These cells are not individual grains, as earlier SLM studies referred to them as. Instead, they form a solidification substructure inside the grains, which typically results in improved mechanical properties in the absence of structural defects (pores or cracks). The SLM grain structure can be defined by a group of adjacent cells having the same crystallographic growth direction, which can be visualized using EBSD (Figure 3.30) [39].

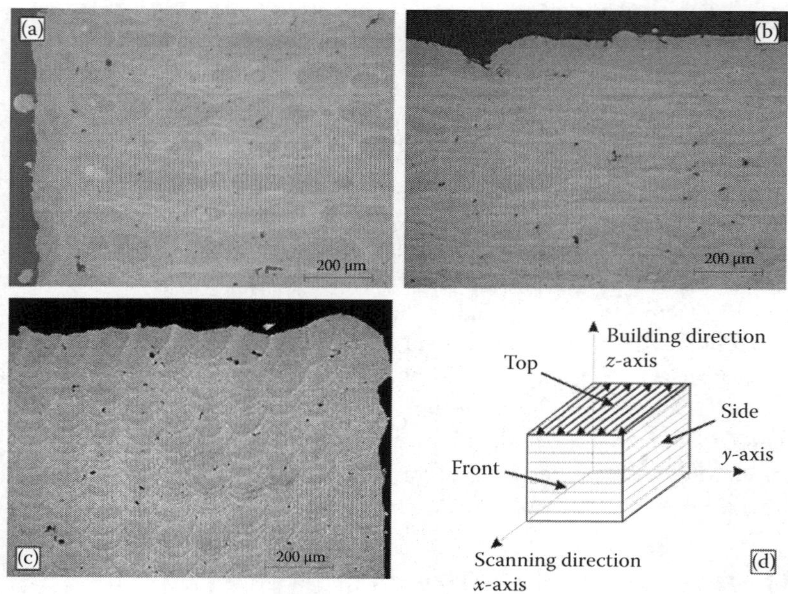

Figure 3.28 The melt-pool structure in SLMed AlSi10Mg alloy showing (a) the side plane, (b) front plane along the laser scanning direction, and (c) cross section (*y*–*z*) plane, according to (d). (From Thijs, L. et al., *Acta Mater.*, 61, 1809–1819, 2013.)

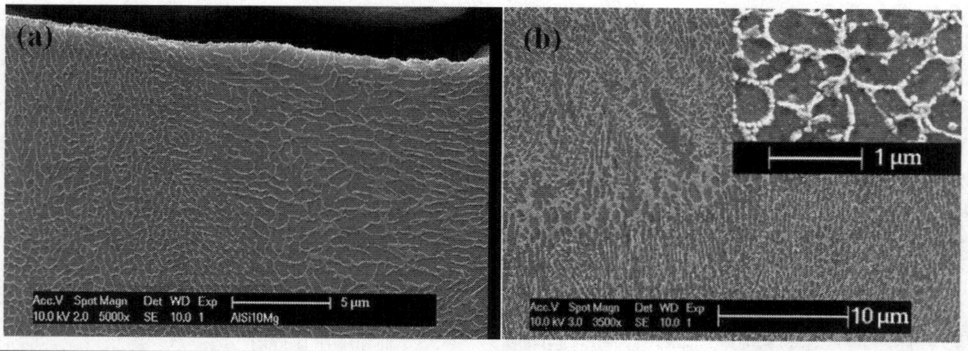

Figure 3.29 SEM micrographs of SLMed AlSi10Mg showing (a) a fine dendritic structure [40] and (b) the cellular solidification microstructures of AlSi10Mg. The red arrow indicates the melt-pool boundary [41].

Several studies investigated the grain and precipitate structures' development in SLM of Al–Si and Al–Si–Mg alloys, where it was reported that Si particles segregate to the Al-matrix at the grain boundaries due to Si solute rejection during solidification, suggesting grain sizes range of ~0.5–2 μm (Figure 3.31) [29,42–45]. However, Thijs et al. [39] observed using EBSD coarser and a wider grain size ranging from 3 to 12 μm depending on the laser scan strategies in AlSi10Mg SLM, suggesting that the previously reported grain size measurements were for the dendritic cells

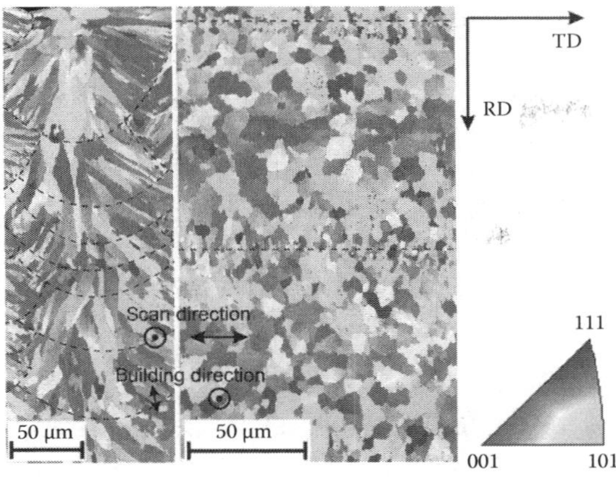

Figure 3.30 EBSD maps for SLMed AlSi10Mg: along the build direction (left) and top view of the build (right). (From Thijs, L. et al., *Acta Mater.*, 61, 1809–1819, 2013.)

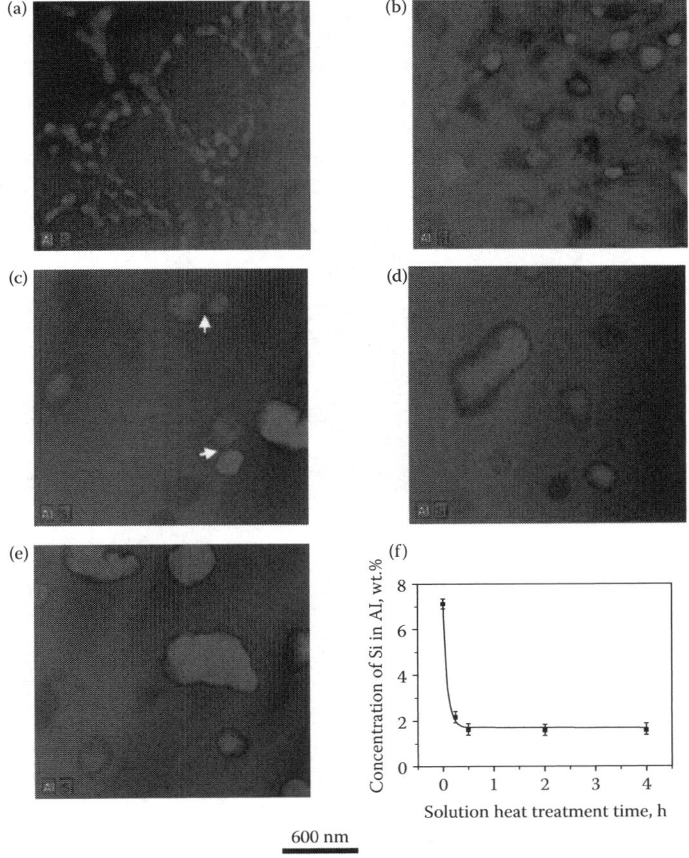

Figure 3.31 STEM-EDX images of the Al and Si distribution in Al–12Si alloy: (a) as-built and after solution treatment for (b) 15 min, (c) 30 min, d) 2 h, and (e) 4 h, and (f) concentration of Si in Al as a function of the solution treatment time. (From Li, X.P. et al., *Acta Mater.*, 95, 74–82, 2015.)

rather than the grain size. Furthermore, several reports showed that the grain growth direction is perpendicular to the melt-pool boundary, heading toward the center of the melt-pool. The grain morphology appears equiaxed at the top plane of the build. The grains are generally oriented along the preferred growth direction for cubic material. On the top planes, coarse grains oriented along [001] were observed along the center of the laser track, with finer grains of random orientation along the boundaries of neighboring tracks (Figure 3.30).

Prashanth et al. [46] also investigated the texture development in SLMed Al–12Si. Samples built using different angles of inclination were studied using x-ray powder diffraction (XRD), quantifying the macrotexture strength using the peak intensity of the (111) and (200) Al-planes (Figure 3.32) and correlating it to the tensile YS. The results suggest that although the texture intensity does vary as a function of the inclination angle, the influence of the texture intensity (and hence the inclination angle) on the strength is negligible in the 30°–90° range.

Due to their importance in understanding the laser–material interaction, melt-pool observation studies were studied by White et al. [38] along the build direction in SLMed AlSi10Mg. Figure 3.33 shows the melt-pool tracks for two different process parameters combinations. The melt-pool geometry is half ellipsoidal-to-cylindrical. Using these images, the size of the melt-pool in relation to the process parameters was correlated. The melt-pool shape can be changed using the laser scanning strategy, which depends on the used SLM platform. Some SLM systems use

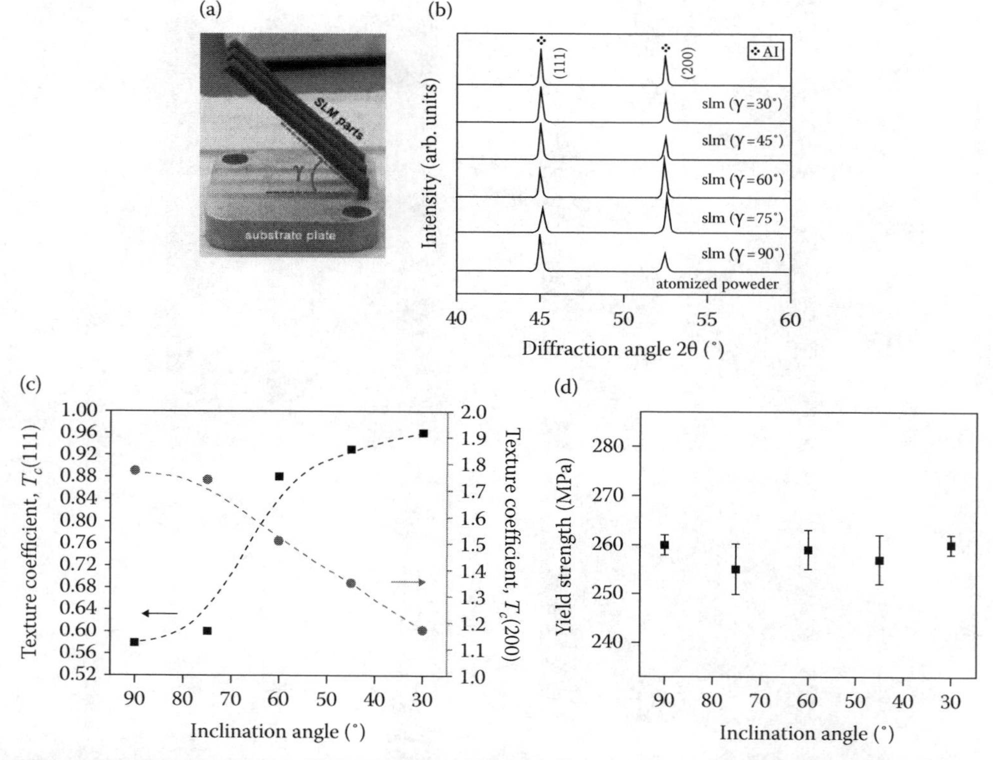

Figure 3.32 Influence of the build inclination angle on the texture in SLMed Al–12Si SLM showing (a) the sample orientation, (b) XRD patterns for the various inclination angles, (c) texture coefficient of Al (111) and (200) planes, and (d) the yield strength as a function of inclination angle. (Prashanth, K.G. et al., *Mater. Sci. Eng. A*, 590, 153–160, 2014.)

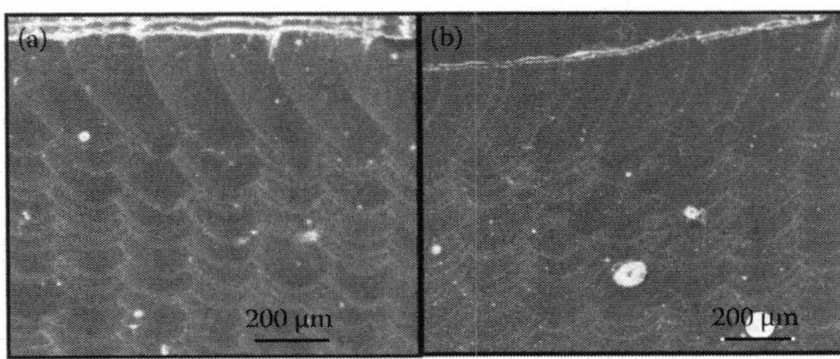

Figure 3.33 Optical micrograph for the melt-pool tracks in SLMed AlSi10Mg: (a) sample built using 300 W, 3500 mm/s, and 60 µm hatch spacing and (b) sample built using 150 W, 500 mm/s, and 60 µm hatch spacing. (From White, J.P. et al., *Prediction of Melt-pool Profiles for Selective Laser Melting of AlSi10Mg Alloy*, MS&T, Pittsburgh, PA, 2014.)

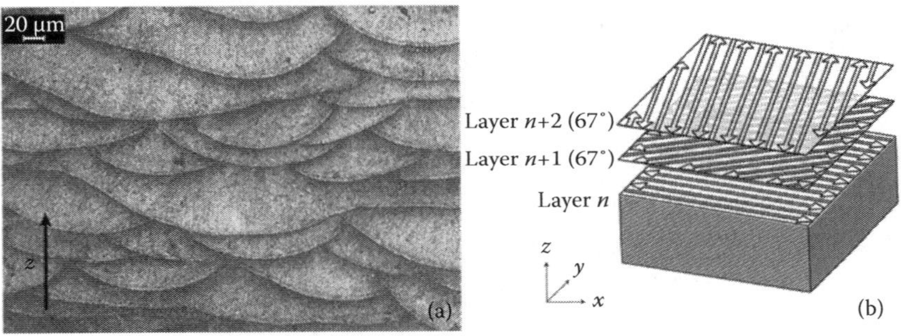

Figure 3.34 Optical micrograph for the laser scan track in SLMed AlSi10Mg with a rotating rater strategy [43], compared to the simple scan strategy in Figure 3.32.

the raster (simple) scan, whereas other systems use a "chessboard" or island scanning strategy, where each layer is divided in the small square section and each square sections are laser scanned individually, or use a gradually rotating scanning strategy (Figure 3.34), resulting in a changing melt-pool profile [43].

3.3.2.2 Influence of the Process Parameters on the Microstructural Development

Several studies investigated the influence of process parameters on the microstructural and porosity development in SLMed Al-alloys through the use of the laser energy density parameter [47–50]. Olakanmi et al. reported the influence of laser energy input on the sintered density in AlSi12 alloy (Figure 3.35). This shows the sample density increases up to a peak at 67 J/mm^3 then becomes constant density above 1.90 g/cm^3. It is important to indicate that this low density likely suggests the use of an inappropriate process window, as other studies have achieved densities approaching 99% in Al-alloys. Read et al. also studied the influence of the energy density on the porosity content in the SLMed AlSi10Mg alloy by varying the laser power, scan speed, and hatch spacing (Figure 3.36). It was found that the porosity content decreased with the increase in energy density, with a narrow window for full consolidation at ~60 J/mm^3. However, the porosity seemed to show an inconsistent trend above that window, which was suggested to be due to the initiation of

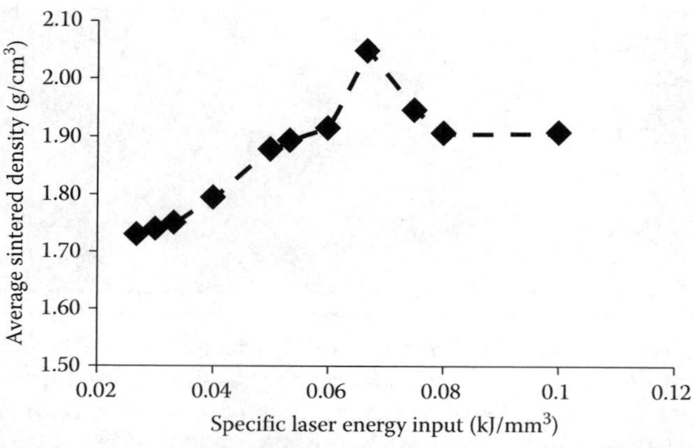

Figure 3.35 The relationship between the sintered density and energy input in SLMed Al-Si12 alloy. (From Olakanmi, E.O. et al., *J. Mater. Process. Technol.*, 211, 113–121, 2011.)

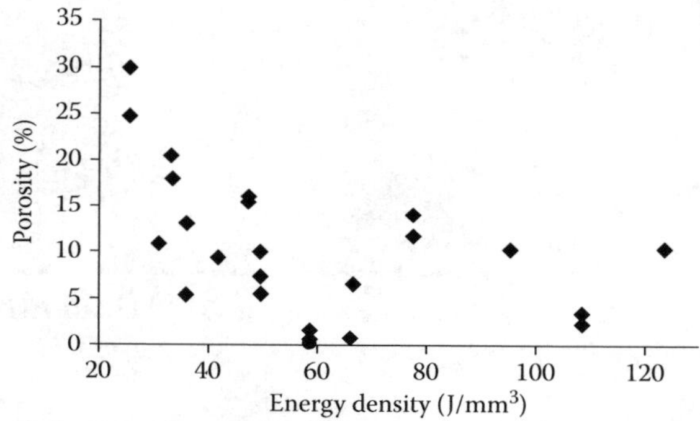

Figure 3.36 Influence of the energy density on the porosity content on the SLMed AlSi10Mg alloy. (From Read, N. et al., *Mater. Des.*, 65, 417–425, 2015.)

evaporation [51]. Similarly, Aboulkhair et al. reported the influence of the scanning speed, hatch spacing, and scan strategy on the density in AlSi10Mg builds, suggesting that increasing the hatch spacing resulted in an increase in the build porosity (Figure 3.37).

Some studies investigated the use of presintering laser remelting, through scanning each layer twice, to reduce the porosity in the builds. Nonetheless, the porosity appeared to increase with the increase in the scan speed. Obviously, remelting results in an increase in the process time, but this cannot be mitigated by increasing the scan speed as the porosity content ends up increasing again [53,54].

In non-AlSi-based alloys, Louvis et al. investigated the influence of the scan speed on the relative density for various hatch spacings in SLMed AA6061 alloy (which is known to be difficult-to-weld among Al-alloys) (Figure 3.38). Their results show that increasing the laser power resulted in an overall increase in the relative density. The highest densities have been achieved at slow scanning speeds (100–200 mm/s). It is important to add that the used laser power was generally lower than what would typically be employed for Al-alloys (100–200 W).

One of the tools to improve the microstructural homogeneity of the builds is to perform in-process or postprocess thermal treatment. Figure 3.39 shows the microstructure evolution in

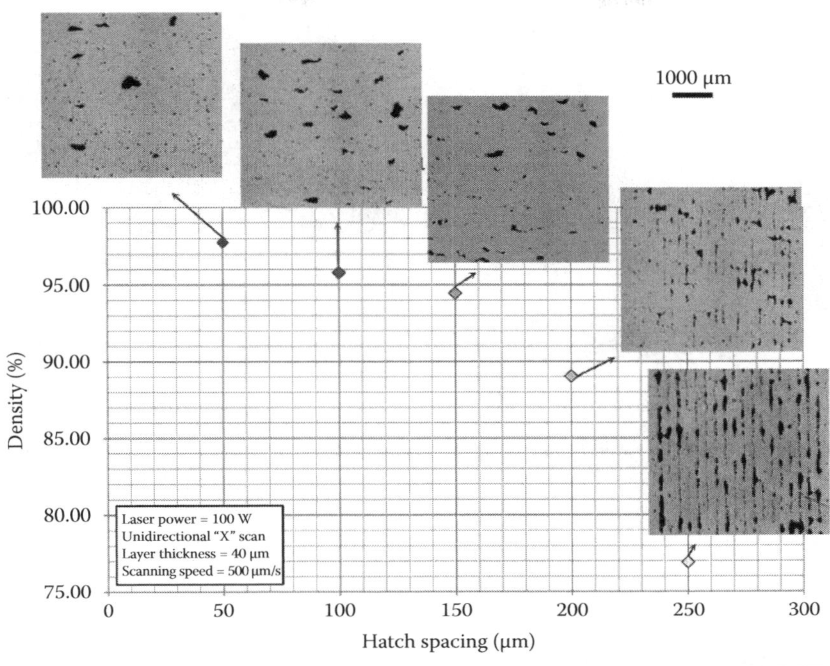

Figure 3.37 Effect of the hatch spacing on the density of SLMed AlSi10Mg alloy. (From Aboulkhair, N.T. et al., *Addit. Manufacturing*, 1, 77–86, 2014.)

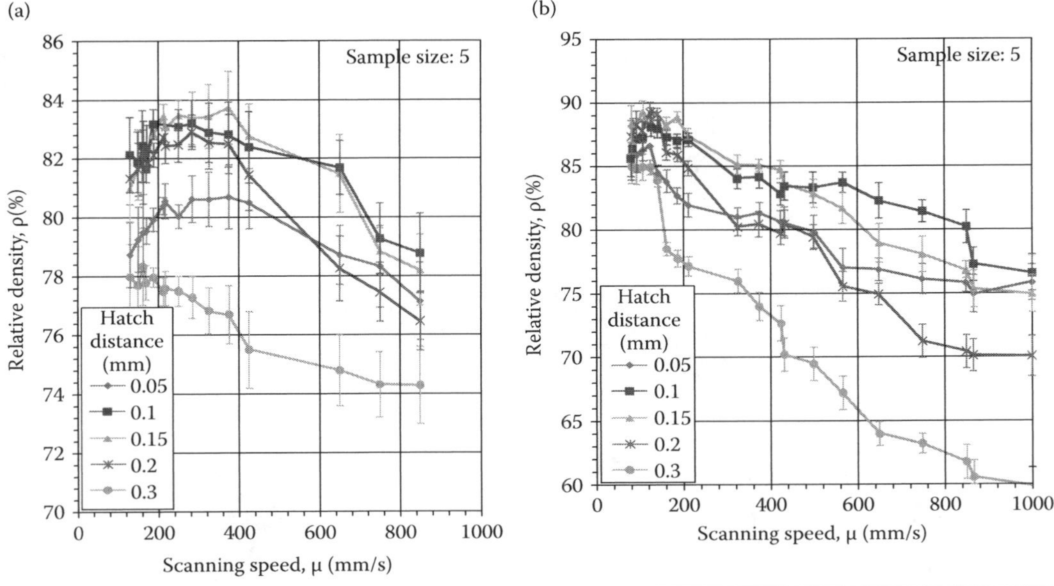

Figure 3.38 Influence of the scan speed on the density for various hatch distances in SLMed AA6061 alloy at (a) 50 W and (b) 100 W. (From Louvis, E. et al., *J. Mater. Process. Technol.*, 211, 275–284, 2011.)

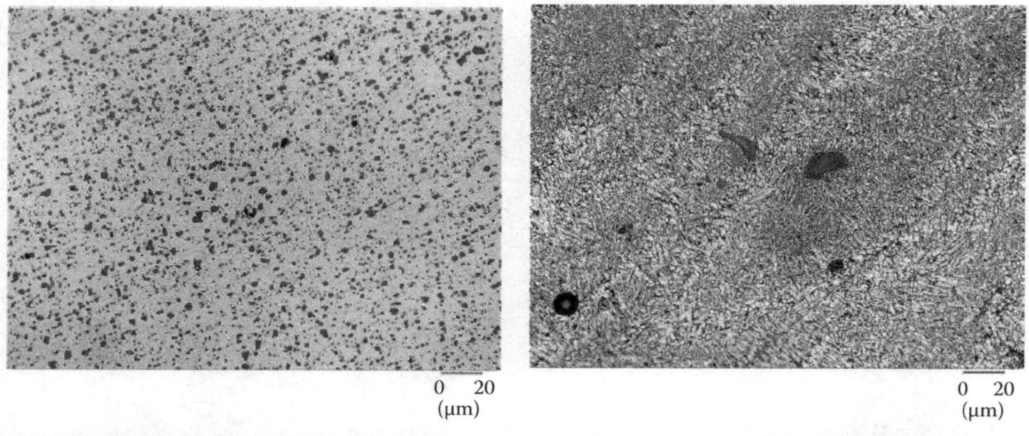

Figure 3.39 Microstructure change due to post-SLM heat treatments and *in situ* heating in SLMed AlSi10Mg alloy, showing the microstructure of the T6 heat treated (left) and the as-built on the heated bed (right). (From Brandl, E. et al., *Mater. Des.*, 34, 159–169, 2012.)

AlSi10Mg builds created using a heated bed (Figure 3.39a) and postprocess peak hardened samples (Figure 3.39b) [56]. It can be seen that the post-SLM heat treatment was more effective in generating homogeneously scattered coarse Si particles, compared to the obvious laser track lines and melt-pool boundaries observed in the in-process heated samples.

Energy-dispersive x-ray (EDX) mapping was used to assess the homogeneity of the Si-containing precipitates in SLMed AlSi12 samples (Figure 3.31) [44]. The distribution of Al and Si appeared to depend on the solution treatment time, with the Si content in the matrix dropping rapidly, following the solution treatment. After long aging cycles, the Si particles started to grow in size via Ostwald ripening and the coalescence of adjacent fine Si particles. Figure 3.40 also shows the influence of the heat treatment temperature on the microstructure in SLMed AlSi12 alloy [46]. As the temperature increases, the microstructure becomes coarser, with Si particles tending to agglomerate along the cell boundaries, as well increasing in size.

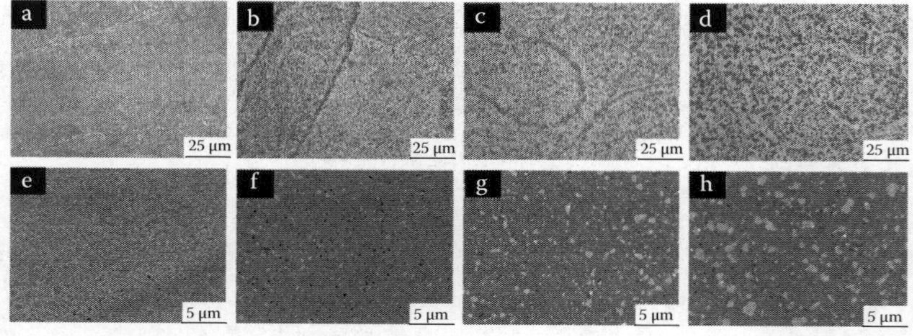

Figure 3.40 The influence of heat treatment temperature on the microstructure in SLMed AlSi12 alloy annealed for 6 h at (a, e) 473 K, (b, f) 573 K, (c, g) 673 K, and (d, h) 723 K. (From Prashanth, K.G. et al., *Mater. Sci. Eng. A*, 590, 153–160, 2014.)

3.3.3 Defect Formation in Al SLM Parts and Its Mechanisms

Macro- and microscale internal defects that are observed in Al-alloys SLMed parts can be classified into three main categories: pores (gas and shrinkage), oxide films, and cracks. Pores can be further classified into two types based on shape: round pores and irregular pores. The round pores, which are formed during the sintering process, can have either rough or smooth internal surfaces. The irregularly shape pores are formed during the powder consolidation process (e.g., lack of fusion defects), and they are of more concern than the round pores since they give rise to higher stress concentrations.

3.3.3.1 Round Shape Pores

Round (or elongated) pores with rough surfaces can originate from dendrite shrinkage porosity. The high coefficient of thermal expansion and volume shrinkage during solidification of Al-alloys, as well as the drastic volume change in the powder bed after melting, can promote the formation of shrinkage porosity. Furthermore, for alloys with large freezing ranges, shrinkage porosity can form at the late stage of the solidification in the mushy zone that contains the liquid-enclosing dendrites. However, due to the small melt-pool and the fine dendrite arm spacing, the shrinkage pores present in SLM parts are at the micron-scale, as shown in Figure 3.41 [39].

The round pores with smooth walls are generally believed to be gas porosity, which is formed at the end of the solidification. Gas porosity can be divided into the common gas porosity induced by dissolved and trapped gas (e.g., hydrogen-induced porosity in Al-alloys) and keyhole porosity induced by the vaporization of material, as shown in Figure 3.42 [51]. Gas porosity is formed as a result of the dissolved gas being rejected at the liquid–solid interface. The gas pores are also visible in Figure 3.41, and are smaller than 10 µm. The possible sources of the enclosed gas porosity may come from three types. First, they can be hydrogen-induced porosity, with the hydrogen coming from the moisture absorbed in surface of the powder particles and the ambient atmosphere. Second, they can also form from the residual gas pores in the gas atomized powder. Finally, they can be related to keyhole porosity. Keyhole pores can be found at the end of the laser scan track. The principle of laser beam–induced keyhole formation is shown in Figure 3.42. When a laser beam

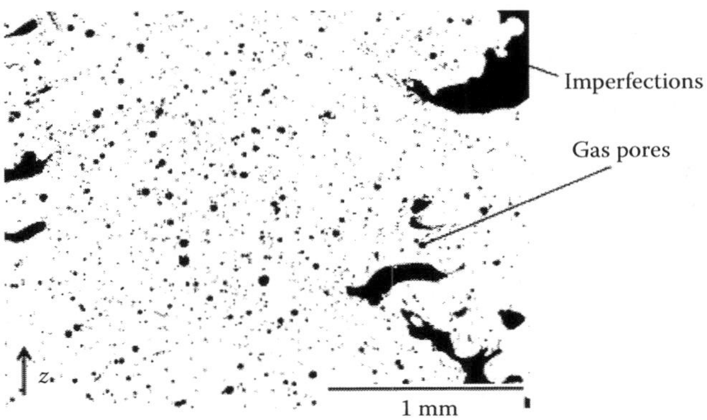

Figure 3.41 Optical micrograph for the cross section (*XY*) of SLMed AlSi10Mg showing round and irregularly shaped shrinkage cavities. (From Weingarten, C. et al., *J. Mater. Process. Technol.*, 221, 112–120, 2015.)

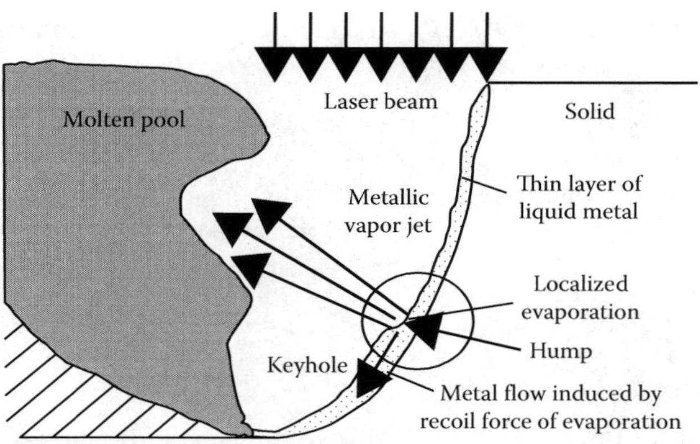

Figure 3.42 Schematic illustration for keyhole formation in laser processing. (From Steen, W., Mazumder, J., *Laser Material Processing*, Springer, London, 2010.)

with high intensity illuminates the build, part of the molten material is vaporized, which can lead to production of high recoil pressure; therefore, a keyhole pore is induced and filled with a partially ionized plume of vapor and the surrounding gas. The high thermal conductivity and laser reflectivity of Al-alloys mean that the power density required for keyhole formation is not easily achieved. However, when some elements such as Zn and Mg that have low vaporization temperatures are added to the alloy, the tendency to form keyhole pores becomes more likely.

Irregular pores, similar to those referred to as imperfections in Figure 3.41, may contain metallic particles on the pore surface, which can be attributed to the so-called balling effect. Balling occurs when the molten material has bad wetting characteristics, resulting in the formation of a coarse bead-shaped surface, instead of a smooth and flat solid surface. Once the coarse balls are formed, fresh powder is hard to fill in the pore channel. Though some small-sized powder particles could fill into the deep pore channel during powder leveling, the laser energy may not be high enough to penetrate and melt the powder in the deep pore channel. In current SLM of various Al, Al–Si, Al–Si–Mg alloys, balling can be present in both excess high and insufficient energy input situations, and/or in presence of oxide and/or low powder packing density.

Excess energy input (high laser power coupled with slow scan speed) induces slow solidification rate, increases lifetime of the molten pool, and results in an uncontrollable large molten pool, which has greater total surface than that of a sphere with the same volume. This causes dendrites to grow coarser, increases the interagglomerate pore size, and reduces the sample density.

On the contrary, the low input energy in combination with low laser power, with extremely fast scan speed, and with low or high scan spacing and/or too thick layer thickness can lead to the formation of a discontinuous track. Under these conditions, the powder cannot absorb sufficient energy to be melted completely and/or create a molten pool at an appropriate temperature and good wettability. A powder layer that is too thick may introduce a larger molten pool far away from the substrate, leading to poor binding between the volume, which is solidifying and the substrate due to the small contact area, as shown in Figure 3.43.

Oxide defects can also be observed in the irregularly shaped porosity in SLMed Al parts, similar to the imperfections shown in Figure 3.42. Figure 3.44 shows a Scanning Electron Microscope (SEM) micrograph of an oxide defect in SLMed AlSi10Mg. Oxides act as barriers restraining the melt flow and fusion, resulting in defects. It has been found that, even in an argon atmosphere with oxygen level of 0.1%–0.3%, oxidation cannot be avoided during SLM

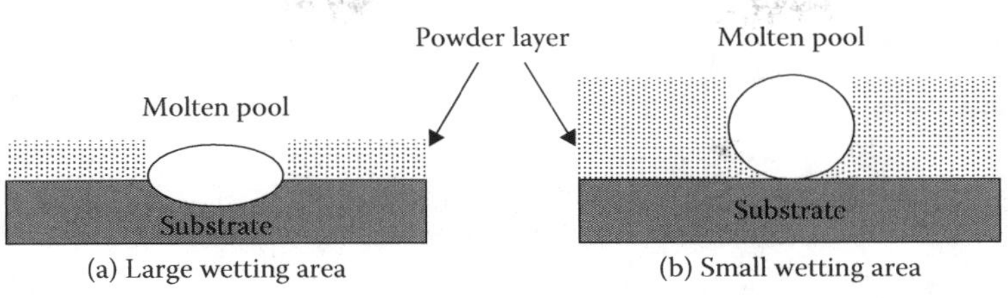

Figure 3.43 Schematic diagram showing the effect of the powder layer thickness on the wetting condition of the melt-pool. (From Li, J. et al., *Int. J. Adv. Manuf. Technol.*, 59, 1025–1035, 2012.)

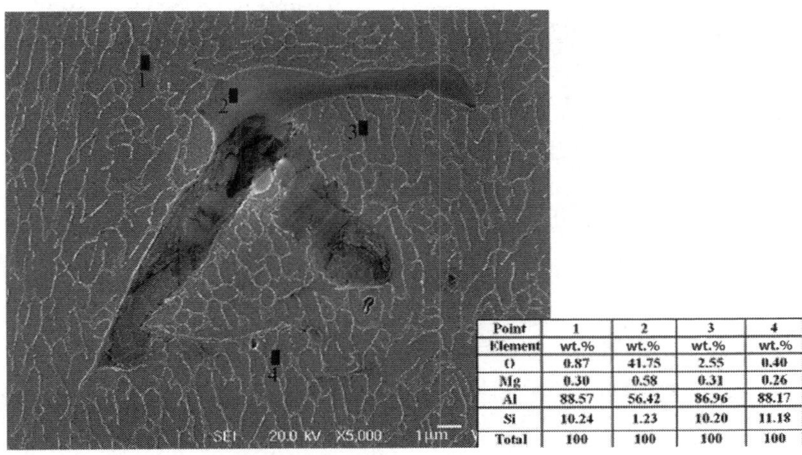

Point	1	2	3	4
Element	wt.%	wt.%	wt.%	wt.%
O	0.87	41.75	2.55	0.40
Mg	0.30	0.58	0.31	0.26
Al	88.57	56.42	86.96	88.17
Si	10.24	1.23	10.20	11.18
Total	100	100	100	100

Figure 3.44 Oxide film defect in SLM AlSi10Mg alloy. (From Read, N. et al., *Mater. Des.*, 65, 417–425, 2015.)

of Al-alloys. Because of the high melting point of alumina, once it is formed, it is hard to be remelted. Moreover, as the Al oxide layer is brittle, it can be cracked if it is thermally stressed. The oxidation behavior is affected by many factors, such as the ambient atmosphere, the turbulence of molten material (which is influenced by the scan speed), the melt temperature, and the chemical composition of the alloy. The thickness of oxide films depends on the oxidation intensity, whereas its size depends on whether it also entraps gas. In casting, the large gas bubbles have sufficient time and buoyancy to break the oxide film. However, in SLM, the size of the gas bubbles is much smaller, making it difficult for the gas to escape during the rapid melting and solidification process.

Finally, cracking during SLM of Al-alloys has been barely discussed. This is probably because the majority of the work in the literature focused on Al-alloys with very low sensitivity to solidification cracking so far, such as AlSi12 and AlSi10Mg alloys.

3.3.4 Mechanical Properties

The majority of the published mechanical properties are for SLMed AlSi10Mg and AlSi12 alloys, focusing on tensile properties, with few on fatigue and creep [52,56].

3.3.4.1 Tensile Strength of As-Fabricated Al-Alloys

Figure 3.45 shows the UTS, 0.2% YS (proof-strength PS), and % elongation of SLMed AlSi10Mg alloy for various building directions from various studies [40,43,52,56,59,60]. Note that in the horizontal (0°) sample, the length of the sample is perpendicular to the building direction, whereas the vertical (90°) sample is parallel to the building direction, and the 45°sample positioned between. For the UTS, large variations are observed in the 0° sample, with UTS varying between 330 and 420 MPa. Similar trend is observed for the 90° samples. However, the 0.2% PS shows consistent levels among all builds, showing no influence for the building direction. For the elongation, the data is scattering widely. For the 0° samples, elongation varies from 1% to 5.8%, and for 90° samples, 1%–3%. Although the data points are few, the data points scatter widely at UTS and elongation. The reason for the inconsistency in the data reported in the literature is due to the complexity of the factors associated with the SLM, notably the machine type, process parameters, and powder quality. The powder quality is possibly the paramount parameter in controlling the mechanical properties, due to the high oxide formation susceptibility in Al-alloys.

3.3.4.2 Influence of Postprocessing on the Tensile Strength

Brandl et al. assessed the influence of using a hot powder bed (300°C) on the SLM process and post-SLMT6 heat treatment on the UTS of AlSi10Mg alloy. It was reported that the as-built

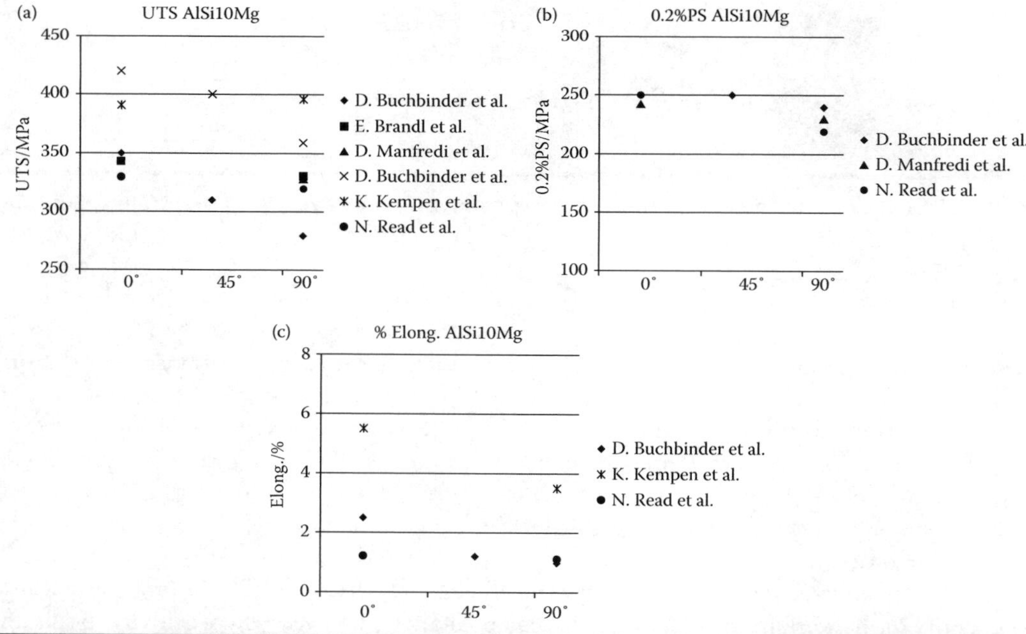

Figure 3.45 RT tensile properties of SLMd AlSi10Mg alloy: (a) UTS, (b) 0.2%PS, and (c) % elongation. (From Kempen, K. et al., *Phys. Proc.*, 39, 439–446, 2012; Manfredi, F.C.D. et al., *Materials*, 6, 856–869, 2013; Read, N. et al., *Mater. Des.*, 65, 417–425, 2015; Brandl, E. et al., *Mater. Des.*, 34, 159–169, 2012; Buchbinder, D., Meiners, W., *Generative Fertigung von Aluminiumbauteilen für die Serienproduktion*, Förderkennzeichen, Aachen, Germany, 2010; Buchbinder, H.S.D. et al., *Phys. Proc.*, 12, 271–278, 2011.)

samples UTS on the heat bed shows the least UTS (250 MPa), whereas T6-treated samples showed the similar as heat bed + T6 condition (~350 MPa) [56], suggesting that although the hot bed may be useful in removing moisture from the powder, postprocess SLM is essential to achieve the required strength. This was also shown in the work of Siddique et al. [61] in AlSi12 alloy. They reported that as-SLMed condition showed the best UTS and PS, a stress-relief treatment, with the hot-bed samples also showing the lowest UTS and PS. For elongation, as build + stress relief shows worst elongation, and as-build on the heat bed shows the best elongation.

The influence of heat treatment temperature on the SLM Al–12Si alloy was studied by Prashanth et al. [46]. Their work showed that increasing of heat treatment temperature results in a dramatic decrease in the PS, while increasing the elongation from 3% to 15% (Figure 3.46).

3.4 Titanium Alloys

Learning Objectives

- Explain the microstructural development of ALMed Ti based
- Understand the influence of laser processing condition and postprocess heat treatment or HIPping on microstructural and mechanical property development.

3.4.1 Microstructure of Ti ALM

Generally, Ti-alloys can be classified into three main categories: $\alpha\beta$, $\alpha+\beta$, and β (metastable and stable) alloys. This classification is relatively qualitative, as it is based on the alloy location in the pseudo-binary β-isomorphous phase diagram. It also identifies the alloy class based on the phases that form following rapid cooling from the β-phase field [62]. On rapid cooling from the β phase field, three potential phases are formed, which are as follows [63]:

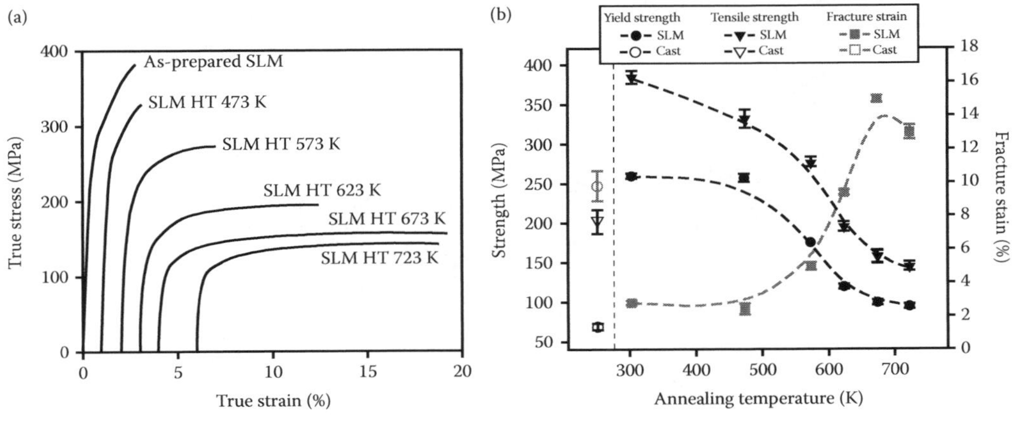

Figure 3.46 Influence of the heat treatment temperature on SLMed AlSi12 alloy. (From Prashanth, K.G. et al., *Mater. Sci. Eng. A*, 590, 153–160, 2014.)

1. α' *(hexagonal) martensite*: which forms either in pure Ti or very dilute alloy in a lath or packet (massive) martensite morphology, or in alloys with higher α-stabilizer solute content, forming acicular martensite.
2. α'' *(orthorhombic) martensite*: which occurs in alloys with β-stabilizing alloying additions, resulting in supersaturation and hence distortion of the hexagonal structure upon rapid cooling from the $\beta 0$ phase field [63–65].
3. $\omega_{athermal}$: which occurs in β-stabilized alloys, creating a fine nano-sized phase.

3.4.1.1 SLM Microstructure

The typical as-fabricated microstructure of SLMed Ti–6Al–4V shows columnar prior β grains, extending along the build direction, with acicular martensitic α' or fine α needles, as shown in Figure 3.47. The columnar grains in the previous layers act as nuclei for the solidification of the following layer, leading to epitaxial growth of the β grains, normally along the same crystallographically favorable orientation. Due to the lack of pinning phases and the favorable growth direction along the maximum head flux, development of equiaxed grains is very much suppressed during SLM. The columnar grains usually grow over many layers of building and can easily reach 3 mm in height with a width of 0.2–0.3 mm [66]. Despite the epitaxial growth of the prior-β grains, the development of α' needles during solid-phase transformation does not necessarily follow the orientation of the prior-β grains. According to Thijs et al. [67], their alignments are more to do with the local heat conduction. In fact, the growing direction is generally parallel to the local conductive heat transfer direction, as shown in Figure 3.48. Vranken et al. [68] suggested that the β phase transforms into α' or α phase according to the Burgers relation given by Equation 3.1.

$$\begin{aligned} (110)_\beta &\leftrightarrow (0001)_\alpha \\ \langle 1\text{-}11\rangle_\beta &\leftrightarrow \langle 11\text{-}20\rangle_\alpha . \end{aligned} \tag{3.1}$$

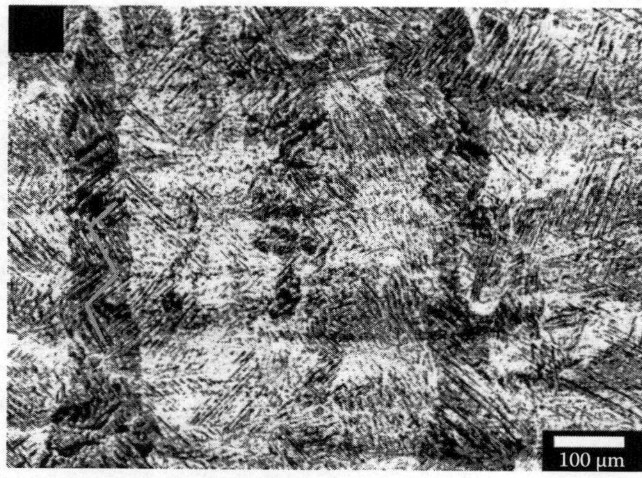

Figure 3.47 Optical micrograph showing the as-SLMEd microstructure of Ti–6Al–4V. (From Qiu, C.L. et al., *Mater. Sci. Eng. A*, 578, 230–239, 2013.)

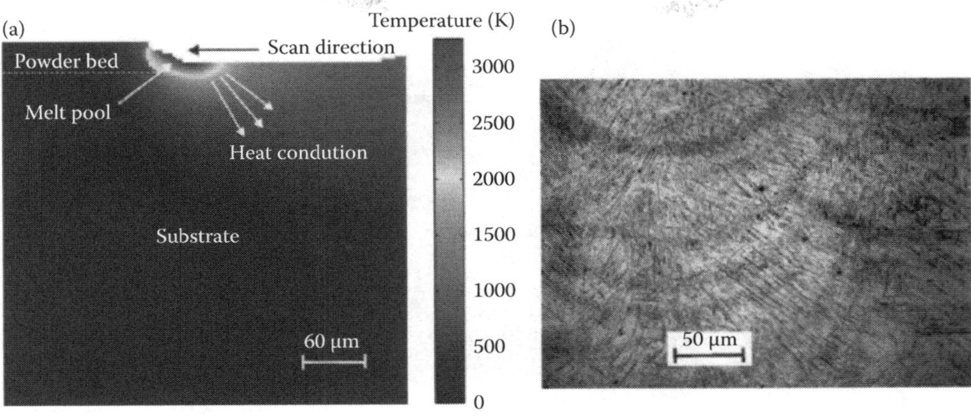

Figure 3.48 (a) Temperature profile in *xz*-plane for one Ti–6Al–4V powder layer on top of a Ti–6Al–4V substrate, scanned from right to left; (b) micrograph of a melt-pool after solidification and cooling. (From Thijs, L. et al., *Acta Mater.*, 58, 3303–3312, 2010.)

Simonelli et al. [69] further proved through extensive EBSD analysis that the β → α′ transformation is governed by the Burgers orientation relationship: most of the α′ laths belonging to the same parent β grain have the (0001)α′ reflection 60° misoriented around the ⟨01-10⟩α′ axes.

With the presence of columnar grains, texture development is unavoidable. According to Simonelli et al. [69], β grain grows epitaxially extending through several fabricated layers and shows a dominant {100} texture component in the direction of grain growth *g* that tends to develop because of the preferential {100} growth direction of cubic crystals during solidification. The overall α′ texture appears random because of the relatively high number of α′ variants within each prior-β grain. Typically, there are 5–6 variants that formed inside every parent β grain. The (0001)α′ plane of each α′ lath is parallel to one of the (110)β planes of the reconstructed parent phase and at least one of the (11-20)α′ directions is parallel to one of the (111)β.

Although the typical microstructure of SLM-processed Ti–6Al–4V is columnar prior-β grains combined with acicular α′, their detailed microstructural features could be affected by a number of factors. The work by Xu et al. [70] suggested that the width of columnar prior-β grains could be changed by powder layer thickness with the thinnest powder layer (30 μm) leading to the finest columnar prior-β grains (Figure 3.50). At a constant powder layer and energy density, focal offset distance (FOD) could also greatly influence microstructural development, primarily affecting the nature of the microstructure from α′ dominant into fine lamellar α + β microstructure. The formation of the acicular α′ martensite or ultrafine lamellar (α+β) structure was also found to be dependent on the energy density. Accordingly, to achieve an ultrafine lamellar (α+β) structure, it is necessary to control the energy input during SLM.

Apart from SLM processing conditions, post-SLM heat treatments and HIPping are commonly used to homogenize the microstructure of SLMed Ti-alloys and relieve the high residual stresses that form due to the rapid cooling during SLM. Generally, the SLMed Ti–6Al–4V builds are associated with high residual stress, together with martensitic needle structure, giving rise to high strength but poor ductility, making them unsuitable for most applications. As a result, postprocessing is usually required to optimize the microstructure and to achieve the required mechanical properties. Qiu et al. [66] firstly investigated the influence of postprocess annealing

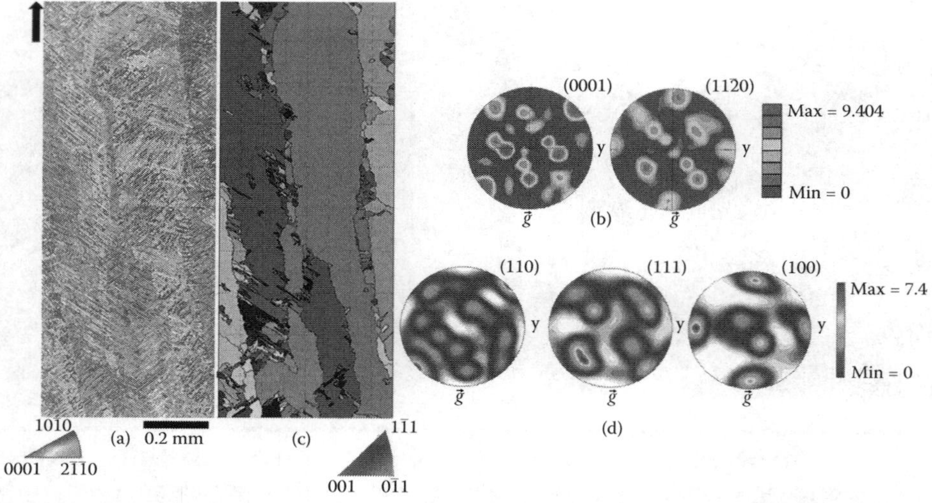

Figure 3.49 (a) EBSD α′ orientation map and the corresponding color scheme of a specimen taken from the lateral *yz*-plane of as-deposited Ti–6Al–4V (the black arrow indicates the building direction); (b) corresponding (0001)α′ and (11-20)α′ pole figures; (c) orientation map of the reconstructed β phase. The black grain boundaries represent β grains misoriented equal or larger than 7°; (d) the corresponding (110), (111), (100) contour pole figures. (From Simonelli, M. et al., *Metall. Mater. Trans.* A, 45, 2863–2872, 2014.)

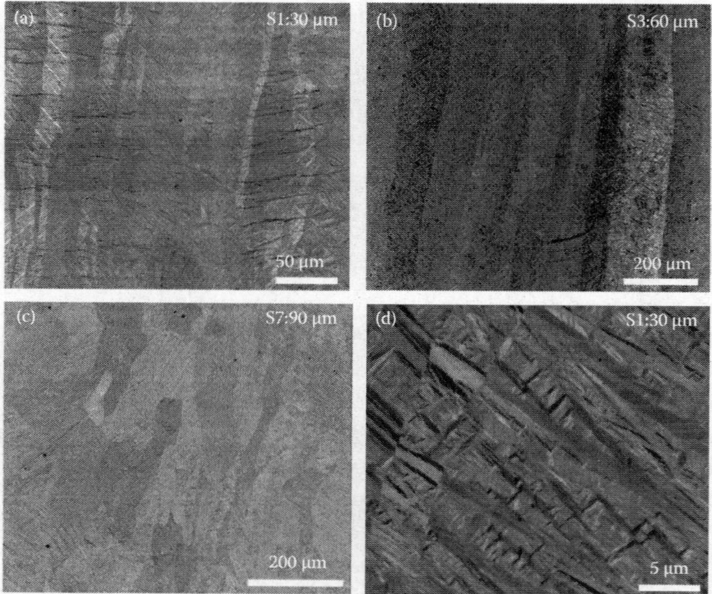

Figure 3.50 SEM BSE micrographs showing typical microstructures of SLMed Ti–6Al–4V. Columnar prior-β grain structure using a layer thickness of (a) 30 µm, (b) 60 µm, and (c) 90 µm. Micrograph (d) showing the α′ martensite formed inside the prior-β grains in S1. (From Xu, W. et al., *Acta Mater.*, 85, 74–84, 2015.)

and HIPping on microstructural development of SLMed Ti–6Al–4V, suggesting that the needle-like structure (whether α or α′) remains predominant when the sample was stress-relief annealed at 600°C (Figure 3.51a). However, after annealing at 700°C, α plates started to form from the martensitic structure (Figure 3.51b), alongside some β phase, as confirmed by XRD. Furthermore, HIPping at 920°C/103 MPa/4 h was found to not only transform the martensitic structure into α+β lamellar structure (Figure 3.51c) but also collapse the majority of porosity present in the as-fabricated samples (Figure 3.52).

According to Vilaro et al. [71] and Simonelli et al. [72], heat treatment at 730°C has allowed the α′ → α+β to take place, leading to a significant increase in α plate width. However, no α colonies are present in the microstructure similar to that observed for SLM-processed Ti–6Al–4V in the as-built condition. In addition, α texture is still weak as a result of the multiple variants that have formed within the β columnar grains. The prior-β grains still formed epitaxially through successive fabricated layers and have a dominant ⟨100⟩ solidification texture along the grain growth direction consistent with that reported for SLMed Ti–6Al–4V [69]. Heat treatment at higher temperatures but below β transus (995°C) was found to lead to increase the volume fraction of the β phase and coarsen α plates, as shown in Figure 3.53 [68]. When heating above the β transus, a fully homogeneous 100% β phase microstructure exists at high temperature, and during subsequent furnace cooling, a lamellar α+β mixture will be formed.

The heat treatment duration was also reported to show a great influence on the microstructural development. Increasing the dwell duration of heat treatment below the β transus could lead to increased development of globularized α grains and to increase α plate width, as shown in

Figure 3.51 Optical micrographs showing the microstructure of (a) SLMed sample that was annealed at 600°C, (b) SLMed sample that was annealed at 700°C, showing that alpha plates formed at the boundaries of the original martensitic plates, and (c) SLMed and HIPed sample. (From Qiu, C.L. et al., *Mater. Sci. Eng. A*, 578, 230–239, 2013.)

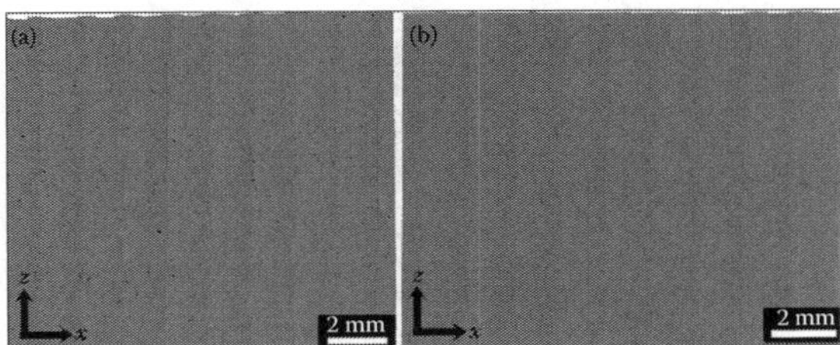

Figure 3.52 Optical micrographs for (a) as-fabricated Ti–6Al–4V, porosity area fraction A_f = 0.35%; (b) SLMed + HIPed sample, A_f < 0.01%. (From Qiu, C.L. et al., *Mater. Sci. Eng. A*, 578, 230–239, 2013.)

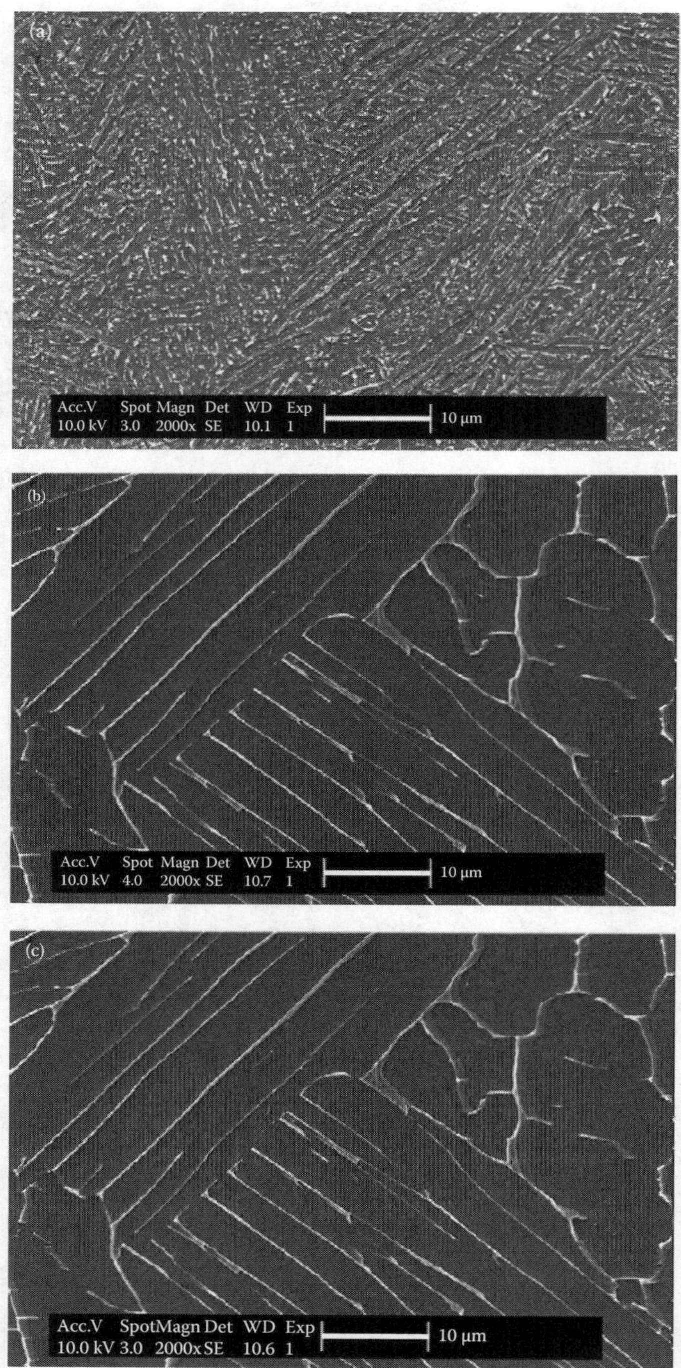

Figure 3.53 Microstructure of Ti6Al4V produced by SLM after heat treating at different temperatures for 2 h, followed by FC. (a) 780°C and (b) 843°C below the β transus, (c) 1015°C above the β transus. Lighter zones are β phase, and the dark phase is the α phase. (From Vrancken, B. et al., *J. Alloys Compd.*, 541, 177–185, 2012.)

Figure 3.54 [68]. The α plates are 2.23±0.12 μm wide after 2 h heat treatment at 940°C and have coarsened to an average width of 2.80±0.16 μm after 20 h. Prolonged dwelling above the β transus was found to yield larger α colonies.

3.4.2.2 DLD Microstructure

DLD includes powder-and wire-based DLD. The blown powder DLD generally gives more flexibility in tool-path design and processing of complicated structures as well as better surface finish and geometrical control, whereas the wire-based DLD usually gives higher build rates, but it is usually restricted to the fabrication of relatively simple geometries. In addition, the surface finish of the wire DLD is much poorer, with surface roughness R_a ~ 1 mm.

The microstructure of blown powder DLDed (α+β) Ti-based alloys is characterized by columnar prior-β grains and α/α′ structure depending on processing condition. Figure 3.55 shows a typical as-DLDed microstructure of Ti–6Al–4V samples, which contain columnar prior-β grains

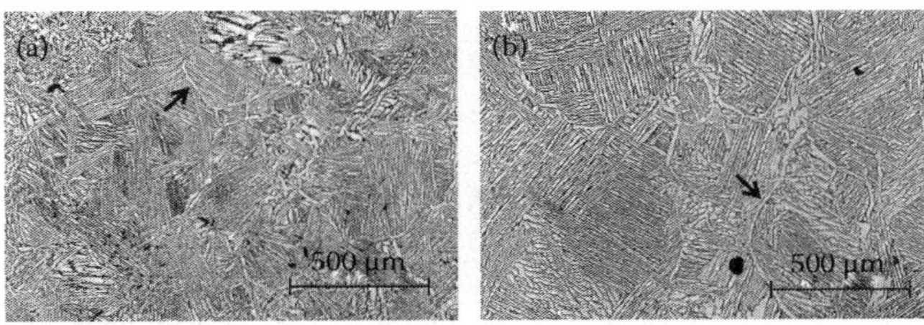

Figure 3.54 Illustration of the smaller α plate and colony size after (a) 2 h at 1020°C compared to after (b) 20 h at 1040°C, followed by furnace cooling. An α phase is light, and a β is dark. The arrows indicate grain boundary α. (From Vrancken, B. et al., *J. Alloys Compd.*, 541, 177–185, 2012.)

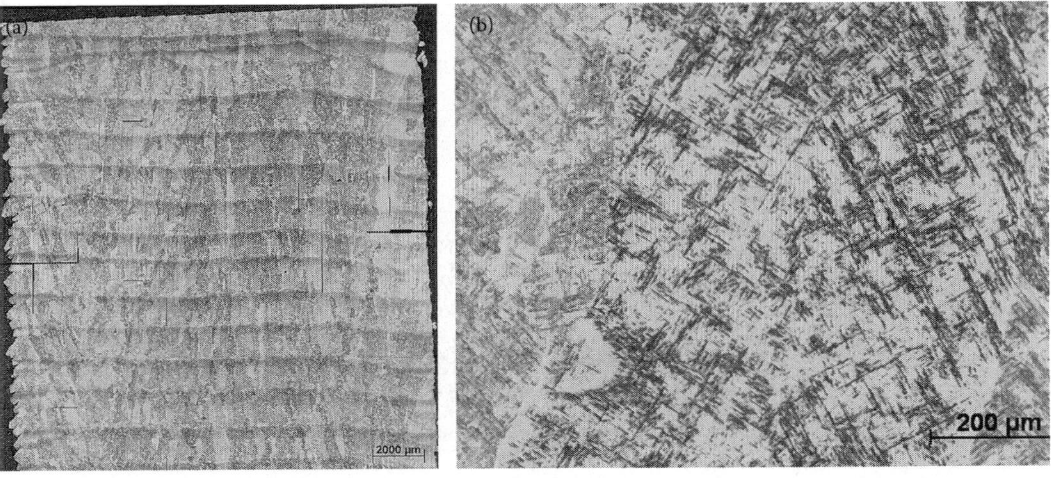

Figure 3.55 Optical micrographs for the *X–Z* plane showing (a) the predominantly columnar prior-β grain structure in as-DLDed samples and (b) the needle-like microstructure inside the prior-β grans. (From Qiu, C.L. et al., *J. Alloys Compd.*, 629, 351–361, 2015.)

together with α' martensitic needles [73]. In other reports [74], a basketweave Widmanstätten α structure has been observed, suggesting that it can be tailored by changing the DLD processing parameters (e.g., heat input). Similar to SLMed structures, post-DLD HIPping can also be used to homogenize the microstructure, to generate the typical $\alpha + \beta$ microstructure (Figure 3.56).

Similar to SLM, the size of the prior-β c grains can be controlled using the laser power and scan speed, as shown in Figure 3.57. Increasing the laser power is believed to keep the substrate at elevated temperatures, and thus the temperature gradient via the substrate is reduced, which leads to the disappearance of columnar grains and large equiaxed grains are formed instead. Furthermore, the influence of the scan speed on the morphology and size of the columnar grains is believed to begin from the effect on the substrate; at a higher scan speed, the energy density input into the substrate during deposition is small, which leads to more rapid cooling and thus to the formation of more nuclei and finer grains. These fine grains become nuclei in subsequent addition of layers to form long and narrow columnar grains. In terms of α/α' structure, the size of α laths could be also changed by changing processing parameters such as laser power, scanning speed, and powder feed rate. The width of α laths also increases with both laser power and powder feeding rate (Figure 3.58) [74].

Given that DLD and the other direct energy depositions methods result in the formation of large prior columnar β grain, which results in anisotropy in mechanical properties; refinement of the grain structure has been one of the major challenges. A number of methods have been suggested to refine the grain structure, including changing of the process parameters, change of laser mode (continuous vs. pulsed mode), and introduction of complimentary mechanical processing

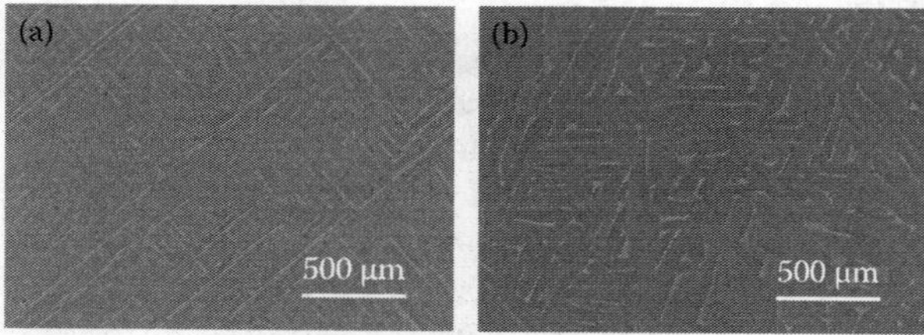

Figure 3.56 The influence of post-DLD HIPping, showing (a) the as-DLDed microstructure and (b) the post-HIPping $\alpha + \beta$ microstructure. (From Qiu, C.L. et al., *J. Alloys Compd.*, 629, 351–361, 2015.)

Figure 3.57 Optical morphologies of DLDed Ti–6Al–4V samples fabricated at 300 mm/min, 12 g/min, and at laser powers of (a) 390 W, (b) 474 W, and (c) 516 W. (From Wu, X. et al., *Mater. Des.*, 25, 137–144, 2004.)

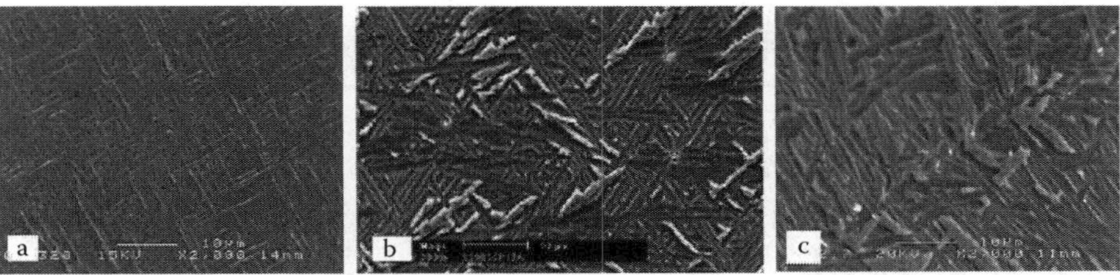

Figure 3.58 (a–c) SEM micrographs showing the increased size of α and β laths with increase of the laser power: (a) 264 W, (b) 390 W, and (c) 516 W. Scan speed = 300 mm/min and powder feed rate = 12 g/min, ΔZ = 0:3 mm. (From Wu, X. et al., *Mater. Des.*, 25, 137–144, 2004.)

method such as cold rolling. Grain refinement through processing conditions is generally very limited, given that a processing condition has to meet the requirement of minimum structural defects, as well as geometrical defects. As such, the flexibility that is offered by changing the process parameters is rather limited. So far, the most effective way to control the microstructure is through mechanical cold rolling after several layers of deposition, which was applied to WAAM process (Figure 3.59a) [75]. The introduction of cold rolling under 50 and 75 kN after several layers of deposition not only changed the columnar grains into equiaxed grains (Figure 3.59b) but also reduced the texture intensity (Figure 3.59c), and the residual stress and distortion (Figure 3.59d) when the samples were unclamped. The use of too high load such as 100 kN, however, was found to lead to fracture of builds. A similar concept has been suggested using impact treatments to reduce the residual stress development in DLD.

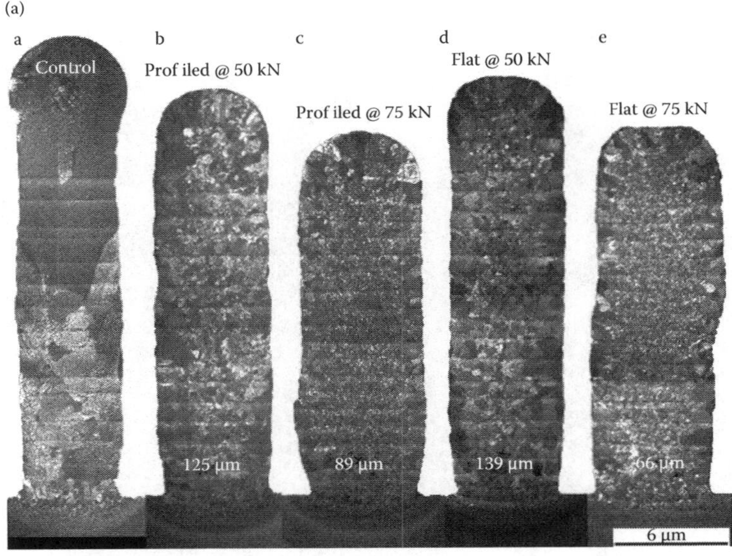

Figure 3.59 The influence of (a) cold rolling during WAAM process on (b) the microstructural and texture development, and (c) residual stress development in Ti–6Al–4V deposits. (From Martina, L. et al., *Metall. Mater. Trans. A*, 46, 6103–6118, 2015.)

(*Continued*)

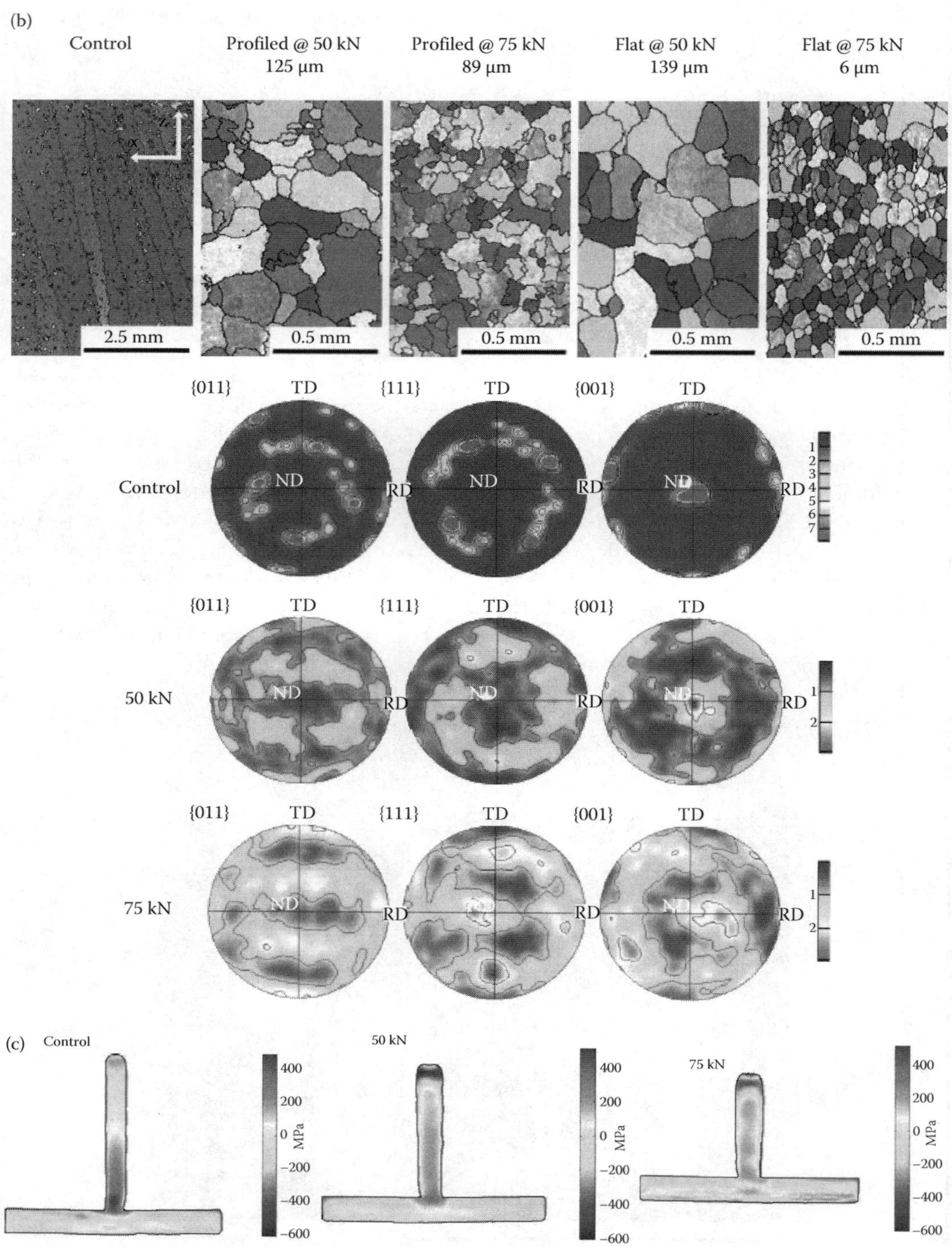

Figure 3.59 (CONTINUED) **The influence of (a) cold rolling during WAAM process on (b) the microstructural and texture development, and (c) residual stress development in Ti–6Al–4V deposits. (From Martina, L. et al., *Metall. Mater. Trans. A*, 46, 6103–6118, 2015.)**

The microstructure of as-DLDed β Ti-based alloys is very different from that of as-DLDed α + β alloys (e.g., Ti–6Al–4V). The as-DLDed Ti-5553 was found to show a mixture of both columnar and equiaxed β grains [76]. Normally, each weld bead contains equiaxed grains in the build center and columnar grains at its peripheral regions. The columnar grains also grow epitaxially across the boundaries between two layers. The fact that the columnar grains did not extend through a whole bead suggests that the thermal gradient between the remaining melt and the columnar grains might have become small (which would impede the directional growth of columnar grains), and according to the classic solidification theory, there may be a turbulent convection current in the liquid (which would help melt off tips or arms of the columnar grains and bring them into the liquid to act as seed crystals), both of which are essential for the formation of equiaxed grains in the bead center [77]. The turbulent Marangoni convection current in the melt-pool is also possible during DLD, given that the powder was blown by argon into the melt-pool, bringing momentum, and turbulence to the pool. Moreover, with increased build height, the fraction and aspect ratio of columnar grains was found to decrease while the number of equiaxed grains increased. This is probably due to the fact that with increased build height and increased building time, both the substrate and build were getting hotter and hotter and thus the thermal gradient between the build and substrate, and that within the build itself were getting increasingly reduced. This would impede the directional growth of columnar grains and promote the development of equiaxed grains. The influence of processing condition on grain structure was also noted, similar to the observations noted in the case of α + β alloys, with respect to grain coarsening with the increase in heat input. The results also suggest that other processing parameters such as powder flow rate could also affect the microstructural development. It is, therefore, concluded that the input energy density (which could be defined by laser power, scanning speed, powder flow rate, etc.) and the thermal history may have determined the microstructural development, and thus any processing parameter that would affect the energy density and thermal history may show influence on microstructure. Similarly, Liu et al. [78] investigated DLD of another β Ti-based alloy (Ti–5Al–5Mo–5V–1Cr–1Fe), revealing that the as-DLDed microstructure contained a mixture of columnar grains and equiaxed grains, which laid out in the form of a sandwich structure.

3.4.2 Mechanical Properties

The evaluation of mechanical properties of SLM-processed Ti–6Al–4V samples has been mainly focused on tensile and fatigue properties. The SLMed Ti–6Al–4V samples generally show high tensile strengths, but poor ductility [66,68,71,79]. The 0.2% YS could reach up to over 1 GPa, and the UTS up to 1.2 GPa, whereas the elongation is generally below 10%, as shown in Figure 3.60. The high strength in the as-SLMed condition is attributed to the fine martensitic microstructure. The as-SLMed samples also show anisotropy in tensile properties, with horizontally built samples generally showing higher tensile strengths than the vertically built, but lower elongation than the vertically built samples. The anisotropy was attributed to the microstructure because of the presence of columnar prior-β grains in the samples and the loading direction relative to the orientation of the columnar grains. In the vertically built samples, the loading direction is nearly parallel to the orientation of columnar grains and the fracture usually happens in a transgranular mode, whereas for the horizontally built samples the loading direction is normal to columnar grains and the fracture occurs in a more brittle intergranular mode, showing entrapped pores on the fracture surface [66,71] (Figure 3.61).

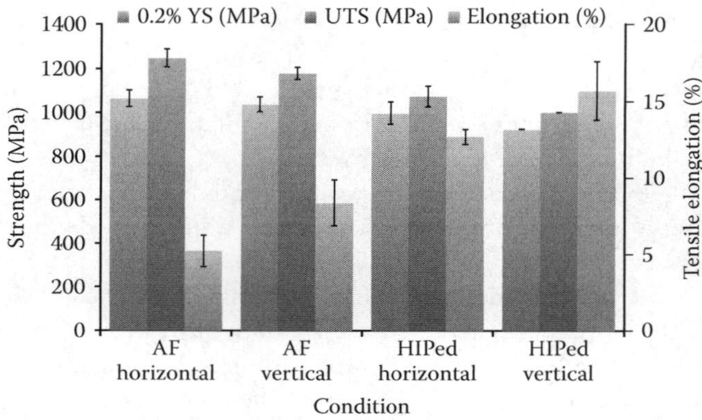

Figure 3.60 Tensile properties of as-fabricated and SLMed+HIPed samples with different orientations; the horizontally and vertically built samples were tested with tensile axis along their longitudinal directions. (From Qiu, C.L. et al., *Mater. Sci. Eng. A*, 578, 230–239, 2013.)

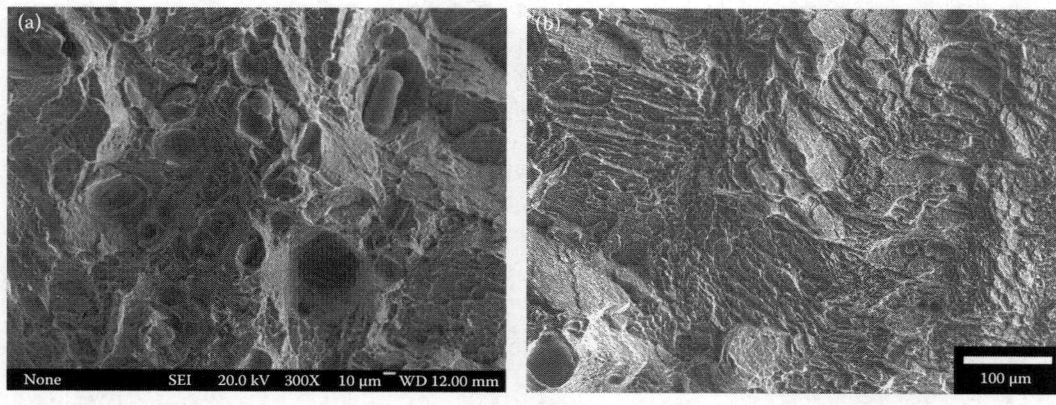

Figure 3.61 SEM fractographs for the horizontally as-SLMed showing (a) the gas pores on the fracture surface and (b) the transgranular fracture morphology. (From Qiu, C.L. et al., *Mater. Sci. Eng. A*, 578, 230–239, 2013.)

As mentioned previously, the microstructure of as-SLMed components could be modified by changing processing conditions. One example is shown in Figure 3.62 where different processing conditions have given rise to varied microstructures and thus to change of tensile properties to varied extents [70]. Some of the processing conditions that yielded ultrafine lamellar $(\alpha+\beta)$ was found to give a combination of high strengths and good elongation (>10%). The influence of postprocess heat treatment and HIPping on tensile properties has also been extensively investigated [66,68,71,73]. Generally speaking, both heat treatment and HIPping at high temperatures lead to improved ductility accompanied by a certain reduction in tensile strengths, as shown in Figure 3.60. The reduction in strengths is due to the coarsening of the fine α/α' microstructure after heat treatment. Nonetheless, the tensile properties after heat treatments or HIPping are comparable to or even better than wrought Ti–6Al–4V. More results on the mechanical properties of ALMed Ti–6Al–4V (including EBSM) can be found elsewhere [68,70].

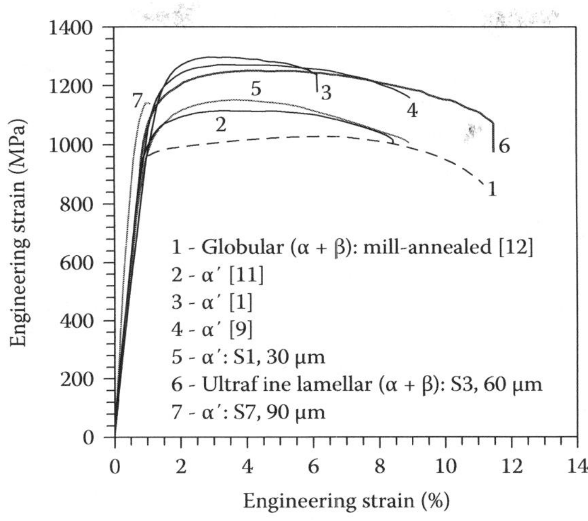

Figure 3.62 Engineering tensile stress–strain curves of SLM-fabricated Ti–6Al–4V comprising ultrafine lamellar (α + β) structure and α′ martensitic structure. (From Xu, W. et al., *Acta Mater.*, 85, 74–84, 2015.)

3.5 Summary

This chapter addressed the microstructural and mechanical properties development in ALMed metallic materials, focusing mainly on Ni-superalloys, Al-alloys, and Ti-alloys. Despite the differences between the three alloy classes, some similarities can be identified.

The microstructural development during ALM technologies shows a combination of melting and rapid solidification, epitaxial growth, and generally unstable microstructure. The as-fabricated microstructure is likely to require a postprocessing thermal treatment, to homogenize the microstructure, or to maximize the mechanical properties, through the combination of heat treatment and HIPping treatments. Depending on the alloy system, ALM techniques are likely to form various defects and microstructural anisotropy, including solidification or solid-state cracking, residual stresses, strong textures, and porosity. As a result, it is unlikely that the as-fabricated properties will be suitable to meet the performance requirements, making it necessary to utilize a postprocessing treatment. Still, using postprocessing treatments, it is possible in some instances to achieve mechanical properties that are comparable to the wrought products, but usually that involves multistage postprocessing treatments, which will likely add to the cost of using an expensive manufacturing operation.

3.6 Research Perspective

The existing literature on ALM of metallic materials provides a wealth of information on the microstructural and mechanical properties development. Nonetheless, there are still gaps in the literature that need to be addressed. Despite the similar microstructural observations for the samples built using various ALM platforms, there have not been sufficient studies assessing the transferability of parameters across the ALM platforms or different feedstock conditions. As such, there is always a need to consider the results in the literature based on the platform and the feedstock condition. Furthermore, work is still needed to understand how to tailor the microstructure, and accordingly the mechanical properties, using the process parameters to be able to meet the performance requirements without significant postprocessing.

The following topics are identified as key areas that need to be addressed in future studies:

- *Alloy design for ALM:* the majority of metal ALM research has focused on a limited number of alloys (e.g., AlSi, IN718, Ti-64, and 316 L) due to their availability in powder form. These alloys were designed for casting, forging, rolling, but not for ALM. As such, it is essential that new ALM-specific alloy types need to be designed. The alloys should have low cracking susceptibility, less likelihood for residual stress development, and less prone to porosity formation.
- *Microstructural design:* The flexibility that ALM offers in creating tailored microstructure has been consistently overlooked in the existing applications. Future work should utilize the flexibility in controlling the ALM process parameters (including the scanning strategy) to create structures tailored for the performance required by the application (e.g., high temperature/coarse grained, low temperature/fine grained, etc.).

3.7 Questions

1. Process optimization of ALM can rely on the energy density to understand the impact of the process parameters on defect formation and microstructural development. Explain.
2. Discuss the significance of the scanning strategy on the microstructural development, in terms of the microstructural homogeneity.
3. Identify the key differences between the SLM, EBSM, and DLD microstructures in Ni-base superalloys.
4. Classify the different types of defects that are associated with SLM of Al-alloys.
5. Discuss the impact of postprocessing using HIPping on the microstructural and mechanical properties development in Ti-alloys processed using SLM and DLD.
6. Outline the benefits of using cold rolling during the WAAM process on the microstructural and residual stress development in Ti-alloys.

References

1. W.F. Smith. *Principles of Materials Science and Engineering*, 3rd edition, McGraw Hill, New York, 1996.
2. M.J. Donachie, S.J. Donachie. *Superalloys: A Technical Guide*, 2nd edition, ASM International, Materials Park, OH, 2002.
3. S. Das. Physical aspects of process control in selective laser sintering of metals, *Advanced Engineering Materials* 5 (2003):701–711.
4. T. Vilaro, C. Colin, J.D. Bartout, L. Naze, M. Sennour. Microstructural and mechanical approaches of the selective laser melting process applied to a nickel-base superalloy, *Materials Science and Engineering A* 534 (2012):446–451.
5. K.N. Amato, S.M. Gaytan, L.E. Murr, E. Martinez, P.W. Shindo, J. Hernandez, S. Collins, F. Medina. Microstructures and mechanical behavior of Inconel 718 fabricated by selective laser melting, *Acta Materialia* 60 (2012):2229–2239.

6. X. Wu, F. Wang, D. Clark. On direct laser deposited Hastelloy X: Dimension, surface finish, microstructure and mechanical properties, *Materials Science and Technology* 27 (2011):344–356.
7. M.A. Anam, J. Dilip, D. Pal, B. Stucker. Effect of scan pattern on the microstructural evolution of Inconel 625 during selective laser melting. *Proceedings of 25th Annual International Solid Freeform Fabrication Symposium*, Louisville, KY, 2014.
8. J. Qingbo, G. Dongdong. Selective laser melting additive manufacturing of Inconel 718 superalloy parts: Densification, microstructure and properties, *Journal of Alloys and Compounds* 585 (2014):713–721.
9. S. Li, Q. Wei, Y. Shi, Z. Zhu, D. Zhang. Microstructure characteristics of Inconel 625 superalloy manufactured by selective laser melting, *Journal of Materials Science & Technology* 31(2015): 946–952.
10. L. Rickenbacher, T. Etter, S. Hovel, K. Wegener. High temperature material properties of IN738LC processed by selective laser melting (SLM) technology, *Rapid Prototyping Journal* 19 (2013):282–290.
11. F. Liu, X. Lin, C. Huang, M. Song, G. Yang, J. Chen, W. Huang. The effect of laser scanning path on microstructures and mechanical properties of laser solid formed nickel-base superalloy Inconel 718, *Journal of Alloys and Compounds* 509 (2011):4505–4509.
12. L.N. Carter, C. Martin, P.J. Withers, M.M. Attallah. The influence of the laser scan strategy on grain structure and cracking behaviour in SLM powder-bed fabricated nickel superalloy, *Journal of Alloys and Compounds* 615 (2014):338–347.
13. R.C. Reed. *The Superalloys: Fundamentals and Applications*, Cambridge University Press, Cambridge, UK, 2006.
14. L.N. Carter, M.M. Attallah, R.C. Reed. Laser powder bed fabrication of nickel-base superalloys: Influence of parameters; characterisation, quantification and mitigation of cracking. *12th International Symposium on Superalloys*, Minerals, Metals and Materials Society, Seven Springs, PA, September 9, 2012–September 13, 2012, 577–586.
15. G.P. Dinda, A.K. Dasgupta, J. Mazumder. Laser aided direct metal deposition of Inconel 625 superalloy: Microstructural evolution and thermal stability, *Materials Science and Engineering A* 509 (2009):98–104.
16. F. Liu, X. Lin, G. Yang, M. Song, J. Chen, W. Huang. Microstructure and residual stress of laser rapid formed Inconel 718 nickel-base superalloy, *Optics and Laser Technology* 43 (2011):208–213.
17. L.L. Parimi, G. Ravi, D. Clark, M.M. Attallah. Microstructural and texture development in direct laser fabricated IN718, *Materials Characterization* 89 (2014):102–111.
18. J. Chen, L. Xue. Process-induced microstructural characteristics of laser consolidated IN-738 superalloy, *Materials Science and Engineering A* 527 (2010):7318–7328.
19. X. Zhao, J. Chen, X. Lin, W. Huang. Study on microstructure and mechanical properties of laser rapid forming Inconel 718, *Materials Science and Engineering A* 478 (2008):119–124.
20. T. Yuan, D. McAllister, H. Colijn, M. Mills, D. Farson, M. Nordin, S. Babu. Rationalization of microstructure heterogeneity in Inconel 718 builds made by the direct laser additive manufacturing process, *Metallurgical and Materials Transactions A* 45 (2014):4470–4483.
21. H. Qi, M. Azer, A. Ritter. Studies of standard heat treatment effects on microstructure and mechanical properties of laser net shape manufactured Inconel 718, *Metallurgical and Materials Transactions A (Physical Metallurgy and Materials Science)* 40 (2009):2410–2422.
22. A. Odabası, N. Unlu, G. Goller, M.N. Eruslu. A study on laser beam welding (LBW) technique: Effect of heat input on the microstructural evolution of superalloy Inconel 718, *Metallurgical and Materials Transactions A (Physical Metallurgy and Materials Science)* 41 (2010):2357–2365.
23. L.E. Murr, E. Martinez, X.M. Pan, S.M. Gaytan, J.A. Castro, C.A. Terrazas, F. Medina, R.B. Wicker, D.H. Abbott. Microstructures of Rene 142 nickel-based superalloy fabricated by electron beam melting, *Acta Materialia* 61 (2013):4289–4296.
24. P. Ganesh, R. Kaul, C.P. Paul, P. Tiwari, S.K. Rai, R.C. Prasad, L.M. Kukreja. Fatigue and fracture toughness characteristics of laser rapid manufactured Inconel 625 structures, *Materials Science and Engineering A* 527 (2010):7490–7497.
25. X. Zhao, X. Lin, J. Chen, L. Xue, W. Huang. The effect of hot isostatic pressing on crack healing, microstructure, and mechanical properties of Rene88DT superalloy prepared by laser solid forming, *Materials Science and Engineering A* 504 (2009):129–134.

26. P.L. Blackwell. The mechanical and microstructural characteristics of laser-deposited IN718, *Journal of Materials Processing Technology* 170 (2005):240–246.
27. P.C.J. Ding, F. Martina, S. Williams, R. Wiktorowicz, M.R. Palt. Development of a laminar flow local shielding device for wire+arc additive manufacture, *Journal of Materials Processing Technology* 226 (2015): 99–105
28. B. Ahuja, M. Karg, K.Y. Nagulin, M. Schmidt. Fabrication and characterization of high strength Al-Cu alloys processed using laser beam melting in metal powder bed, *Physics Procedia* 56 (2014):135–146.
29. D. Manfredi, F. Calignano, M. Krishnan, R. Canali, E.P. Ambrosio, S. Biamino, D. Ugues, M. Pavese, P. Fino. Additive manufacturing of Al alloys and aluminium matrix composites (AMCs), *Light Metal Alloys Applications* 11 (2014): 3–34. https://www.intechopen.com/books/light-metal-alloys-applications additive-manufacturing-of-al-alloys-and-aluminium-matrix-composites-amcs-.
30. L.H.S. Dadbakhsh. Effect of Al alloys on selective laser melting behaviour and microstructure of in situ formed particle reinforced composites, *Journal of Alloys and Compounds* 541 (2012):328–334.
31. J. Ion. *Laser Processing of Engineering Materials: Principles, Procedure and Industrial Application*, Butterworth-Heinemann, Boston, MA, 2005. https://www.elsevier.com/books/laser-processing-of-engineering-materials/ion/978-0-7506-6079-2.
32. F.C. Campbell. *Manufacturing Technology for Aerospace Structural Materials*, Elsevier, 2006. https://www.elsevier.com/books/manufacturing-technology-for-aerospace-structural-materials/campbell-jr/978-1-85617-495-4.
33. J.R. Davis, (Ed.) *Aluminum and Aluminum Alloys*, ASM International, Materials Park, OH, 1993.
34. J.M. Sánchez-Amaya, T. Delgado, J.J. De Damborenea, V. Lopez, F.J. Botana. Laser welding of AA 5083 samples by high power diode laser, *Science and Technology of Welding & Joining* 14 (2009):78–86.
35. G. Mathers. *The Welding of Aluminium and Its Alloys*, Woodhead Publishing Limited, Cambridge, UK, 2002.
36. E. Anselmino, A. Miroux, S. van der Zwaag. Dispersoid quantification and size distribution in hot and cold processed AA3103, *Materials Characterization* 52 (2004):289–300.
37. I.J. Polmear. *Light Alloys: Metallurgy of the Light Metals*, St. Edmundsbury Press Ltd, Bristol, UK, 1995.
38. J.P. White, N. Read, R.M. Ward, R. Mellor, M.M. Attallah. Prediction of melt-pool profiles for selective laser melting of AlSi10Mg alloy. *Materials Science and Technology Conference and Exhibition 2014*, MS&T, Pittsburgh, PA, 2014, pp. 1985–1992.
39. L. Thijs, K. Kempen, J.-P. Kruth, J. Van Humbeeck. Fine-structured aluminium products with controllable texture by selective laser melting of pre-alloyed AlSi10Mg powder, *Acta Materialia* 61 (2013):1809–1819.
40. K. Kempen, L. Thijs, J. Van Humbeeck, J.-P. Kruth. Mechanical properties of AlSi10Mg produced by selective laser melting. *Physics Procedia* 39 (2012):439–446
41. R.W.B. Vrancken, J.P. Kruth, J. Van Humbeeck. Study of the influence of material properties on residual in selective laser melting. *Proceedings of Solid Freeform Fabrication Symposium*, Austin, TX, 2013.
42. K. Bartkowiak, S. Ullrich, T. Frick, M. Schmidt. New developments of laser processing aluminium alloys via additive manufacturing technique. *6th International WLT Conference on Lasers in Manufacturing*, LiM, Elsevier, Munich, Germany, 393–401, May 23, 2011–May 26, 2011.
43. F.C.D. Manfredi, M. Krishnan, R. Canali, E.P. Ambrosio, E. Atzeni. From powders to dense metal parts: Characterization of a commercial AlSiMg alloy processed through direct metal laser sintering, *Materials* 6 (2013):856–869.
44. X.P. Li, X.J. Wang, M. Saunders, A. Suvorova, L.C. Zhang, Y.J. Liu, M.H. Fang, Z.H. Huang, T. B. Sercombe. A selective laser melting and solution heat treatment refined Al–12Si alloy with a controllable ultrafine eutectic microstructure and 25% tensile ductility, *Acta Materialia* 95 (2015):74–82.
45. N.T. Aboulkhair, C. Tuck, I. Ashcroft, I. Maskery, N. M. Everitt. On the precipitation hardening of selective laser melted AlSi10Mg, *Metallurgical and Materials Transactions A* 46 (2015):3337–3341.
46. K.G. Prashanth, S. Scudino, H.J. Klauss, K.B. Surreddi, L. Löber, Z. Wang, A.K. Chaubey, U. Kühn, J. Eckert. Microstructure and mechanical properties of Al–12Si produced by selective laser melting: Effect of heat treatment, *Materials Science & Engineering A* 590 (2014):153–160.

47. E.O. Olakanmi, K.W. Dalgarno, R.F. Cochrane. Laser sintering of blended Al–Si powders, *Rapid Prototyping Journal* 18 (2012):109–119.
48. E.O. Olakanmi, R.F. Cochrane, K.W. Dalgarno. Densification mechanism and microstructural evolution in selective laser sintering of Al–12Si powders, *Journal of Materials Processing Technology* 211 (2011):113–121.
49. E.O. Olakanmi. Selective laser sintering/melting (SLS/SLM) of pure Al, Al–Mg, and Al–Si powders: Effect of processing conditions and powder properties, *Journal of Materials Processing Technology* 213 (2013):1387–1405.
50. X.J. Wang, L.C. Zhang, M.H. Fang, T.B. Sercombe. The effect of atmosphere on the structure and properties of a selective laser melted Al–12Si alloy, *Materials Science & Engineering A* 597 (2014):370–375.
51. C. Weingarten, D. Buchbinder, N. Pirch, W. Meiners, K. Wissenbach, R. Poprawe. Formation and reduction of hydrogen porosity during selective lasermelting of AlSi10Mg, *Journal of Materials Processing Technology* 221 (2015):112–120.
52. N. Read, W. Wang, K. Essa, M.M. Attallah. Selective laser melting of AlSi10Mg alloy: Process optimisation and mechanical properties development, *Materials and Design* 65 (2015):417–425.
53. N.T. Aboulkhair, N. Everitt, I. Ashcroft, C. Tuck. Reducing porosity in AlSi10Mg parts processed by selective laser melting, *Additive Manufacturing* 1 (2014):77–86.
54. E. Yasa, J.P. Kruth. Microstructural investigation of selective laser melting 316 stainless steel parts exposed to laser re-melting, *Procedia Engineering* 19 (2011) 389–395.
55. E. Louvis, P. Fox, C.J. Sutcliffe. Selective laser melting of aluminium components, *Journal of Materials Processing Technology* 211 (2011): 275–284.
56. E. Brandl, U. Heckenberger, V. Holzinger, D. Buchbinder. Additive manufactured AlSi10Mg samples using selective laser melting (SLM): Microstructure, high cycle fatigue, and fracture behavior, *Materials and Design* 34 (2012):159–169.
57. W. Steen, J. Mazumder. *Laser Material Processing*, Springer, London, 2010.
58. J. Li, J. Liu, Y. Shi, L. Wang, W. Jiang. Balling behavior of stainless steel and nickel powder during selective laser melting process. *The International Journal of Advanced Manufacturing Technology* 59 (2012):1025–1035.
59. D. Buchbinder, W. Meiners. *Generative Fertigung von Aluminiumbauteilen für die Serienproduktion*, Förderkennzeichen. Aachen, Germany, 2010. http://edok01.tib.uni-hannover.de/edoks/e01fb11/667761012.pdf; http://publica.fraunhofer.de/dokumente/N-203765.html.
60. H.S.D. Buchbinder, S. Heidrich, W. Meiners, J. Bültmann. High power selective laser melting (HP SLM) of aluminum parts, *Physics Procedia* 12 (2011):271–278.
61. S. Siddiquea, M. Imran, E. Wycisk, C. Emmelmann, F. Walther. Influence of process-induced microstructure and imperfections onmechanical properties of AlSi12 processed by selective laser melting, *Journal of Materials Processing Technology* 221 (2015):205–213.
62. G. Lutjering, J.C. Williams. *Titanium*, 2nd edition, Springer, Berlin, 2007.
63. S. Banerjee, P. Mukhopadhyay. *Phase Transformations Examples from Titanium and Zirconium Alloys*, Elsevier, Amsterdam, 2007.
64. J.C. Williams, B.S. Hickman. Tempering behavior of orthorhombic martensite in titanium alloys, *Metallurgical Transactions* 1 (1970):2648–2650.
65. M. Young, E. Levine, H. Margolin. The aging behaviour of orthorhombic martensite in Ti-6-2-4-6, *Metallurgical Transactions A (Physical Metallurgy and Materials Science)* 5 (1974):1891–1898.
66. C.L. Qiu, N.J.E. Adkins, M.M. Attallah. Microstructure and tensile properties of selectively laser-melted and of HIPed laser-melted Ti–6Al–4V, *Materials Science & Engineering A* 578 (2013):230–239.
67. L. Thijs, F. Verhaeghe, T. Craeghs, J.V. Humbeeck, J.-P. Kruth. A study of the microstructural evolution during selective laser melting of Ti–6Al–4V, *Acta Materialia* 58 (2010):3303–3312.
68. B. Vrancken, L. Thijs, J.-P. Kruth, J.V. Humbeeck. Heat treatment of Ti6Al4V produced by selective laser melting: Microstructure and mechanical properties, *Journal of Alloys and Compounds* 541 (2012): 177–185.

69. M. Simonelli, Y.Y. Tse, C. Tuck. On the texture formation of selective laser melted Ti–6Al–4V, *Metallurgical and Materials Transactions A*45 (2014):2863–2872.
70. W. Xu, M. Brandt, S. Sun, J. Elambasseril, Q. Liu, K. Latham, K. Xia, M. Qian. Additive manufacturing of strong and ductile Ti–6Al–4V by selective laser melting via *in situ* martensite decomposition, *Acta Materialia* 85 (2015): 74–84.
71. T. Vilaro, C. Colin, J.D. Bartout. As-fabricated and heat-treated microstructures of the Ti–6Al–4V alloy processed by selective laser melting, *Metallurgical and Materials Transactions A* 45 (2014):3190–3199.
72. M. Simonelli, Y.Y. Tse, C. Tuck. Effect of the build orientation on the mechanical properties and fracture modes of SLM Ti–6Al–4V, *Materials Science & Engineering A* 616 (2014): 1–11.
73. C.L. Qiu, G.A. Ravi, C. Dance, A. Ranson, S. Dilworth, M.M. Attallah. Fabrication of large Ti–6Al–4V structures by direct laser deposition, *Journal of Alloys and Compounds* 629 (2015): 351–361.
74. X. Wu, J. Liang, J.F. Mei, C. Mitchell, P.S. Goodwin, W. Voice. Microstructures of laser-deposited Ti–6Al–4V, *Materials and Design* 25 (2004): 137–144.
75. F. Martina, P.A. Colegrove, S.W. Williams, J. Meyer. Microstructure of interpass rolled wire+arc additive manufacturing Ti–6Al–4V components, *Metallurgical andMaterials Transaction A* 46 (2015):6103–6118.
76. C.L. Qiu, G.A. Ravi, M.M. Attallah. Microstructural control during direct laser deposition of a β-titanium alloy, *Materials and Design* 81 (2015): 21–30.
77. P.A. Porter, K.E. Easterling. *Phase Transformations in Metals and Alloys*, 2nd edition, Chapman & Hall, London, 1992.
78. C.M. Liu, X.J. Tian, H.B. Tang, H.M. Wang. Microstructural characterization of laser melting deposited Ti–5Al–5Mo–5V–1Cr–1Fe near β titanium alloy, *Journal of Alloys and Compounds* 572 (2013): 17–24.
79. L. Facchini, A. Molinari, S. Höges, K. Wissenbach. Ductility of a Ti–6Al–4V alloy produced by selective laser melting of prealloyed powders, *Rapid Prototyping Journal* 16 (2010):450–459.

Chapter 4

Structural Integrity of Additive Manufactured Parts

Steve R. Daniewicz
University of Alabama

Alexander Johnson
Mississippi State University

Scott M. Thompson and Nima Shamsaei
Auburn University

Contents

CHAPTER OUTLINE

In this chapter, we will discuss the following:

- Transport phenomena in the laser-based additive manufacturing (LBAM) process
- Fatigue loading characterization
- Factors that affect stress–number of cycles–probability (S–N–P) curves
- Nonzero mean fatigue stresses
- Spectrum loading and cumulative damage
- Fatigue crack growth
- Fatigue behavior of LBAM parts
- Fatigue crack growth in LBAM parts
- Some related examples for LBAM

4.1 Basics of Laser-Based Additive Manufacturing

Learning Objectives

- Understand what transfer mechanisms are involved during common forms of LBAM
- Understand the various modes of heat transfer during LBAM
- Know how design and process parameters impact part quality
- Methods for predicting and monitoring part temperature during LBAM

Note that many excerpts have been used from (Thompson et al. 2015) Elsevier © 2015; all permission has been obtained for reuse.

4.1.1 The Laser-Based Additive Manufacturing Process

To fully understand and predict the mechanical response and structural integrity of an additively manufactured part, one must be cognizant of the details surrounding its manufacture. In fact, for any part utilized in the field, it is absolutely necessary to understand how the part was sourced for ensuring its quality and safety to the end customer. The manufacturing process can introduce faults or anomalies in parts, making them more prone to failure. Hence, we first look into some of the details regarding the LBAM process, in order to better understand and appreciate the susceptibility of manufactured parts to controllable process and design features.

One can classify AM by its material and energy delivery methods, which can vary considerably for the various AM processes. For LBAM of metals, the two most common types of processes are Direct Laser Deposition (DLD) and Powder-Bed Fusion-Laser (PBF-L). For each of these processes, there is a distinct means of delivering and processing material at a specific working location/region—to accomplish the layer wise 3D model generation. The deposition and energy delivery method can either occur simultaneously, like DLD, or subsequently, like PBF-L. When the material and energy delivery processes occur at separate times, the AM process is said to be "selective," whereas if the processes are concurrent, the process is said to be "direct." As the process changes, then so can the material "preform" or "feedstock" that exists prior to AM, which may be powder, wire, sheets, etc. For LBAM, powder metal is commonly used, although LBAM systems that use wire preforms are also available.

For any AM process, the material preform is altered in shape or composition via a specific energy delivery method, which is typically thermal or chemical in nature. Chemical reactions are often reserved for processing material that "cures" in response to ultraviolet electromagnetic (EM) radiation. This is seen via the polymerization of liquid preforms used in material jetting and vat photopolymerization. Chemical processes can also be accomplished through the use of binders (binder jetting), which allow for layer-to-layer adhesion. For the case of metals, chemical reactions are not feasible for processing many preforms. Instead, a change in phase of metallic preform is sought via directed thermal energy. Similar to focusing sunlight for starting a fire, directed thermal energy can result in magnificent material responses. Thermal processing of material allows for easier operations due to elevated temperatures or altered phase. During thermal processing, the heat transfer may either be sensible (no phase change) or latent (phase change), and this results in elevated temperatures and melting (and any subsequent superheating), respectively, in order to more easily build a layer while in liquid form.

For LBAM, thermal energy is focused and delivered to a relatively small area by using a laser. A laser is a coherent form of light that carries energy in the form of EM waves. For the process engineer, this form of energy transfer is often referred to as a thermal irradiation, since the EM results in a thermal response in the material. For most LBAM systems, a single neodymium-doped yttrium aluminum garnet (Nd:YAG) laser is utilized, with typical power ranging between 1 and 5 kW. However, CO_2-type and pulsed-wave lasers may also be used. The Nd:YAG laser has been used widely for many laser welding and cutting processes due to its incident, spectral radiation being less reflective for most common metals (e.g., at a wavelength of 1.067 μm) (Thompson et al. 2015). CO_2-type lasers have wavelengths an order of magnitude higher at 10.6 μm and can result in a less energy-efficient LBAM process, since higher laser powers are required. For most applications, the diameter of the beam is typically on the order of millimeters; hence, very extreme heat fluxes are experienced given the laser beam area and power. For example, an LBAM process operating with a laser at 200 W total power and a beam size of 1 mm provides an average heat flux in excess of 20,000 W/cm^2. For comparison, common central processing units (CPUs) dissipate approximately 10–50 W/cm^2. These high heat fluxes are required in LBAM for overcoming the high latent heat of many metals in a short period of time. The thermal irradiation "felt" by processed metallic powder used in LBAM depends on many factors, including laser wavelength, surface cleanliness, surface morphology, chamber operating gas, material type, laser-to-surface angle, and more. Only a "blackbody" can absorb all incident thermal irradiation. The metallic powders used in LBAM will only absorb a portion of the incident thermal energy due to their specific surface properties and reflectivity.

PBF-L is used to generate metallic parts via the incremental heightwise movement of a table consisting of a compact, uniformly distributed layer of metallic powder that is selectively melted by a focused laser beam. In Figure 4.1, a uniform bed of powder is first deposited and then specific regions of the bed are melted by the laser beam in order to build a single layer of the part. The melting pattern, or laser scanning pattern, can be continuous lines or near-random pulses. Upon

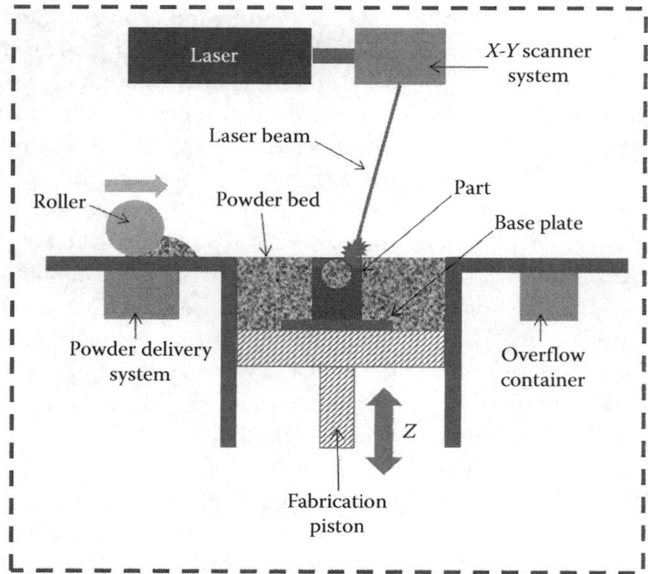

Figure 4.1 The powder-bed fusion-laser process. (From Thompson, S.M. et al., *Addit. Manuf.*, 8, 36–62, 2015. With permission.)

the completion of a single layer, the powder bed is lowered by the height of the deposited layer, a new bed of powder is deposited via a roller (or blade), and the process is repeated. This repetitious process results in excess metal powder, which can help in supporting the part during the build and can also result in powder remaining in the part if it consists of passages/channels in its design. The use of PBF-L for generating parts with "overhanging" structures is advantageous, as these sections can be supported by the unmelted powder bed, thus reducing residual stress formation and the potential of part collapse during the build. The PBF-L and DLD processes typically occur in an enclosed, inert-gas atmosphere in order to reduce the oxidization rate of the part during the build. The part is built upon a base plate (i.e., build plate, substrate, platen), and the finished part must be sheared off from the base plate after the AM process. This is typically accomplished by Electric Discharge Machining (EDM). Heat transfer to or from the base plate can influence part-to-plate adhesion, as well as part microstructure.

Like PBF-L, DLD is a means to build metallic parts. However, instead of a separate material and selective energy delivery process, DLD combines the material/energy delivery for simultaneous deposition and part forming within a similar region, as shown in Figure 4.2. The metal preform can be wire or powder. Wire-fed DLD provides better control on the deposition efficiency, while powder preform is typically blown through nozzles and can result in nonused powder accumulation in the machine. Earlier DLD systems consisted of a single, coaxial nozzle (coaxial with laser beam) in atmosphere, while current DLD machines may have up to four (or more) nozzles and utilize inert gas as to minimize high oxidization rates inherent for elevated temperature metal processing (Griffith et al. 1998). The Laser Engineered Net Shaping (LENS) system is a common, multinozzle/blown-powder DLD system. Unlike PBF-L, DLD can be used to repair precious components by laser cladding additional material to affected areas. In addition, DLD can be used for *in situ* alloying/mixing of materials by using two (or more) powder sources during operation.

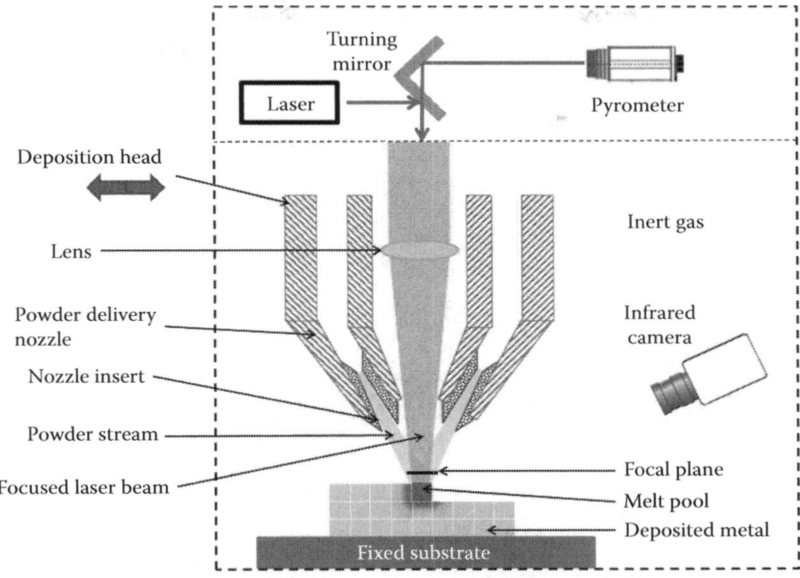

Figure 4.2 Blown powder direct laser deposition (DLD) process with thermal monitoring. (From Thompson, S.M. et al., *Addit. Manuf.*, 8, 36–62, 2015. With permission.)

4.1.2 Manufacturing Process and Design Parameters

LBAM consists of many process parameters that are at the end-user's discretion. The major process parameters include laser power, laser speed (relative to base plate), powder feed rate (for DLD) and hatch spacing. For PBF-L, the powder layer thickness is also controllable and important. Note that any combination of these process parameters, coupled with the chosen "design parameters"—which can include part orientation during the build, base plate operating temperature, laser scan strategy, and more—will result in a unique thermal response, thus unique heating/cooling rates and encumbered microstructure. Nonoptimal process parameter selections can result in parts that fail to densify completely or that contain significant defects and undesired microstructure. Process parameters can be nondimensionalized or reduced in dimension in order to make their utility broader, and this is a common practice in modeling heat transfer during manufacturing processes.

In addition to process/design parameter selection, the feedstock type, form, and condition are other important factors influencing the LBAM process. Powders can vary in size and shape and in terms of how they are produced. For most DLD applications, the powders are larger in shape as compared to those used in PBF-L processes. For LBAM, it is common to see powders ranging between 10 and 100 μm and being spherical in shape. Powders for alloyed materials will consist of all alloyed constituents as they are produced from the alloyed metal in its molten state. Gas atomization, water atomization, and plasma rotating electrode processes (PREP) are typical means for producing powders effective for LBAM application (Davis 2000). For DLD, the powder feed rate is the average powder mass leaving DLD nozzles with respect to time and can range between 1 and 10 g/min (grams per minute). The interlayer idle time is the finite time elapsed between successive material/energy deposits or layers and can range, e.g., from 0 to 1000 s (Costa et al. 2005; Yin et al. 2008; Zheng et al. 2008; Manvatkar et al. 2011). For PBF-L, the surface roughness and compactness of the powder bed, although difficult to control, will also influence part manufacture. Process parameters can be variable with time or constant during the manufacturing process. For instance, if an LBAM machine employs some form of control system, the laser power can be increased or decreased throughout fabrication of an individual layer.

4.1.3 Momentum and Energy Transfer during LBAM

For a given instant in time, and per a selected set of process parameters, there are several possible paths for momentum and energy transfer. Figure 4.3 shows that this energy and/or momentum transfer can be categorized as to occur either subsequently or in parallel. The principal "participants" in LBAM heat transfer include the laser, feedstock (e.g., powder), melt pool, heat-affected

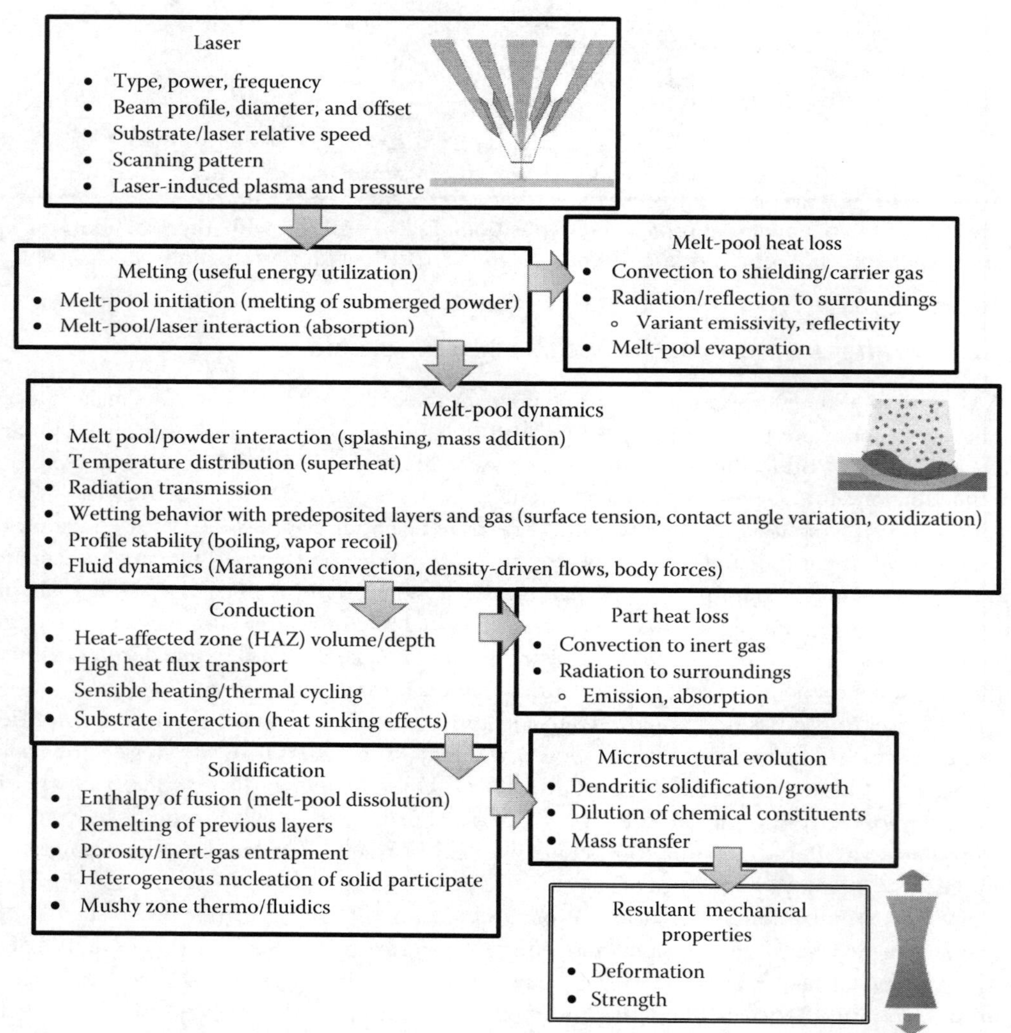

Figure 4.3 Physical events occurring during LBAM for given instant in time. (Adapted from Thompson, S.M. et al., *Addit. Manuf.*, 8, 36–62, 2015. With permission.)

zone (HAZ), bulk part, and surrounding environment. All participating media (except the laser) possess a thermal capacity, which dictates the measurable temperature rise within it, per unit heat transfer. By conservation of energy (i.e., the First Law of Thermodynamics), thermal energy delivered from the laser (via thermal radiation) is utilized for forming the melt pool via melting and then exchanged with the bulk part (i.e., previous, solidified layers and base plate) and working environment—which may include the shielding gas and enclosure or a powder bed for the case of PBF-L. For the case of powder-based LBAM, the injected or surrounding particles can exchange momentum with each other and with the melt pool and solid part. In addition to thermal energy and momentum, one can also discuss/model the mechanical response of the part during its manufacture, meaning "residual stress" can form within the part due to local regions being in either tension or compression.

The melt pool is a very important aspect of the LBAM process and is the bridge between powder deposition and part. It is a volume of molten material at the interface of the solidified part and laser, with a characteristic size of ~1 mm. The HAZ is the region in vicinity to the melt pool that consists of elevated temperature and thermal gradients. Its temperature distribution and fluidic/wetting behavior dictates solidification heat transfer and thus encumbered microstructure. Therefore, its modeling and diagnostics are of immediate interest for DLD quality control (Thompson et al. 2015). Note that the fluid within the melt pool, consisting of its own momentum, is actually moving for the time it exists. This means, due to temperature variation along the surface and volume of the melt pool, properties such as surface tension and density are nonconstant in space, and this gives rise to Marangoni and free-convection flows, respectively.

All modes of heat transfer exist during LBAM. Energy delivery from the laser occurs via EM radiation, and when a temperature response occurs due to EM radiation, it is commonly referred to as thermal radiation. The energy leaving the laser source will not equal the energy absorbed for powder processing due to laser attenuation, reflection, scattering and surface optical properties, and surface condition. It is for these reasons that powders with surface oxide layers will absorb more thermal radiation, since oxygen is a better absorber of thermal radiation than many cleaned powder surfaces. Thermal radiation will depend on many parameters, including the surface emissivity, absorptivity, reflectivity, morphology, and cleanliness. The surface absorptivity and reflectivity will depend on the temperature of the irradiated powder, as well as the wavelength and angle-of-attack of the employed laser. In many cases, powder "receives" only 10%–50% of the original laser power (Pinkerton and Li 2004). Powder injected during DLD will undergo sensible heating prior to being introduced to the melt pool, and the shielding gas used in the chamber will introduce a drag force. The powder delivery for a DLD process can be quantified via powder efficiency, which is the ratio of powder utilized for final part formation over the actual amount of powder delivered/blown by the system over a given time interval. This powder (or deposition) efficiency depends on the nozzle geometry, angle, and the size distribution of the particles utilized (Thompson et al. 2015).

Upon the irradiation of the LBAM feedstock to its liquidus temperature, thermal energy is then used to break molecular metallic bonds in the material for forming the melt pool. The energy with observed temperature response is often referred to as "sensible," while the energy used for changing a material's phase, which will not have a discernable temperature response, is often referred to as "latent." For materials consisting of multiple chemical elements, such as metallic alloys, the chemical constituents can diffuse while in the molten, and even solid state due to mass gradients, i.e., mass diffusion/transport. The remaining thermal energy is transported via conduction throughout the HAZ and bulk part, as well as by convection and thermal radiation with the surroundings. For the case of PBF-L, conduction with the surroundings is also achievable due to the presence of the powder bed.

Convection heat transfer is typically quantified using a heat transfer coefficient, the local temperature of the part, and the bulk temperature of the surrounding gas. Conduction heat transfer is quantified by solving a form of the heat equation, which is the governing equation for a time-dependent temperature field, and by employing Fourier's law. Quantification of all heat transfer during LBAM allows one to better select processing parameters and to determine "process–property" relationships. This means, one can better select laser power, laser speed and hatch distance, and more, based on preliminary prediction of the associated heat transfer and microstructural evolution.

Some of the major physical events occurring during LBAM, in order of approximate occurrence, include laser delivery, powder interaction (energy and momentum exchange), laser/powder/gas interactions, melt-pool initiation (melting), melt-pool energy/stability/morphology, heat loss to environment via thermal radiation and convection, solidification, intrapart conduction, and part-to-substrate conduction. Some of the detailed physical "subevents" for each category are provided in Figure 4.3. Many of these subevents within the LBAM process are research topics in of themselves, and their understanding and/or modeling is critical for the engineers responsible for designing, improving, or monitoring, the LBAM process.

4.1.4 Solidification

As the laser beam moves away from a particular region, the melt pool cools and solidifies very quickly. This solidification of the melt pool is governed by the net heat transfer through the melt pool and can be classified into three major events: (1) heterogeneous nucleation, (2) mushy-zone heat/mass transfer, and (3) microstructural evolution via heat treatment. The mushy zone is the region between the melt pool and the solid material. It supersedes or is colocated with the HAZ. The mushy zone is a two-phase mixture—with remnants of solid particulate and molten metals. Natural convection is the means of heat transfer in the mushy zone, as the density variation is significant. The mushy zone is a result of melting and solidification occurring over a finite temperature range over a spatial domain—as observed by the coexistence of liquid and solid metals. This temperature range is in between the solidus and liquidus temperature.

The phase-change heat transfer and momentum at the solid/liquid boundary of the melt pool is perhaps the most technically sophisticated physical process inherent to LBAM. The powder interfaces lead to melt-pool instability and transient morphology; while the trailing edge of the melt pool surrenders its latent heat of fusion to form the solid deposited layer. This complex diffusion process is further coupled with the temperature-dependent microstructure evolution that depends strongly on the material type. Solidification for carbon steels is different than that of Ti–6Al–4V, and this is due to the various chemical elements in the alloy.

4.1.5 Modeling the LBAM Process

The LBAM process consists of many complex physical phenomena that are all coupled to some degree. Predicting momentum, energy, and mass diffusion in the melt pool, HAZ, and bulk region is difficult to accomplish via pure analytical methods. However, the LBAM heat transfer has many similarities to the welding problems encountered and mastered in the past. There is a moving heat source, the formation/solidification of a moving melt pool, and a finite HAZ. The primary difference is the existence of mass addition into the melt pool via blown metallic powders and more prevalent bulk heating affects—due to repetitious laser passes. Unlike laser welding models, a solution is sought in which the spatial domain is increasing with time—in which new layers are required—a concept referred to as massification or consolidation. Although the Rosenthal (1946) and Mazumder and Steen (1980) closed-form solution to the heat equation provides a good approximation for the moving heat source problem on semi-infinite, single-phase mediums for conduction and deep-penetration (keyhole) welding, respectively—it does not incorporate mass addition (due to powder injection) or phase-change heat transfer due to melting and solidification. Regardless, the quasi-steady-state temperature distribution of an HAZ typically has the form of Equation 4.1:

$$T - T_o \cong \frac{q}{4R\pi k} \exp \frac{-v(x+R)}{2\alpha}, \tag{4.1}$$

where T is local temperature, T_o is a relative temperature, q is the heat rate, k is the part thermal conductivity, ν is the traverse speed of the moving heat source, x is the position along the coordinate axis parallel with laser motion, R is the absolute distance from the heat source, and α is the thermal diffusivity. This solution demonstrates that the HAZ temperature decreases exponentially as the traverse speed increases.

The finite element method (FEM) is commonly used as a means to discretize the energy equation for determining the temperature of many isothermal elements (i.e., the mesh). However, when using FEM for the LBAM problem, one must be creative in respecting the increasing volume with respect to time—a direct result of the material deposition. A three-dimensional FEM has been demonstrated as a convenient means for estimating the temperature field in and around a part during its LBAM. For instance, Wang et al. (2008) investigated the temperature response and microstructure evolution in a thin-wall build of stainless steel 410 during multinozzle, powder-based DLD. The phase transformations were accounted for by using continuous cooling transformation (CCT) diagrams. CCT diagrams are frequently used in heat treatment of steels and represent which types of phase changes will occur in a material for various cooling rates. Because of the strong thermal gradients encountered during DLD, temperature-dependent properties should be employed for better accuracy. High-performance computing may be required if high-resolution solutions are sought, since the scale of heat transfer is relatively small in time and space.

Based on the importance of heat transfer during LBAM, real-time nondestructive evaluation (NDE) via thermal diagnostics/monitoring continues to aid in predicting postfabrication properties and/or existence of part defects. These evaluation methods can also be used for validating analytical and numerical models. During LBAM, and for many common metals for engineering applications, the melt pool achieves temperatures that emit EM waves primarily in the infrared wavelength spectrum. This measurable thermal signature of the melt pool, as well as the temperature distribution along the part, can be correlated with final part properties so that closed-loop control algorithms, which can be integrated into modern LBAM systems, can tailor a part for optimal functionality. Residual stress within the part can also be controlled by monitoring and controlling the inherent LBAM temperatures.

4.2 Introduction to Part Fatigue

Learning Objectives

- Explore fatigue design methodologies used for conventional wrought materials
- Introduce the concept of damage-tolerant design

The heat transfer inherent to the laser-based additive manufacturing (LBAM) process will dictate the final part quality. As will be discussed, the presence of pores, impurities, and/or unmelted regions within parts fabricated via LBAM can negatively affect their mechanical performance and lifetime. The structural integrity of metallic components fabricated using LBAM is not well understood and can be significantly lower than components manufactured using conventional methods. There is a level of conservatism in the immediate use of additively manufactured parts. Although AM is an enabling and appealing technology for fabricating new materials with complex geometries not practical with conventional manufacturing, many industries are hesitant to utilize AM if metal fatigue is a potential failure mode. One must be aware of the mechanical "loading," or forces, experienced by parts in application. Static loading, which is time-invariant in nature, is rarely observed in modern engineering practice, making it essential to consider the consequences of time-dependent cyclic loading. Such loading results in cyclic stresses that may result in fatigue failure.

Extensive fatigue-related research over many years has led to the observation that the fatigue process is actually comprised of two types, or domains, of cyclic loading that are significantly different in character. One domain of cyclic loading consists of significant plastic strain that occurs during each cycle. This domain is associated with high loads and short lives, or low numbers of cycles to produce failure, and is commonly referred to as low-cycle fatigue. The other domain of cyclic loading consists of strain cycles that are largely confined to the elastic range. This domain is associated with lower loads and longer lives, or high numbers of cycles to produce fatigue failure, and is commonly referred to as high-cycle fatigue. Low-cycle fatigue is typically associated with fatigue lives ranging from 1 to 10^4 or 10^5 cycles. In this chapter, the focus is on high-cycle fatigue.

Fatigue may be characterized as a progressive failure phenomenon that proceeds by the initiation and slow, continued propagation of cracks to a sufficient size that final fracture occurs. In traditionally manufactured metallic machine components, fatigue cracks typically initiate from surface damage, inclusions, particles, or pores. An initiated crack will continue to grow in size if the cyclic loading is of a sufficient magnitude, until the crack length reaches a critical size and complete failure occurs.

The fatigue resistance of components fabricated using LBAM will differ from that of wrought materials fabricated using traditional machining methods. Despite the significance of the topic; to date, the fatigue resistance of components fabricated using LBAM has received relatively little attention. The unique thermal histories imposed when using LBAM result in unique microstructures, additional fatigue crack initiation sites, and significant residual stress and distortion. The relatively rough surface finishes resulting from some AM processes will also negatively affect part fatigue resistance.

4.2.1 Fatigue Loading Characterization

The simplest fatigue stress spectrum, or stress versus time, a machine element may be subjected to is a sinusoidal stress–time pattern of constant amplitude and frequency, applied for a defined number of cycles (Collins and Daniewicz 2005). Such a constant amplitude stress–time pattern is illustrated in Figure 4.4, which we now use to define several useful terms and symbols:

σ_{max} = maximum stress in the cycle
σ_m = mean stress = $(\sigma_{max} + \sigma_{min})/2$
σ_{min} = minimum stress in the cycle

σ_a = alternating stress amplitude = $(\sigma_{max} - \sigma_{min})/2$
$\Delta\sigma$ = range of stress = $\sigma_{max} - \sigma_{min}$
R = stress ratio = $\sigma_{min}/\sigma_{max}$

There are many possible constant-amplitude stress–time patterns. To aid their description, one resorts to describing two components of the pattern, including the (1) completely reversed with zero mean stress (R=−1) and (2) released tension or pulsating loading with a zero minimum stress (R=0). Any two of these quantities, except the combinations σ_a and $\Delta\sigma$, are sufficient for effectively describing a stress–time pattern. Note that units for stress are force, or load, per unit area.

Less trivial stress–time patterns are realized when the stress amplitude or mean stress change during the operational cycle. An example of such loading is shown in Figure 4.5, while a random stress spectrum is provided in Figure 4.6. These types of stress–time patterns may be encountered in, e.g., airframe structural members during flight or in tractors performing agricultural tasks. Obtaining realistic data that is meaningful and useful is challenging. Instrumentation aboard existing machines used in the field include accelerometers, strain gauges, and other transducers, and these devices can provide useful information regarding the stress–time pattern, and allow engineers to better design parts against fatigue failure.

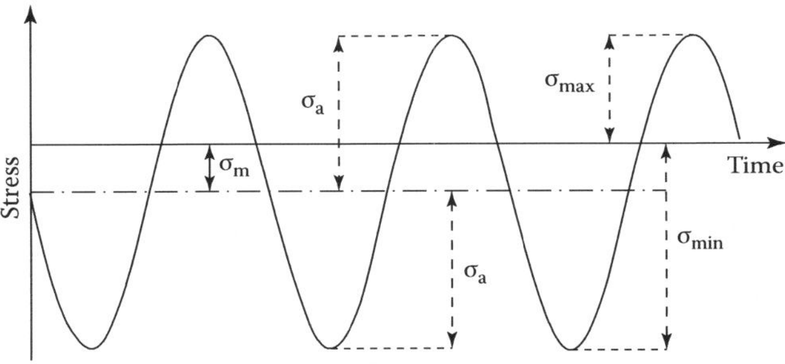

Figure 4.4 An example of constant amplitude cyclic loading. (From Collins, J., *Failure of Materials in Mechanical Design: Analysis, Prediction, Prevention,* John Wiley & Sons, New York, 1993. With permission.)

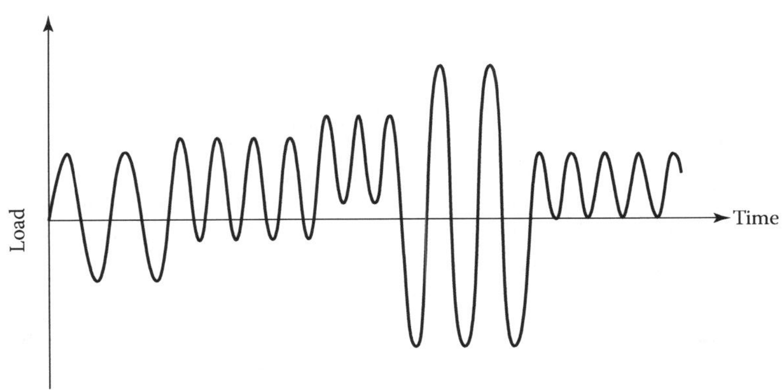

Figure 4.5 An example of variable amplitude cyclic loading.

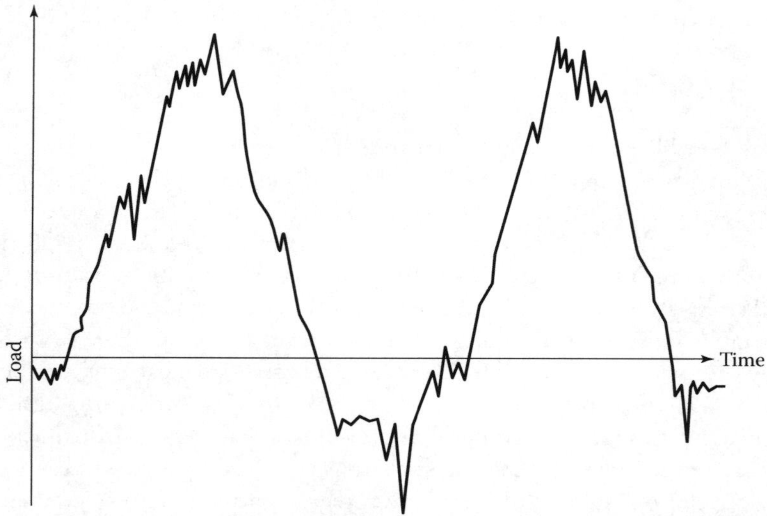

Figure 4.6 An example of a realistic loading condition.

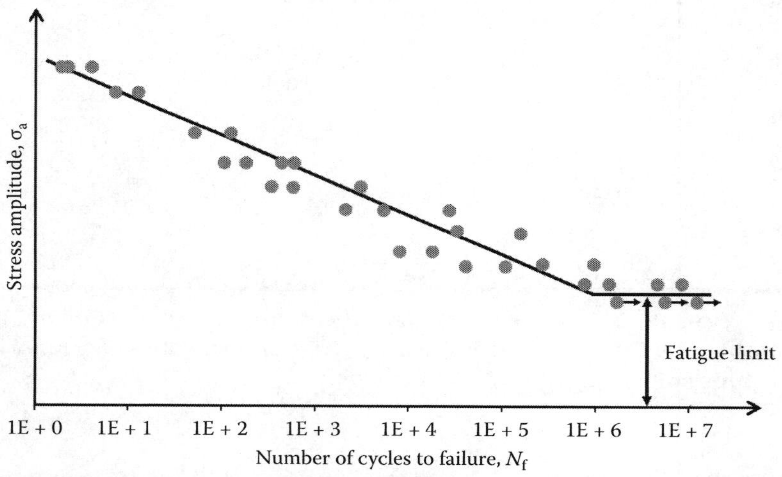

Figure 4.7 Stress–life (S–N) curve showing data scatter and the fatigue limit.

Representative fatigue data in the high-cycle life range are typically displayed on a plot of cyclic stress versus logarithmic life, or alternatively, on a plot of logarithmic stress versus logarithmic life. These plots, called S–N curves, constitute design information of fundamental importance for machine parts subjected to cyclic loading (Collins and Daniewicz 2005). A representative S–N curve is shown in Figure 4.7. Fatigue is a stochastic phenomenon. Because of the significant scatter in fatigue life data that can exist for a particular stress level, it is important to recognize that there is more than one S–N curve for a given material. In fact, a family of S–N curves with the probability of failure as an additional parameter exists. These plots are called S–N–P curves or curves of constant probability of failure on a stress-versus-life plot (Collins and Daniewicz 2005). A representative family of S–N–P curves is illustrated in Figure 4.8. A comprehensive discussion regarding fatigue testing and the experimental generation of S–N–P curves may be found in Juvinall

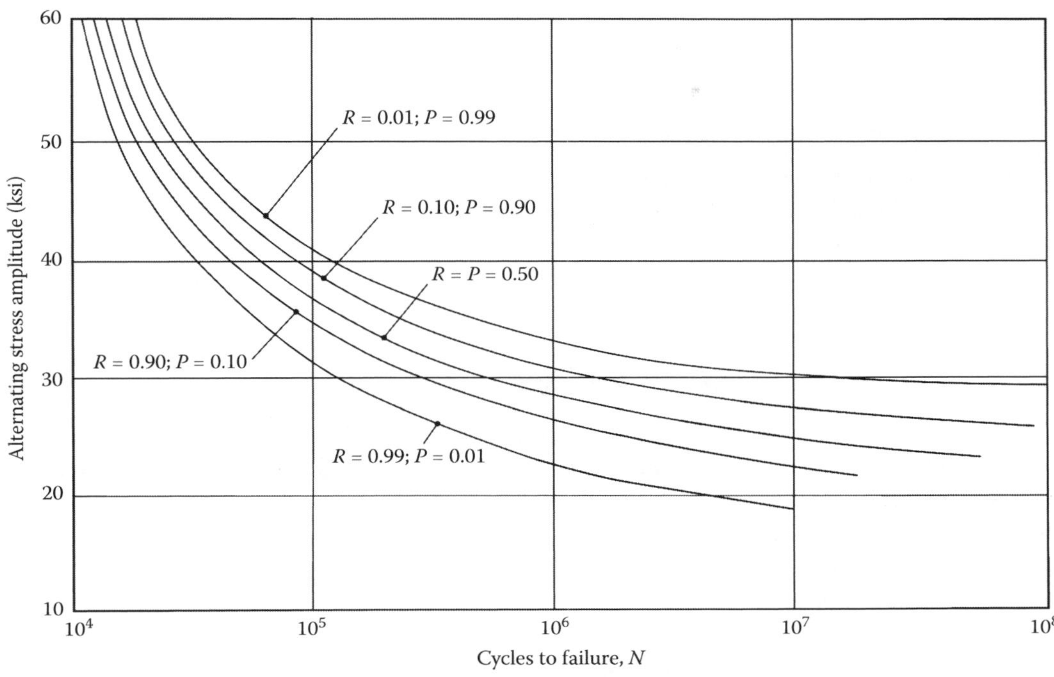

Figure 4.8 S–N–P curves. (From Collins, J., *Failure of Materials in Mechanical Design: Analysis, Prediction, Prevention*, John Wiley & Sons, New York, 1993. With permission.)

(1967). Note that the logarithmic life is typically plotted (in base 10) to allow one to visualize all cycle of fatigue, which will typically range from 1 to 10^5 or more. All fatigue tests associated in producing data for publication or for quality assurance have been standardized and continue to be regulated by international, professional societies such as American Society for Testing and Materials (ASTM) International and the International Organization for Standardization (ISO).

Many metals exhibit a relatively flat curve in the long life range, giving the perception that an infinite life is possible, as shown in Figure 4.7. The term fatigue limit or endurance limit is used to describe the stress level below which an infinite number of cycles can be sustained without failure. The term fatigue strength at a specified life σ_f is used to characterize failure response in the finite life range, where the term *fatigue strength* identifies the stress level at which the failure will occur for a specified life. The specification of fatigue strength without specifying the corresponding life is meaningless. The specification of a fatigue limit always implies an infinite life. Not all materials exhibit an endurance limit. There are fatigue experts who question the existence of a fatigue limit in general, suggesting that if enough cycles of loading were to be applied to a component, fatigue failure would eventually occur at any stress level. As a practical matter, fatigue testing is typically ceased for fatigue lives greater than 10^7 cycles, in order to reduce time and cost investments related to experimentation.

4.2.2 Factors That Affect S–N–P Curves

There are several factors that influence fatigue resistance, including: configuration of the stress–time pattern, nonzero mean stress, material composition, microstructure, geometrical discontinuities, size, surface roughness, residual stress, operating temperature, and prior fatigue damage. It is usually necessary to search the open literature to find the information required for a specific application, and it may be necessary to perform experiments to produce data where they are unavailable (Collins 1993).

Surface finish effects can be significant, and an increase in a part's surface roughness can produce dramatic reductions in its fatigue resistance. Fabrication using LBAM requires careful attention to surface finish, if good fatigue resistance is needed. Residual stresses, which are static compressive or tensile stresses that exist within the part in the absence of external forces, induced during the building process will likely result in build orientation effects. The accompanying distortions may make adherence to tight dimensional tolerances difficult. Significant microstructural texturing and unique microstructures associated with LBAM are varied and strongly depend on the material used and the subsequent cooling rate induced by the employed build parameters.

4.2.3 Nonzero Mean Fatigue Stresses

Fatigue testing of parts in the laboratory can be conducted while holding either the load or strain constant. Most basic fatigue load-controlled data are collected in the laboratory using uniaxial, completely reversed alternating stresses with $R = -1$. Most service applications involve nonzero mean cyclic stresses. Fatigue data produced using different levels of mean and alternating stress are often presented similarly to Figure 4.9. An alternative means of presenting this type of fatigue data is illustrated in Figure 4.10. If data are not available, the influence of nonzero mean stressmay

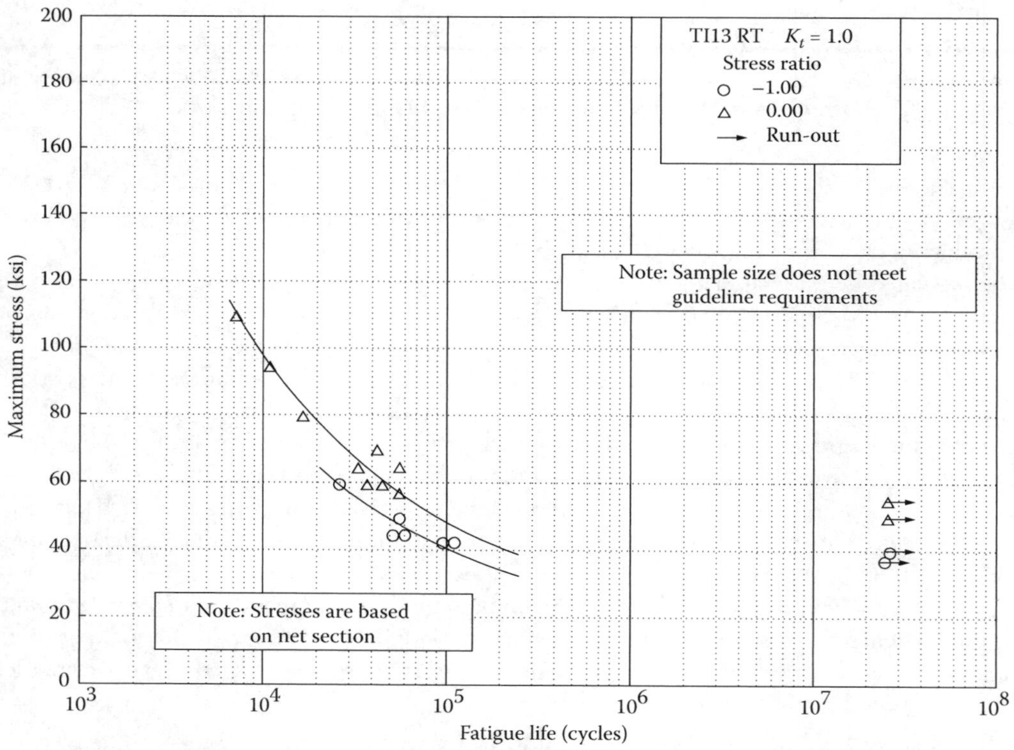

Figure 4.9 Fatigue resistance with mean stress effects for un-notched titanium alloy TI13. (From US Department of Transportation, and Federal Aviation Administration, Metallic materials properties development and standardization (MMPDS), 2003.)

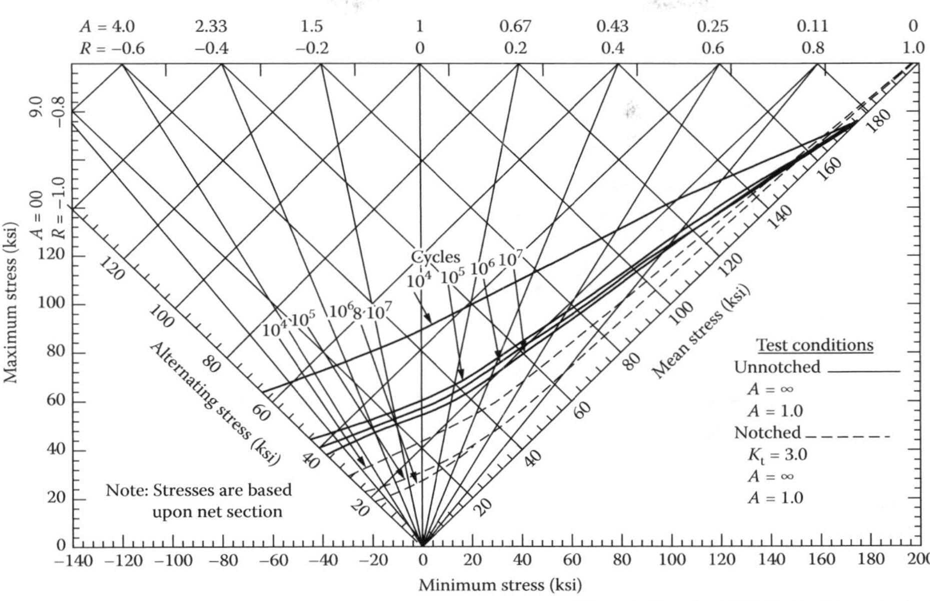

Figure 4.10 Constant life diagram for notched and un-notched Ti–13V–11Cr–3Al alloy sheet at room temperature. (From US Department of Transportation, and Federal Aviation Administration, Metallic materials properties development and standardization (MMPDS), 2003.)

be estimated using the empirical Goodman equation, which is a widely used relationship in experimental fatigue testing. For tensile mean stress ($\sigma_m > 0$), this relationship may be written as:

$$\frac{\sigma_a}{\sigma_f} + \frac{\sigma_m}{\sigma_u} = 1, \tag{4.2}$$

where σ_u is the material ultimate strength and σ_f is the completely reversed ($R=-1$) fatigue strength for a prescribed number of cycles N. For a given alternating stress, compressive mean stresses ($\sigma_m < 0$) may conservatively be assumed to exert no influence on fatigue life. Thus, for $\sigma_m < 0$, the fatigue response would be assumed identical to that for $\sigma_m = 0$ with $\sigma_a = \sigma_f$. The Goodman relationship is illustrated graphically in Figure 4.11. Any cyclic loading that produces an alternating stress and mean stress that exceeds the bounds of the curve is predicted to induce a fatigue failure in fewer than N cycles. Any alternating stress–mean stress combination that lies within the curve is predicted to result in more than N cycles without failure.

4.2.4 Spectrum Loading and Cumulative Damage

In most engineering applications where fatigue is a concern, the alternating stress amplitude may be expected to vary or change in some way during the service life. Such variations in load amplitude, often referred to as spectrum loading, make direct use of standard, constant-amplitude S–N data inapplicable. It thus becomes necessary to employ a methodology that permits good design estimates for parts undergoing spectrum loading based on using standard, constant-amplitude S–N curves.

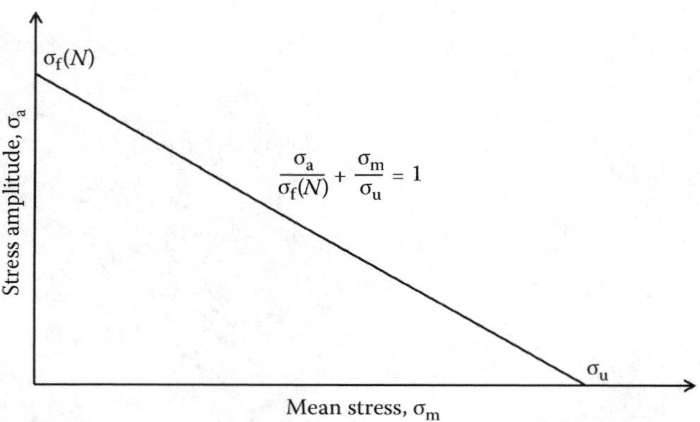

Figure 4.11 Goodman diagram to approximate mean stress effects.

Operations at any cyclic stress amplitude will produce fatigue damage. The severity of this damage will be related to the number of cycles incurred during operation at that stress amplitude and the magnitude of that stress amplitude. The damage incurred is permanent and does not dissipate or diminish with time or inactivity; thus, the term fatigue might be considered misleading. Parts undergoing several different stress amplitudes in sequence will result in an accumulation of total damage equal to the sum of the damage increments accrued at each individual stress level. When the total accumulated damage reaches a critical value, fatigue failure is predicted to occur. Although this concept is simple in principle, difficulty is encountered in engineering practice since the precise assessment of the amount of damage incurred per operation at a given stress level σ_i for a specified number of cycles n_i is not straightforward. The most widely used damage rule is the Palmgren-Miner hypothesis or the linear damage rule.

Consider Figure 4.12; by definition, operation at a constant stress amplitude σ_1 will produce complete damage, or failure, in N_{f1} cycles. Operation at stress amplitude σ_1 for a number of cycles n_1 with $n_1 < N_{f1}$ will produce a smaller fraction of damage known as the damage fraction D_1. Operation over a number of different stress levels results in a damage fraction D_i for each of the different stress levels σ_i. When these damage fractions sum to unity, failure is predicted to occur:

$$D_1 + D_2 + \cdots + D_{i-1} + D_i \geq 1. \tag{4.3}$$

The Palmgren-Miner hypothesis asserts that the damage fraction at any stress level σ_i is linearly proportional to the ratio of the number of cycles during operation (n_i) to the total number of cycles that would produce a failure at that stress level (N_{fi}):

$$D_i = \frac{n_i}{N_{fi}}. \tag{4.4}$$

Hence, failure is predicted to occur if:

$$\frac{n_1}{N_{f1}} + \frac{n_2}{N_{f2}} + \cdots + \frac{n_i}{N_{fi}} \geq 1. \tag{4.5}$$

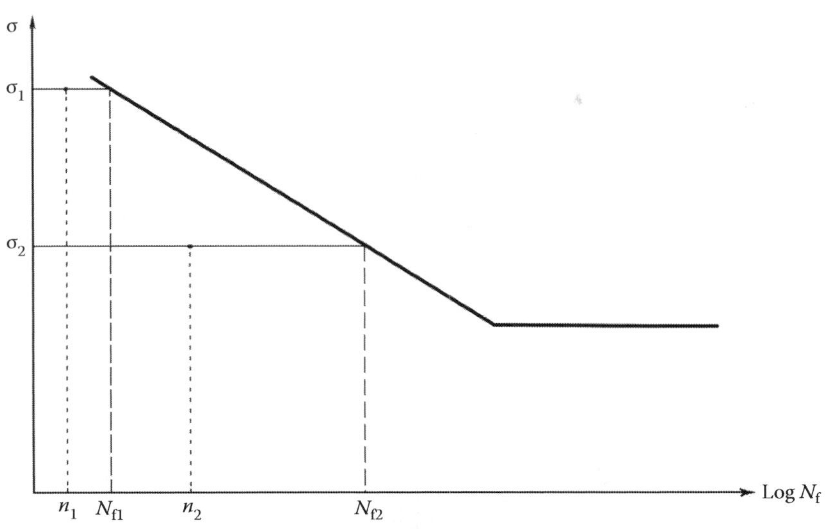

Figure 4.12 Schematic showing the cumulative damage calculation.

The Palmgren-Miner hypothesis has one important virtue—simplicity—and for this reason, it is widely used. It must be recognized, however, that in its simplicity certain significant influences are unaccounted for, and errors may therefore be expected (Collins 1993). The most significant shortcomings of the linear theory are that no influence regarding the order of application of various stress levels is recognized, and damage is assumed to accumulate at the same rate at a given stress level without regard to past loading history. Experimental data indicate that the order in which various stress levels are applied does have a significant influence, and also that the damage rate at a given stress level is a function of prior loading history. Experimental values for the Miner's sum at the time of failure range from about ¼ to 4, depending on the specific order in which the stress levels are applied (Collins 1993). If the cyclic stress amplitudes are mixed in the sequence randomly, the experimental Miner's sum will more closely approach unity at the time of failure, with values of Miner's sums corresponding to failure in the range of about 0.6–1.6. Since many service applications involve random fluctuating stresses, the use of the Palmgren-Miner linear damage rule is often satisfactory.

If the Palmgren-Miner hypothesis is to be employed to predict fatigue under a stress spectrum such as that given in Figure 4.6, the stress spectrum needs to be converted to an equivalent number of constant amplitude cycles with each having an associated stress amplitude and mean stress. This empirical conversion is accomplished with what is known as a cycle counting method. The most commonly used cycle counting method is the rain flow method illustrated in Figure 4.13. The stress–time history is orientated so that the time axis is vertically downward, and the lines connecting the stress peaks are imagined to be a series of roofs. Several rules are imposed on rain flowing down these roofs so that cycles and half-cycles can be calculated. Rain flow begins successively at the inside of each stress peak. The rain flow initiating at each peak is allowed to flow downward and continue, except that, if it initiates at a minimum, it must stop when it comes opposite to a minimum "more negative" than the minimum from which it started. For example, in Figure 4.13, rain flow begins at peak 1 and stops opposite peak 7 since peak 7 is more negative than peak 1. A half-cycle is thus counted between peaks 1 and 6. Similarly, if the rain flow initiates at to a maximum, it must stop when it comes opposite a maximum "more positive" than the maximum from which it initiated.

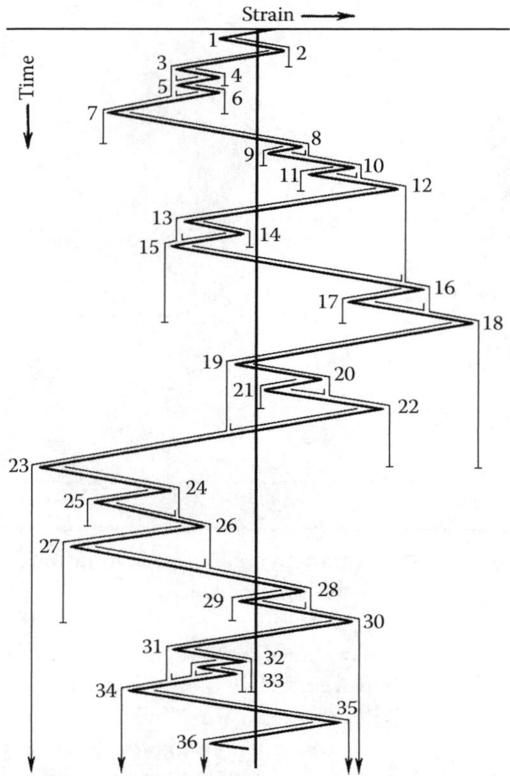

Figure 4.13 Rainflow cycle counting. (From Collins, J., *Failure of Materials in Mechanical Design: Analysis, Prediction, Prevention,* John Wiley & Sons, New York, 1993. With permission.)

For example, in Figure 4.13, rain flow begins at peak 2 and stops opposite peak 4; thus, a half-cycle between peaks 2 and 3 is counted. Rain flow must also stop if it meets the rain flow from a roof above. For example, in Figure 4.13, the half-cycle beginning at peak 3 ends beneath peak 2. Note that every part of the strain–time history is counted once only (Collins 1993).

The nonzero mean stress cycles computed from the cycle counting method are then converted to be equivalent to the reversed cycles via the Goodman equation or another comparable empirical expression based on specific material data. Using Equation 4.2, for a tensile mean stress, the equivalent completely reversed stress σ_{eq} is written as:

$$\sigma_{eq} = \frac{\sigma_a}{1 - \sigma_m / \sigma_u}, \quad \sigma_m \geq 0 \tag{4.6}$$

and for a compressive mean stress as:

$$\sigma_{eq} = \sigma_a, \quad \sigma_m \leq 0. \tag{4.7}$$

The fatigue damage fraction from each counted cycle with associated equivalent stress σ_{eq} is then found using the existing standard constant amplitude S–N curve generated under $R = -1$.

4.2.5 Fatigue Crack Growth

An alternative to the stress-based fatigue design methodology previously discussed is a damage tolerance approach. In this relatively new engineering approach to fatigue design, if a crack with length a is present in the component, then the question becomes: "How many cycles of loading can be applied before the crack slowly grows to a sufficient length such that fracture occurs?" For many structures or machine elements, the time required for a fatigue-initiated crack or preexisting flaw to grow to critical size is a significant portion of its useful life. A damage tolerance approach to fatigue allows the incorporation of nondestructive testing technologies that are employed to detect cracks during maintenance in existing hardware under service and makes decisions regarding the further use of the cracked hardware. In addition, damage tolerance can be used during the conceptual design of new hardware. Here, the presence of crack-like flaws is assumed to exist in select critical locations, and the designer incorporates features that are more damage tolerant and not severely weakened by the presence of the flaw.

The foundation of the damage-tolerant perspective is the stress intensity factor K. For a given remote applied stress σ and crack length a:

$$K = F(a)\sigma\sqrt{\pi a}, \tag{4.8}$$

where $F(a)$ is dependent on the type of loading and the geometry away from the crack. Much work has been completed in determining the values of F for a wide variety of geometries and loading conditions (see e.g., Tada and Paris 2000). The *stress intensity factor* is a single-parameter measure of the intensity of the stress near the crack tip. The stress intensity factor is also a similitude parameter. Thus, if the stress intensity factors are equal, different geometries under different applied stresses will have identical crack tip stress fields. Therefore, crack growth data generated from simple laboratory specimens may be utilized for crack growth predictions in more complex machine elements or structures.

Most commercial finite element analysis (FEA) software packages have special crack tip elements allowing for the numerical computation of stress intensity factors. A discussion of some of the techniques employed within these software packages is given by Anderson (1995). Through the use of weight functions (Wu and Carlsson 1991), stress intensity factors may also be computed using numerical integration. The fatigue crack growth rate $\mathrm{d}a/\mathrm{d}N$ under cyclic loading has been found to correlate with the crack-tip stress intensity factor range such that:

$$\frac{\mathrm{d}a}{\mathrm{d}N} = g(\Delta K), \tag{4.9}$$

where ΔK is the stress intensity factor range computed using the maximum and minimum applied stresses with $\Delta K = K_{max} - K_{min}$. Most commonly, fatigue crack growth is characterized using a power law:

$$\frac{\mathrm{d}a}{\mathrm{d}N} = C(\Delta K)^n, \tag{4.10}$$

where the empirical parameters C and n are functions of material type, R ratio, thickness, temperature, and environment. Standard methods exist for conducting fatigue crack growth tests in a laboratory setting ("ASTM E647-08, Standard Test Method for Measurement of Fatigue Crack Growth Rates" 2008).

Given an initial crack length a_i, corresponding to a crack either initiated by cyclic loading or initially present as a flaw. Equation 4.10 may be solved to give the number of cycles N required to grow the crack to a size a_N such that

$$N = \int_{a_i}^{a_N} \frac{da}{C(\Delta K)^n}. \tag{4.11}$$

Given that ΔK is a function of crack length, numerical integration techniques are typically required to compute N. An approximate procedure and several idealized examples are presented by Parker (1981). The fatigue crack propagation life N will also be influenced by the presence of residual stresses, which may exist due to welding, heat treatment, carburizing, grinding, or shot-peening. Compressive residual stresses are beneficial, decreasing the rate of fatigue crack growth and increasing life. While approximate methodologies exist for incorporating the effects of residual stress within fatigue crack growth predictions (Rice 1997), residual stress distributions are often difficult to characterize. In addition, these stresses could evolve or change over time, while the machine component is in service.

4.3 Fatigue Behavior of Additive Manufactured Parts

Learning Objectives

- Survey the fatigue behavior of LBAM specimens
- Appreciate unique fatigue crack growth behavior of LBAM parts
- Understand the opportunities and challenges related to LBAM

The fatigue resistance of machine components fabricated using LBAM is not well understood and can be significantly lower than the components manufactured using conventional methods. LBAM components experience a thermal history dramatically different than wrought material, and this thermal history is controllable by altering the user-selected build parameters. This thermal history alters the metallurgical structure; therefore, the subsequent fatigue strength of the material is also affected. Residual stresses are also generated as a consequence of the nonuniform plastic deformations induced by the severe thermal strain variations. Columnar grains produced during solidification also lead to microstructural anisotropy, such that build orientation becomes a factor. LBAM components may also exhibit rough surface finishing, porosity, regions of unmelted powder, and lack of fusion between layers, all of which will reduce fatigue resistance of the part.

The extreme thermal gradients and histories associated with LBAM represent both a challenge and an opportunity. If the LBAM process is properly controlled with optimal build parameters, it is theoretically possible to produce a desired microstructure in a prescribed location within a given component to enhance its fatigue resistance. In addition, it is possible that the anisotropic nature of components built using AM can be exploited, aligning component materials axes with axes of loading.

4.3.1 Fatigue

Wang et al. (2013) evaluated the high-cycle fatigue behavior of double annealed TC18 titanium alloy (Ti–5Al–5Mo–5V–1Cr–1Fe) fabricated using LBAM. A thick plate-like specimen of TC18 titanium alloy was fabricated using LBAM. The plate was stress relieved at 750°C for 1 h, followed by air cooling. It was then annealed at 880°C for 1 h, followed by furnace cooling to 750°C for another 2 h, and air cooled to room temperature. Finally, it was aged at 580°C for 4 h and again air cooled. Fatigue specimens were removed from the plate using EDM according to the GB/T-15248 standard ("GB/T 15248: The Test Method for Axial Loading Constant-Amplitude Low-Cycle Fatigue of Metallic Materials" 2008) in a horizontal direction. The surfaces of all specimens were mechanically polished, and fatigue testing was conducted at a stress ratio of $R = 0.10$. The fatigue strength of the LBAM TC18 with Widmanstätten (e.g., lamellar) microstructure and its comparison to wrought TC18 with lamellar/bimodal microstructures is presented in Figure 4.14 (Wang et al. 2013; Wu et al. 2013). It may be seen that LBAM TC18 shows significantly lower fatigue strength at 10^7 cycles. This is primarily due to the presence of micropores in the LBAM specimens. This leads to the question of whether the LBAM process employed was optimized for generating fully dense TC18 and if this is typical for any reported mechanical data corresponding to any parts fabricated via AM. Standardization of the LBAM process, and the subsequent testing methods, is still ongoing.

Pores may greatly affect the fatigue behavior of AM parts by creating stress concentrations. Porosity and unmelted feedstock can accelerate the nucleation of cracks and decrease the fatigue life. The micropore size and its distance from the material surface both strongly affect the HCF life of titanium alloys (Lin et al. 2005). As reported by Wang et al. (2013), the life of LBAM TC18 decreases with large pore sizes and with distance from the surface. Moreover, pore location is more influential on the fatigue resistance than its size at lower stress levels.

The HCF performance of Inconel 718 fabricated using LBAM is presented in Figure 4.15 (Amsterdam and Kool 2009). Uniaxial fatigue tests were performed at a stress ratio of $R = 0.10$. Before machining and polishing specimens to a surface roughness $R_a < 0.2\,\mu m$, a heat treatment schedule was performed. Results indicate that the fatigue strength for specimens in the mid-life

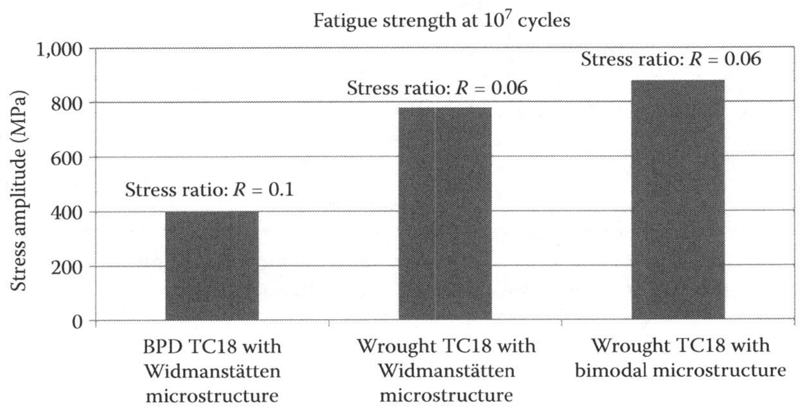

Figure 4.14 Comparison of fatigue strength of LBAM TC18 specimens (Wang et al. 2013) to wrought TC18 with Widmanstätten (lamellar) and bimodal microstructures (Wu et al. 2013) at 10^7 cycles. (From Shamsaei, N. et al., *Addit. Manuf.*, 8, 12–35, 2015. With permission.)

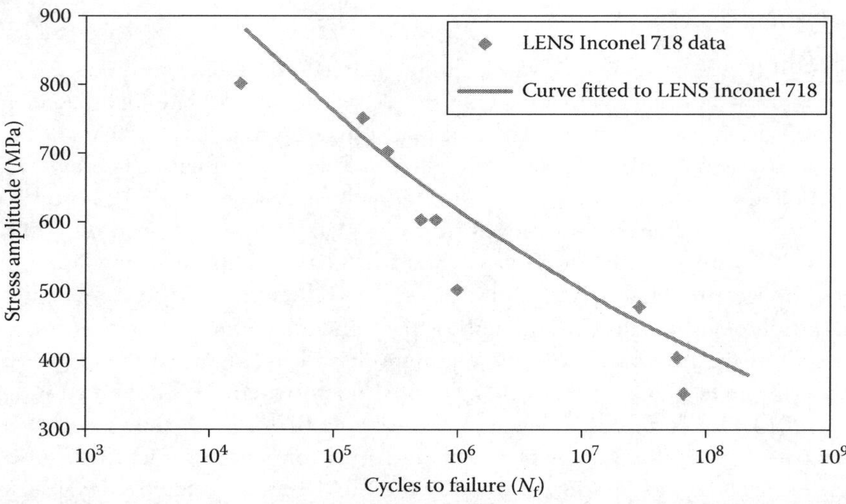

Figure 4.15 S–N data for Inconel 718 specimens fabricated with LBAM (Amsterdam and Kool 2009). (From Shamsaei, N. et al., *Addit. Manuf.*, 8, 12–35, 2015. With permission.)

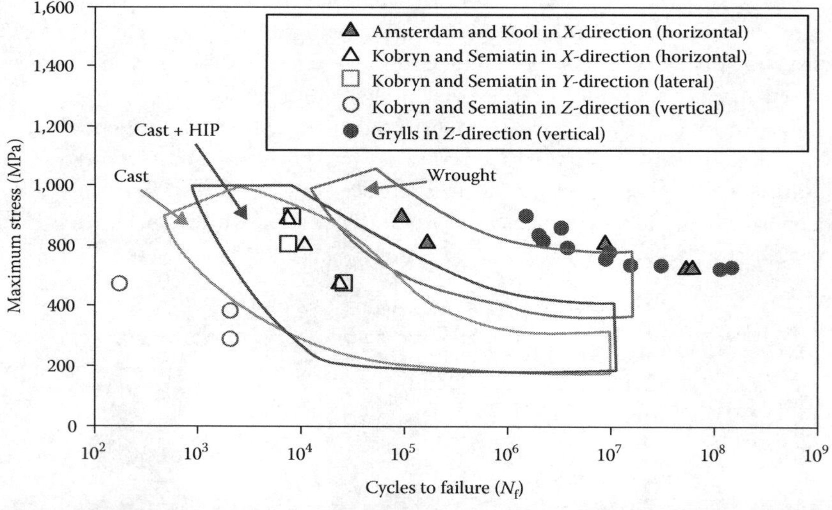

Figure 4.16 Comparison of S–N data of different study for the LENS Ti–6Al–4V specimens in different directions (Kobryn and Semiatin 2001; Grylls 2005) to the cast, cast plus HIP, and wrought materials (Donachie 2000). (From Shamsaei, N. et al., *Addit. Manuf.*, 8, 12–35, 2015. With permission.)

and long life regimes may be comparable with that of wrought Inconel 718. However, a significant drop in the fatigue strength was seen in the very high-cycle fatigue regime ($N_f > 10^7$), as seen in Figure 4.15. Lack of fusion or bonding between layers and the presence of defects were reported as the reason for this behavior (Amsterdam and Kool 2009).

LBAM 316 L stainless steel was fatigue tested by Spierings et al. (2013) in its as-built, machined, and polished condition to investigate the role of surface finish. The as-built condition corresponds to the condition of the part immediately upon removal of the AM machine and utilized substrate. It was found that the endurance limit was strongly impacted by the surface finish, but for 1.0 E5 $< N <$ 3.0 E6, surface finish showed little effect. The as-built surface exhibited a surface roughness $R_a = 10\,\mu m$, while the machined specimens were measured to give $R_a = 0.4\,\mu m$. Polishing resulted in a surface with $R_a = 0.2\,\mu m$.

Contradictory results have been reported in the literature regarding the fatigue behavior of LBAM Ti–6Al–4V specimens. Fatigue data from various studies on LBAM Ti–6Al–4V, printed in different directions, are summarized in Figure 4.16 (Kobryn and Semiatin 2001; Grylls 2005; Amsterdam and Kool 2009). Some results indicate that LBAM Ti–6Al–4V has comparable fatigue behavior to its wrought form (Grylls 2005; Amsterdam and Kool 2009). However, Kobryn and Semiatin (2001) report significantly lower fatigue strength for LBAM Ti–6Al–4V in the midlife regime. Variation in test setups, manufacturing parameters, and postmanufacturing processes can easily contribute to discrepancies in published fatigue data. These variations are a result of there being insufficient process and testing standardization.

Grylls (2005) and Amsterdam and Kool (2009) used a stress ratio of $R = 0.10$ in their uniaxial fatigue tests; however, the stress ratio used in Kobryn and Semiatin's study was not reported. Specimen surface condition and size may also contribute to the different fatigue lives observed in these studies. In Amsterdam and Kools' study, fatigue specimens were machined and polished to a roughness $R_a < 0.2\ \mu m$. The other fatigue studies do not report in detail on the specimens' surface condition. The LBAM process parameters greatly impact a part's thermal history, microstructure, bonding between layers, pore size, and shape; thus, the fatigue resistance of the fabricated part will depend on such parameters as well.

Residual stresses accumulated in parts also depend on manufacturing thermal history and can greatly affect the fatigue behavior of LBAM products (Costa and Vilar 2009). Postmanufacturing process parameters, such as heat treatment, are also influential factors

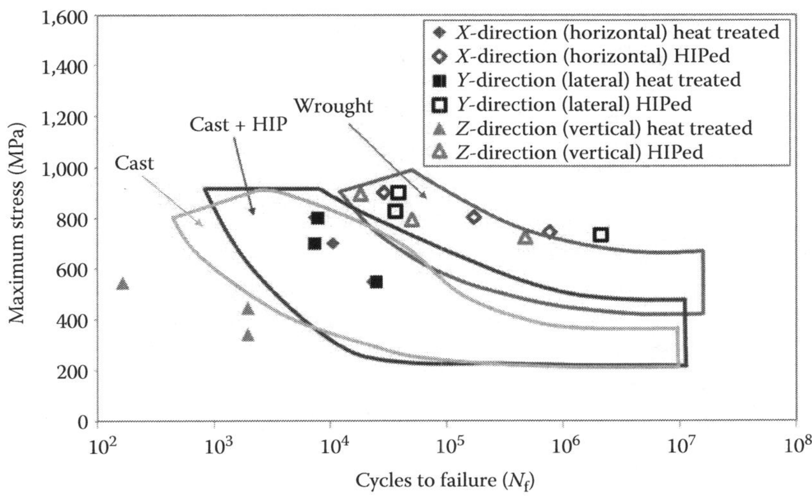

Figure 4.17 Comparison of fatigue strengths for LBAM Ti–6Al–4V in different building orientations for both heat treated and HIPed specimens (Kobryn and Semiatin 2001). (From Shamsaei, N. et al., *Addit. Manuf.*, 8, 12–35, 2015. With permission.)

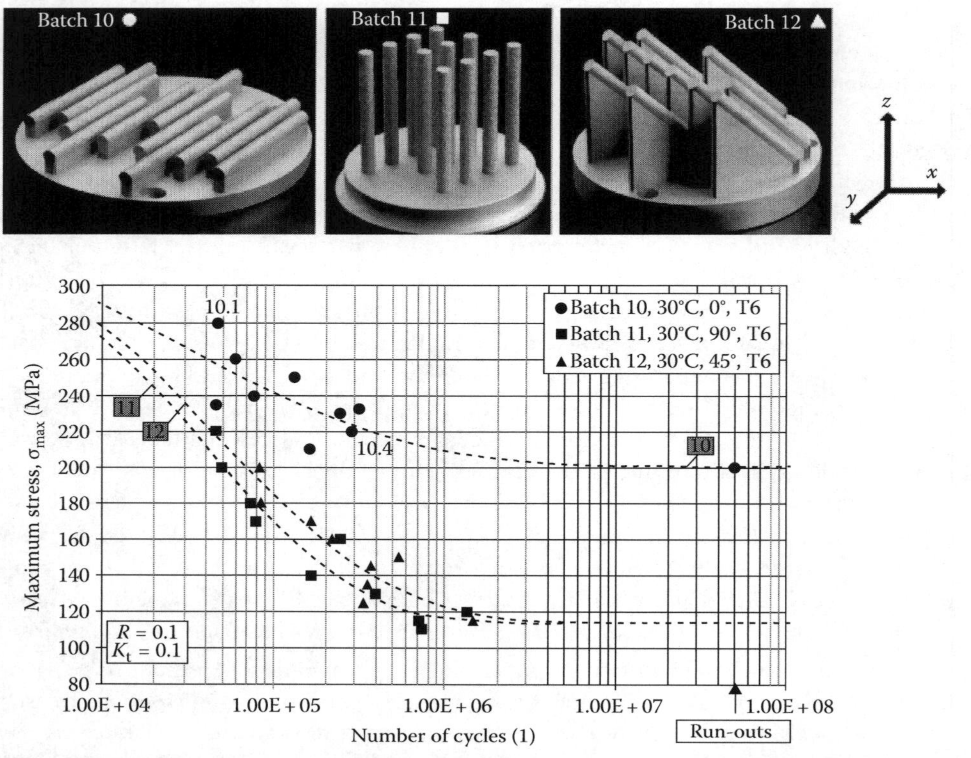

Figure 4.18 Effect of build orientation on S–N behavior for LBAM AlSi10Mg-T6. (Adapted from Brandl, E. et al., *Mater. Des.,* 34: 159–69, 2012.)

on the fatigue behavior. In the Amsterdam and Kool's study (2009), Ti–6Al–4V specimens were heat treated at 970°C for 1 h, water quenched and aged at 538°C for 4 h, and then air cooled, while Kobryn and Semiatin's (2001) specimens were stress relieved in vacuum for 2 h at 700°C–730°C. Grylls (2005) does not report any heat treatment details employed for their investigated specimens. Research focused on the effects of process parameters, part orientations, and postmanufacturing treatments on the final fatigue behavior of LBAM parts is still needed.

Anisotropic fatigue resistance of LBAM parts has been reported (Kobryn and Semiatin 2001). As shown in Figure 4.17, the fatigue behavior of LBAM Ti–6Al–4V specimens depends on the fabrication/deposition direction. Specimens vertically oriented (*Z*-direction) during fabrication display significantly lower fatigue strength than those either horizontally oriented (*X*-direction) or laterally oriented (*Y*-direction) during fabrication (Kobryn and Semiatin 2001). This can be attributed to the laser scanning directions changing with and along each layer, inducing nonuniform cooling rates, resulting in different microstructurural features, residual stresses, and consequently fatigue behavior. Specimens fabricated in the *Z*-direction exhibited extensive porosity compared to those built in the *X*- and *Y*-directions (Kobryn and Semiatin 2001). In addition, parts fabricated using an alternating laser scanning pattern, that rotated 90° with each new layer, possessed similar fatigue resistance as specimens fabricated in the *X*- and *Y*-directions. Postbuild processes, such as hot isostatic pressing (HIP), can reduce or diminish unmelted regions and pores and thus improve the fatigue strength of laser-deposited parts, potentially giving properties comparable to wrought, especially in the long life regime. In addition, specimens treated with HIP exhibit less anisotropic behavior. All of these findings suggest that porosity may greatly influence the anisotropic behavior of LBAM parts. In general, the effects of different manufacturing parameters on the anisotropic behavior of LBAM parts are not fully understood.

Brandl et al. (2012) evaluated the HCF performance of LBAM AlSi10Mg. Specimens were manufactured using different build directions with and without a build platform preheat of 300°C and fatigue tested with R=0.10. Cylindrical specimens built horizontally gave a much higher fatigue resistance than those built vertically or at an angle of 45° when no platform preheat was used, as shown in Figure 4.18. This direction-dependent mechanical behavior, or anisotropy, vanished when a 300°C preheat was employed. All specimens received a postbuild heat treatment to induce a T6 condition. Fatigue cracks were observed to initiate from pores or unmelted regions.

4.3.2 Fatigue Crack Growth

Fatigue crack growth in LBAM specimens and components will be influenced by the relationship between the build direction and the plane of crack growth. This anisotropy is likely the result of residual stresses induced during the build, as the effect is most prominent in as-built parts. Postbuild heat treatments diminish the level of anisotropy as discussed by Cain et al. (2015). Conducting the build at an elevated temperature is another way to remove anisotropy as performed by Edwards et al. (2013) for Ti–6Al–4V. Eliminating postbuild heat treatments helps reduce manufacturing costs. Leuders et al. (2013) studied fatigue crack growth in Ti–6Al–4V and found that residual stresses and microstructure impacted fatigue crack growth, while pores did not.

Fatigue crack growth rates for LBAM Inconel 625 and AISI 316 L SS are compared with their corresponding wrought materials (DiMatteo 1996), as shown in Figure 4.19 (DiMatteo 1996; Ganesh et al. 2010, 2014). 25-mm-thick compact tension (CT) specimens were extracted from LBAM samples and tested under ΔK increasing and ΔK decreasing conditions with $R = 0.10$ according to ASTM E647 (ASTM 1993). Crack growth in the CT specimens was directed parallel to laser scan directions and perpendicular to the specimen's build direction. As shown in Figure 4.19a, the fatigue crack growth rate for the LBAM Inconel 625 specimens is lower than that of the wrought Inconel 625 (DiMatteo 1996) for the stress intensity factor range of $\Delta K = 20–25$ MPa √ m. However, the fatigue crack growth behavior for the LBAM AISI 316L SS is similar to that of the wrought material for larger ΔK values, as presented in Figure 4.19b. Monotonic tension tests revealed the LBAM Inconel 625 possessed a higher yield strength and lower ductility when compared with the wrought material.

The fatigue crack growth threshold ΔK_{th}, below which a crack will not grow under cyclic loading, was measured by Wycisk et al. (2014), Leuders et al. (2013), and Seifi et al. (2015) for LBAM, as well as electron beam AM, Ti–6Al–4V. In the as-built condition, Lueders reported a threshold value of 1.4 MPa√m. If a postbuild heat treatment consisting of 800°C for 2 h followed by a furnace cool was used, the threshold was observed to improve to 4.6 MPa√m. Wycisk reported a threshold value of 3.48 MPa√m when using a postbuild heat treatment consisting of 650°C for 3 h followed by argon cooling. In the as-built condition, Seifi reported threshold values varying from 3.4 to 5.7 MPa√m depending on the build orientation.

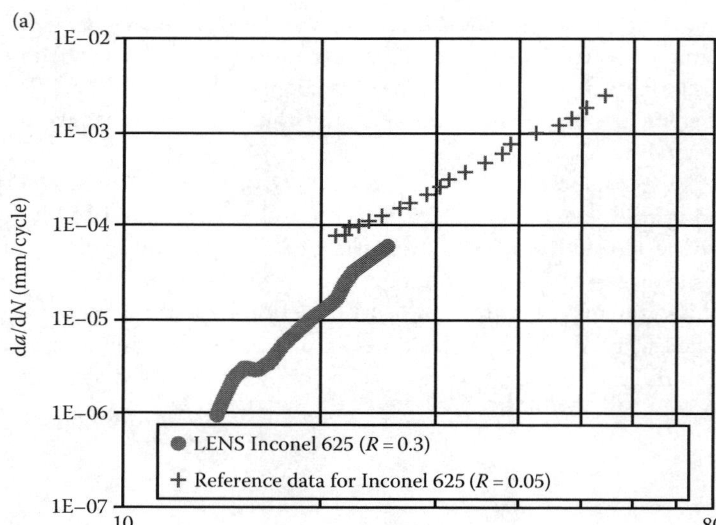

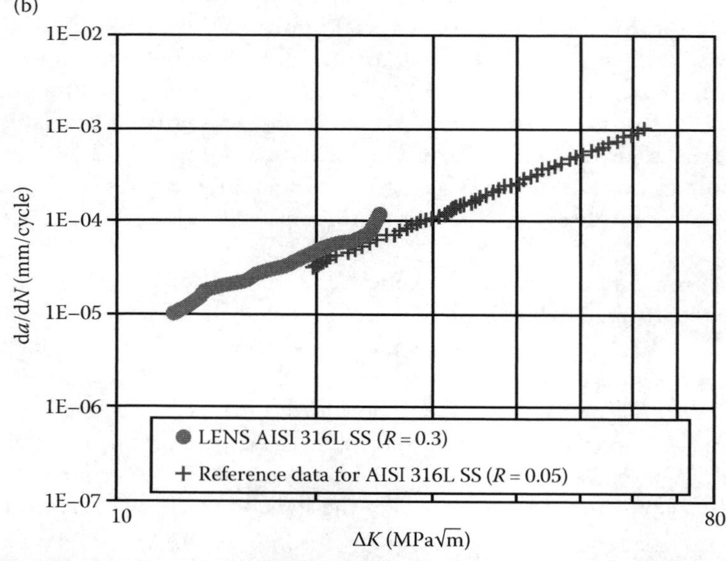

Figure 4.19 Comparison of fatigue crack growth rate d*a*/d*N*, versus intensity factor, ΔK, for LENS (a) Inconel 625 (Ganesh et al. 2010) and (b) AISI 316 L SS (Ganesh et al. 2014) specimens to wrought materials (DiMatteo 1996). (From Shamsaei, N. et al., *Addit. Manuf.*, 8, 12–35, 2015. With permission.)

4.4 Summary

AM, in general, offers the ability to manufacture parts through the repetitious deposition of material layers directly from a digital CAD model (Daniewicz and Shamsaei 2017). Several AM techniques have been developed, providing opportunities to manufacture complex geometries or functionally graded materials, which are unobtainable through traditional manufacturing techniques. However, to fully exploit these capabilities, the structural integrity of parts fabricated using AM technology must be understood and fully characterized. While fabricating geometrically complex parts using AM technology may be possible by employing topological optimization, their mechanical behavior and overall trustworthiness is not yet fully documented.

Heat transfer is the driving force for accomplishing LBAM—directly or indirectly affecting part quality and structural integrity as dictated by material coupled, thermally driven solidification, and microstructural evolution. The two most common types of LBAM processes are DLD, which involves *in situ* laser and feedstock delivery, and PBF-L, which involves a spread layer of powder and a selective laser scan. The successful LBAM of materials with structural integrity requires one to understand and appreciate the energy, momentum, and transfer mechanisms during the process. The LBAM process is not always "plug-and-play." Time must be invested in determining an effective group of process and design parameters for ensuring part quality. If an end-user employs a nonoptimal group of process/design parameters, lack of fusion, and pore formation can occur within the material. These material "defects" can be due to lack of heat penetration, or the realization of an ineffective HAZ, and this results in the part having porosity. Since the melt pool is the initiation of the solid part, its morphology, temperature, and wetting behavior are of paramount interest in quality control. The melt pools inherent to LBAM are not too different from the melt pools generated in more traditional laser and arc welding procedures; however, the powder-coupling complicates its temperature/momentum predictability and also induces free surface instability (e.g., splashing).

The fatigue resistance, i.e., structural integrity of LBAM parts may vary from that of traditionally manufactured ones. This is primarily attributed to the process itself, which consists of highly localized heating and cooling, coupled with powders that generally have varying degrees of quality, porosity, cleanliness, shape/size, etc. The cooling rates are generally much higher than those associated with conventional fusion welding, leading to previously unobserved microstructures evolved under nonequilibrium conditions with significant anisotropy. In addition, the severe temperature gradients experienced during LBAM result in the generation of potentially significant residual stresses and distortion. Powder–heat source interactions and the layer-by-layer nature of LBAM also result in a vast number of deposited tracks or layers that are subject to repetitious heating and remelting during the build, creating susceptibility to defects such as porosity, lack of fusion, and unmelted powder—all of which can serve as fatigue crack initiation sites and influence fatigue behavior dramatically.

4.5 Review Questions

4.5.1 Critical Thinking

1. Why is thermal energy employed, and directed, for the AM of metals? Can you think of an alternate method, which does not use thermal energy, for the AM of metals?
2. Describe an industry where AM technology could provide a pivotal boost in productivity or design that could be strongly influential.
3. Based on research shown at the end of the chapter, which material in your opinion demonstrates the best potential for AM production?

4. Because of how AM is quickly emerging into the main stream of manufacturing techniques, what might be some legal issues the industry may have to deal with in the future?
5. Currently, AM research has only been conducted for the most part on metals. Do you think other materials will be able to be processed in this method? Why or Why not?

4.5.2 Conceptual

1. What is a melt pool? How is it formed? Of what dimension is the melt pool, and how long does it typically exist, for a given point in space, during LBAM?
2. There are three main modes of heat transfer, namely conduction, convection, and radiation. Discuss how each of these modes relate to a specific event encountered in a typical LBAM process.
3. How is the heat transfer during a typical LBAM process eventually related to a part's fatigue behavior?
4. What types of fatigue loading modes (e.g., axial, torsion, bending, combined torsion/bending, pressure) might become present in the following systems: (1) Hip replacement prosthesis, (2) Fighter jet engine turbine blade, (3) Back leg of a desk chair, (4) Motorcycle front axle, and (5) Oil pipeline?
5. Sketch what you believe would be an estimated fatigue load graph for each of the five systems above in Problem 4.
6. What are two domains of cycling stressing? Describe characteristics of each type and sketch a graph.
7. List three mechanical characteristics linked to LBAM thermal histories.
8. List three microstructural factors that initiate fatigue crack growth in additively manufactured metals.
9. List three techniques that could be applied to increase the fatigue life in AM processed metals.
10. What is fatigue strength?
11. What is a stress intensity factor?
12. List and describe two different LBAM techniques.

References

Amsterdam, E., and G. A. Kool. 2009. High cycle fatigue of laser beam deposited Ti–6Al–4V and Inconel 718. *ICAF, Bridging the Gap between Theory and Operational Practice*. Springer, Dordrecht, the Netherlands, 1261–74.

Anderson, T. 1995. *Fracture Mechanics: Fundamentals and Applications*, 2nd ed. Boca Raton, FL: CRC Press.

ASTM Standard E647-15e1. 2015. Standard test method for measurement of fatigue crack growth rates. ASTM International, West Conshohocken, PA.

Brandl, E., U. Heckenberger, V. Holzinger, and D. Buchbinder. 2012. Additive manufactured AlSi10Mg samples using selective laser melting (SLM): Microstructure, high cycle fatigue, and fracture behavior. *Materials & Design* 34: 159–69.

Cain, V., L. Thijs, J. Humbeeck, B. Hooreweder, and R. Knutsen. 2015. Crack propagation and fracture toughness of Ti6Al4V alloy produced by selective laser melting. *Additive Manufacturing* 5: 68–76.

Collins, J. 1993. *Failure of Materials in Mechanical Design: Analysis, Prediction, Prevention*. New York: John Wiley & Sons.

Collins, J. A., and S. R. Daniewicz. 2005. Failure modes: Performance and service requirements for metals. In Myer Kutz (ed.)., *Mechanical Engineers' Handbook: Materials and Mechanical Design*, 860–924. Hoboken, NJ: John Wiley & Sons.

Costa, L., and R. Vilar. 2009. Laser powder deposition. *Rapid Prototyping Journal* 15(4): 264–79. doi:10.1108/13552540910979785.

Costa, L., R. Vilar, T. Reti, and A. M. Deus. 2005. Rapid tooling by laser powder deposition: Process simulation using finite element analysis. *Acta Materialia* 53(14): 3987–99. doi:10.1016/j.actamat.2005.05.003.

Daniewicz, S. R., and N. Shamsaei. 2017. An introduction to the fatigue and fracture behavior of additive manufactured parts. *International Journal of Fatigue* 94: 167.

Davis, J. R. 2000. Powder metallurgy processing of nickel alloys. In Davis, J. R. (ed.) *Nickel, Cobalt, and Their Alloys*, 215. Materials Park, OH: ASM International.

DiMatteo, N. D. 1996. *Fatigue and Fracture*. Materials Park, OH: ASM International.

Donachie, M. J. 2000. Titanium: A technical guide. Materials Park, OH: ASM International.

Edwards, P., A. O'Conner, and M. Ramulu. 2013. Electron beam additive manufacturing of titanium components: Properties and performance. *Journal of Manufacturing Science and Engineering* 135: 061016.

Ganesh, P., R. Kaul, C. P. Paul, P. Tiwari, S. K. Rai, R. C. Prasad, and L. M. Kukreja. 2010. Fatigue and fracture toughness characteristics of laser rapid manufactured Inconel 625 structures. *Materials Science and Engineering: A* 527(29–30): 7490–97. doi:10.1016/j.msea.2010.08.034.

Ganesh, P., R. Kaul, G. Sasikala, H. Kumar, S. Venugopal, P. Tiwari, S. Rai, R. C. Prasad, and L. M. Kukreja. 2014. Fatigue crack propagation and fracture toughness of laser rapid manufactured structures of AISI 316L stainless steel. *Metallography, Microstructure, and Analysis* 3(1): 36–45. doi:10.1007/s13632-013-0115-3.

GB/T 15248-2008 Standard. The test method for axial loading constant-amplitude low-cycle fatigue of metallic materials. Haidian District, Beijing: Standardization Administration of China.

Griffith, M. L., M. E. Schlienger, L. D. Harwell, M. S. Oliver, M. D. Baldwin, M. T. Ensz, J. E. Smugeresky et al. 1998. Thermal behavior in the LENS process. In *Proceedings of the 9th Solid Freeform Fabrication Symposium*, Austin, TX, 89–96.

Grylls, R. 2005. LENS process white paper: Fatigue testing of LENS Ti-6-4. Optomec whitepapers. http://www.optomec.com.

Juvinall, R. 1967. *Engineering Considerations of Stress, Strain, and Strength*. New York City, NY: McGraw Hill.

Kobryn, P. A., and S. L. Semiatin. 2001. Mechanical properties of laser-deposited Ti–6Al–4V. *Solid Freeform Fabrication Proceedings*, Austin, TX, 179–86.

Leuders, S., M. Thöne, A. Riemer, T. Niendorf, T. Tröster, H. Richard, and H. Maier. 2013. On the mechanical behaviour of titanium alloy TiAl6V4 manufactured by selective laser melting: Fatigue resistance and crack growth performance. *International Journal of Fatigue* 48: 300–7.

Lin, C. W., C. P. Ju, and J. H. Chern Lin. 2005. A comparison of the fatigue behavior of cast Ti–7.5Mo with Cp Titanium, Ti–6Al–4V and Ti–13Nb–13Zr alloys. *Biomaterials* 26(16): 2899–2907.

Manvatkar, V. D., A. A. Gokhale, G. Jagan Reddy, A. Venkataramana, and A. De. 2011. Estimation of melt pool dimensions, thermal cycle, and hardness distribution in the laser-engineered net shaping process of austenitic stainless steel. *Metallurgical and Materials Transactions A: Physical Metallurgy and Materials Science* 42(13): 4080–87. doi:10.1007/s11661-011-0787-8.

Mazumder, J., and W. M. Steen. 1980. Heat transfer model for CW laser material processing. *Journal of Applied Physics* 51(2): 941–47. doi:10.1063/1.327672.

Parker, A. 1981. *The Mechanics of Fracture and Fatigue: An Introduction*. London: E. & F.N. Spon.

Pinkerton, A. J., and L. Li. 2004. Modelling the geometry of a moving laser melt pool and deposition track via energy and mass balances. *Journal of Physics D: Applied Physics* 37(3704): 1885–95. doi:10.1088/0022-3727/37/14/003.

Rice, R. 1997. *SAE Fatigue Design Handbook*, 3rd ed. Warrendale, PA: Society of Automotive Engineers.

Rosenthal, D. 1946. The theory of moving sources of heat and its application to metal treatments. *Transactions of the American Society of Mechanical Engineers* 68: 849–66.

Seifi, M., M. Dahar, R. Aman, O. Harryson, J. Beauth, and J. Lewandowski. 2015. Evaluation of orientation dependence of fracture toughness and fatigue crack propagation behavior of as-deposited ARCAM EBM Ti–6Al–4V. *JOM* 67(3): 597–607.

Shamsaei, N., A. Yadollahi, L. Bian, and S. M. Thompson. 2015. An overview of direct laser deposition for additive manufacturing; Part II: Mechanical behavior, process parameter optimization and control. *Additive Manufacturing* 8: 12–35. doi:10.1016/j.addma.2015.07.002.

Spierings, A., T. Starr, and K. Wegener. 2013. Fatigue performance of additive manufactured metallic parts. *Rapid Prototyping Journal* 19: 88–94.

Tada, H., and P. Paris. 2000. *The Stress Analysis of Cracks Handbook*. New York: ASME Press.

Thompson, S. M., L. Bian, N. Shamsaei, and A. Yadollahi. 2015. An overview of direct laser deposition for additive manufacturing; Part I: Transport phenomena, modeling and diagnostics. *Additive Manufacturing* 8: 36–62. doi:10.1016/j.addma.2015.07.001.

US Department of Transportation, and Federal Aviation Administration. 2003. Metallic materials properties development and standardization (MMPDS). http://app.knovel.com/hotlink/toc/id:kpMMPDSMM1/metallic-materials-properties-3/metallic-materials-properties-3.

Wang, L., S. Felicelli, Y. Gooroochurn, P. T. Wang, and M. F. Horstemeyer. 2008. Optimization of the LENS process for steady molten pool size. *Materials Science and Engineering A* 474(1–2): 148–56. doi:10.1016/j.msea.2007.04.119.

Wang, Y., S. Zhang, X. Tian, and H. Wang. 2013. High-cycle fatigue crack initiation and propagation in laser melting deposited TC18 titanium alloy. *International Journal of Minerals, Metallurgy, and Materials* 20(7): 665–70. doi:10.1007/s12613-013-0781-9.

Wu, G. Q., C. L. Shi, W. Sha, A. X. Sha, and H. R. Jiang. 2013. Microstructure and high cycle fatigue fracture surface of a Ti–5Al–5Mo–5V–1Cr–1Fe titanium alloy. *Materials Science and Engineering: A* 575: 111–18. doi:10.1016/j.msea.2013.03.047.

Wu, X., and J. Carlsson. 1991. *Weight Functions and Stress Intensity Factor Solutions*. Oxford, UK: Pergamon Press.

Wycisk, E., A. Solbach, S. Siddique, D. Herzog, F. Walther, and C. Emmelmann. 2014. Effects of defects in laser additive manufactured Ti–6Al–4V on fatigue properties. *Physics Procedia* 56: 371–78.

Yin, H., L. Wang, and S. D. Felicelli. 2008. Comparison of two-dimensional and three-dimensional thermal models of the LENS® process. *Journal of Heat Transfer* 130(10): 102101. doi:10.1115/1.2953236.

Zheng, B., Y. Zhou, J. E. Smugeresky, J. M. Schoenung, and E. J. Lavernia. 2008. Thermal behavior and microstructural evolution during laser deposition with laser-engineered netshaping: Part I. Numerical calculations. *Metallurgical and Materials Transactions A: Physical Metallurgy and Materials Science* 39(9): 2228–36. doi:10.1007/s11661-008-9557-7.

Chapter 5

Optimization of Laser-Based Additive Manufacturing

Amir M. Aboutaleb and Linkan Bian
Mississippi State University

Contents

CHAPTER OUTLINE

In this chapter, we

- Define the problem of laser-based additive manufacturing (LBAM) process optimization
- Discuss the effects of process parameters on LBAM processes
- Overview physics-based methods for LBAM modeling and optimization
- Discuss the use of design of experiments and model fitting in the application of LBAM optimization
- Provide a portfolio of examples of LBAM optimization

5.1 Introduction: Effects of Parameters

Learning Objectives

- Understand the problem of process optimization for LBAM
- Understand the challenge of process optimization resulting from the large number of process parameters and responses
- Understand the interaction between process parameters and why process parameters cannot be optimized individually

The optimization of LBAM processes requires the understanding and characterization of the relation between a vector of process control parameters $\boldsymbol{x}$ and a vector of response $\boldsymbol{y}$, which can be mechanical properties such as porosity, fatigue, and yield strength, or in-process variables, such as melt-pool dimensions and deposit height. Since the mechanical properties of LBAM parts depends on the process parameters, which affect the microstructural distribution via thermal history, it is important to optimize the LBAM process parameters to generate near-net-shaped parts with minimal defects. The significance of LBAM process optimization is twofold: (1) the optimized process parameters can then be utilized for effectively "seeding" a thermally monitored, feedback-controlled LBAM process, which may result in LBAM parts with improved and/or customized properties and (2) it facilitates the research in material science to accumulate suitable AM processing data for various metallic materials, since comprehensive knowledge is involved in AM processes, including laser technology, material science, and solidification.

Despite the recent advances in LBAM technologies, the expense of running a large number of experiments remains nontrivial. Optimal process parameters are typically determined via extensive experiments, which usually require high experimental costs and a significant time investment. Because of the high machine, materials, and operation costs, single experimental studies of LBAM part density could cost thousands of dollars, if not more, and take up to weeks. Moreover, this type of trial-and-error approach may never uncover the functional relationship between the process parameters $\boldsymbol{x}$ and part features $\boldsymbol{y}$ and, therefore, may never estimate the optimal combinations for process control parameters.

Two types of methods/models have been used to address the challenge of LBAM process optimization: (1) physics-based models that seek to characterize the underlying thermophysics deposition process of LBAM and (2) data-driven methods that target identifying the patterns in the existing experiment data through design of experiments and empirical modeling. For both approaches, the goal is to identify the functional relationship:

$$y = f(x) + \varepsilon, \tag{5.1}$$

where f represents the unknown relation between process parameters $\boldsymbol{x}$ and LBAM response $\boldsymbol{y}$, and ε represents the random error and uncertainty associated with the process.

Characterizing the functional form of f, i.e., the underlying phenomena that govern the LBAM process is nontrivial because several LBAM process parameters may affect the response vector of the part, which include, but are not limited to, laser power, laser velocity, laser scanning strategy, etc. For instance, laser power and laser velocity affect the melt-pool shape and incident energy, and consequently, the cooling rate and local thermal gradients, and eventually the mechanical properties of parts.

In what follows, we summarize the effects of a number of key process parameters on the thermal history and mechanical properties of LBAM parts to delineate the complex relationship between process parameters and microstructural/mechanical properties. A more detailed discussion about the effects of various process parameters can be found in Shamsaei et al. (2015) and Thompson et al. (2015).

5.1.1 Laser Power and Laser Velocity

The combination of lower laser power and higher laser velocity results in lower incident energy at the top of the part, which in turn leads to higher cooling rate as well as finer microstructure (Shamsaei et al., 2015). Conversely, a lower cooling rate and coarser microstructure can be achieved by increasing laser power and decreasing laser traverse speed (Shamsaei et al., 2015). Lower incident energy, which can be due to laser attenuation and/or radiation effects, tends to result in finer equiaxed morphologies. By contrast, higher incident energy generally results in a coarser microstructure and columnar grains (Shamsaei et al., 2015). Since the cooling rate (and consequently the solidification rate) increases toward the surface of the deposit, a transition from columnar to mixed equiaxed-columnar microstructure has been observed at the surface of the deposit. Therefore, various combinations of laser power and laser velocity may result in differential microstructure, morphology, and thus mechanical properties.

5.1.2 Layer Height

Layer height influences the microstructure and mechanical properties, as well as the geometric accuracy, of LBAM-fabricated parts. Layer height is affected by different factors for different powder deposition mechanisms: For laser-based powder-bed fusion (PBF-L) processes, also referred to as selective laser melting (SLM), a uniform bed of powder is first spread on the previous layer or substrate by a roller or recoater blade, and then specific regions of the bed are selectively melted by a laser beam in order to build a single layer of the part. Upon the completion a layer, the powder bed is lowered by the height of the deposited layer (layer height), and a new layer of powder is deposited with the roller. In this case, the layer height is mainly determined by the thickness of powder and may be accurately controlled. For direct energy deposition (DED) processes, such as laser-engineered net shaping (LENS), the amount of powder forming a layer is mainly determined by the amount of powder injected into the melt pool (deposited mass flow rate). The resulting powder density distribution in the melt pool is then the most important factor for layer height control. Thus, characterizing the effect of process control parameters on the layer height is equivalent to modeling their effects on the amount of the powder injected into the melt pool and its distribution. In general, the layer height increases as the powder feed rate increases. Powder feed rate interacts with other parameters, such as laser power and traverse velocity. A higher laser power coupled with lower powder feed rate may result in increased porosity and vice versa.

5.1.3 Deposition Pattern

There are four common deposition patterns used in LBAM: raster, bidirectional, offset, and fractal patterns. The offset patterns can be further divided into two types—offset-out and offset-in—depending on the direction and starting point of deposition, as shown in Figure 5.1. The raster pattern is the most commonly used because of its ease of implementation (Shamsaei et al., 2015). The choice of laser scanning path of the raster pattern does not depend on the geometry of the part. However, different deposition patterns significantly affect the geometric and mechanical properties of the fabricated parts (Shamsaei et al., 2015). Choosing the appropriate scanning patterns reduces the incidence of residual stresses and thermal distortions. Moreover, the fractal and offset-out deposition patterns generate the smallest and second smallest substrate deformations, respectively, according to Yu et al. (2011). Despite the

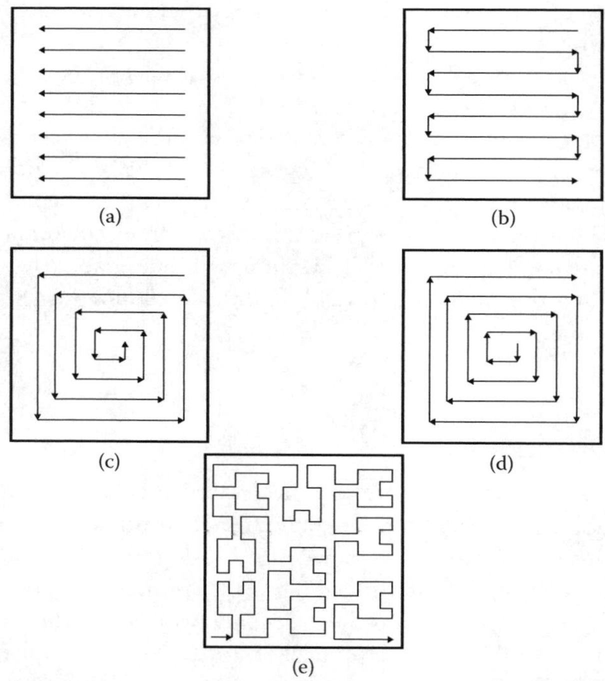

Figure 5.1 Different deposition patterns. (a) Raster, (b) bidirectional, (c) offset-in, (d) offset-out, and (e) fractal postmanufacturing processing. (From Yu et al., *Mater. Sci. Eng. A*, 528(3), 1094–1104, 2011.)

advantageous resultant mechanical properties and geometric accuracy, applying offset or fractal patterns requires additional customization of deposition path per part geometry, which is very challenging for a part with complex geometry.

Postmanufacturing processes, such as hot isostatic pressing (HIP), may be used to enhance to physical and mechanical properties of LBAM parts. HIP is a manufacturing process used for reducing the porosity and increasing the density of materials, by exposing the part to elevated temperature under constant/isostatic gas pressure. As reported by Thompson et al. (2015), adequate densification of LENS Ti–6Al–4V can be accomplished through HIP, leading to a higher ductility and elongation-to-rupture. Furthermore, the high temperatures experienced during the HIP process cause an increase in the alpha-platelet thickness of Ti–6Al–4V, which consequently reduces the strength of the material. In addition, HIPed specimens exhibit less anisotropic behavior than heat treated ones. Such observations indicate that porosity may play a significant role in the anisotropic behavior of laser-deposited parts. In general, the effects of different manufacturing and postmanufacturing parameters on the anisotropic behavior of laser-deposited parts are not yet well understood; thus, further investigations are required.

5.2 Physics-Based Models

Learning Objectives

- Understand the impact of thermal history on the microstructure evolution and mechanical properties during LBAM processes
- Understand the advantages of using dimensionless parameters for process characterization
- Understand the method of process maps: applications and limitations

Because of the layer-by-layer additive nature of LBAM, the complex thermal histories are experienced repeatedly in different regions of the deposited layers, which normally involve melting as well as numerous reheating cycles at a relatively lower temperature. Such complicated, cycling thermal behavior during LBAM results in the complex phase transformations and microstructural developments, and consequently exists significant difficulties in obtaining targeted optimal mechanical properties. On the other hand, the use of a finely focused laser to form a rapidly traversing melt pool may result in considerably high solidification rate and melt instability. Complicated residual stresses tend to be locked into the parts during the building process, due to the thermal transients encountered during solidification. Understanding of the parameter–thermomechanical relationship facilitates the optimization of the LBAM process and thus final mechanical properties. Actually, a series of complex thermomechanical phenomena, including heat transfer, phase changes, mass addition, and fluid flow, are involved in the melt pool during LBAM. Therefore, physics-based models developed based on the knowledge of temperature, velocity, and composition distribution history is essential for an in-depth understanding of the process, and subsequent mechanical properties.

5.2.1 Buckingham's Π Theorem

Determining the effects of process parameters on the LBAM process is challenging because of the high dimensionality of the parameter space and their interaction among each other. In other words, each process parameter may not be singled out and studied individually because its impact on the microstructure and mechanical properties of LBAM parts depends on the values/levels of other process parameters. Buckingham's Π (Pi) theorem provides a key tool in dimensional analysis for reducing the dimensionality of the process parameter space and a guideline in identifying the maximal number of dimensionless parameters, needed to characterize the process. The main idea is that if there is a physically governing equation involving a certain number, say n, of physical variables, and k is the number of fundamental dimensions (time, location, density, etc.) required to describe these n variables, then the original expression is equivalent to an equation involving a set of $p=n-k$ dimensionless parameters constructed from the original variables. Buckingham's Π theory suggests the number of dimensionless parameters that can be constructed but the dimensionless parameters generated via Buckingham's theory are not unique, and there exist multiple choices for the selection of these dimensionless parameters. In most cases, researchers select dimensionless parameters that are useful in understanding the underlying process physics. Possible choices of dimensionless parameters include melting efficiency, deposition efficiency, process efficiency, laser absorptivity, specific energy, and others. The utilization of common thermofluidic dimensionless numbers, such as Re, Pr, and Bo, is also sought as these classical numbers can provide physical insight into various aspects of the LBAM process. As an example, one can find the Reynolds number for the melt pool to help in assessing if laminar or turbulent flow exists.

Example 1: Buckingham's Π Theorem (Kahlen and Kar, 2001)

Determining the effects of LBAM process parameters on the quality of the final parts requires an extensive experimental study for various combinations of process parameters, as listed in Table 5.1. The parameters taken into account include materials, laser, product, process, and environment, resulting in a parameter space with high dimensionality (e.g., 19 parameters). Nevertheless, even such a high-dimensional parameter space only includes a number of major parameters, and does not incorporate all the factors that could affect the LBAM process. For example, the effects of the powder delivery system (e.g., powder-feeding location) and the shielding and carrier gases are not considered. The distribution of the laser irradiance at the focal spot and the position of the focal spot relative to the melt surface are not taken into account. Considering the high dimensionality, fully characterizing the effect of process parameters on the LBAM process would be very challenging. Any possible reduction of the parameter space will significantly improve the efficiency of the study.

Kahlen and Kar (2001) reduced the number of parameters by combining some of the process parameters and by identifying similar parameters using Buckingham's Π theorem. With a basis of five dimensions (i.e., time, length, mass, temperature, and energy), Buckingham's Π theorem groups the original 19 process parameters into 14 (=19−5) dimensionless numbers to characterize the process, as shown in Table 5.2. These resulting dimensionless numbers–which included Reynold number (*Re*), Bond number (*Bo*), Prandtl number (*Pr*), and various process "efficiencies"—aid in constraining the process parameter dimension and help in reducing the number of experiment required to optimize the process.

The advantage of using dimensionless parameters is that these parameters are usually defined based on LBAM processes, instead of materials to be processed or geometries of parts to be fabricated. Thus, dimensionless parameters are less dependent on material properties and part geometries. In practice, when a material is to be processed or when a part is to be built, dimensionless parameters can be used to suggest the range of parameters in which good metal parts are formed. Furthermore, the use of dimensionless parameters can alleviate the confusion of dealing with various units for a single process parameter. For example, the units utilized for traverse speed can vary substantially (e.g., in./min, cm/s, etc.). Utilization of a dimensionless traverse speed allows for easier communication of process parameters among the international community.

Table 5.1 Parameters for Metal Deposition

Material	*Density, ρ*	*Thermal Conductivity, k*	*Solidification Time, t_s*	*Viscosity, μ*	*Heat Capacity, c_p*	*Surface Tension, σ*	*Melting Energy, E_m*
Laser	Beam diameter at focus, d_F	Power, P					
Product	Height of deposit, h_D	Width of deposit, w					
Process	Delivered mass flow rate, m_{del}	Deposited mass flow rate, m_{dep}	Energy loss to environment, E_l	Translation speed, v_s	Powder size, d_p	Initial powder temperature, T_i	Powder stream speed, v_p
Environment	Gravity, g						

Table 5.2 Dimensionless Parameters

Similarity Parameter	*Expression*
Energy loss efficiency	$\eta_L = \frac{E_l}{P}$
Melting efficiency	$\eta_m = \frac{E_m}{P}$
Superheating efficiency	$\eta_x = \frac{E_x}{P}$
Powder delivery efficiency	$\eta_P = \frac{m_{dep}}{m_{del}}$
Bond number	$Bo = \frac{\rho d_F^2 g}{\sigma}$
Froude number	$Fr = \frac{v_P^2}{g d_F}$
Galileo number	$Ga = \frac{g d_F^3 \rho^2}{\mu^2}$
Prandtl number	$Pr = \frac{c_p \mu}{k}$
Reynolds number	$Re = \frac{v_p \rho d_F}{\mu}$
Length scale of the melt-pool	$L_P = \frac{v_s t_s}{d_F}$
Powder particle number density in the melt-pool	$N = \left(\frac{d_P}{d_F}\right)^2$
Melt-pool shape factor	$S_P = \frac{v_P}{v_s}$
Product aspect ratio	$A_S = \frac{h_D}{w}$
Powder dissolution factor	$D_P = \frac{4k(T_a - T_i)d_F}{E_m}$

5.2.2 Process Characterization

One well-established method for developing dimensionless parameters is process maps that can be used to understand the LBAM process parameters and their impact on the thermal history and microstructure of the part. Process maps possess dimensionless ordinates to help in determining the effects of laser velocity, part preheating, laser power, etc.; and are generalized for the moving heat source problem (e.g., Rosenthal's solutions) inherent to LBAM processes. Plots are generated based on analytical, numerical, or experimental results, which can be used as tools to aid LBAM users in ascertaining the appropriate, initial process parameters for a given material for fabrication via LBAM.

5.2.3 Steady-State Thermal Maps

Early research of process maps mainly focuses on predicting the steady-state melt-pool size for various practical combinations of process parameters. During LBAM, the laser beam creates a moving melt pool on the substrate in which powder is melt. The melt-pool size has been identified as a critical parameter for maintaining optimal building conditions. The formation of steady melt pool with a small heat affected zone and an uninterrupted solidification front tends to results in homogeneous part quality. A typical steady-state thermal process map can demonstrate how melt-pool length is affected by normalized height of substrate and melt temperature. These steady-state process maps have been used for determining process parameters that result in desired melt-pool lengths. An advantage of the steady-state process maps is its ease of implementation: the relations between dimensionless process parameters and the melt-pool lengths are contained in the form of a single three-dimensional or two-dimensional plot that process engineers can use directly.

5.2.4 Thermal Maps for Transient Analysis

Thermal process maps have also been developed to conduct transient analysis, accounting for dynamics LBAM processes and their effects on, for example, melt-pool/track morphology. One typical dynamic LBAM process is the so-called "boundary issue," which results in an increase in the melt-pool size as the boundary of each layer is approached. The boundary problem is mainly caused by the fact that the laser velocity reduces, and thus the energy density increases, to accurately deposit materials near the boundary of layers. In addition, a velocity reversal is needed to continue with deposition of the next layer of material. Efficient and effective optimization of the melt-pool size/morphology and other process parameters requires comprehensive understanding of how the melt-pool size changes over a range of process size scales, as well as various laser power or velocity. In other words, a transient analysis allows one to determine how changes in process parameters affect the melt-pool geometries, as well as the cooling rates, thermal gradients, etc.

5.2.5 Process Maps for Different Scales of LBAM Systems

Whenever a new laser-based manufacturing system is developed at a different size scale, engineers must perform a large number of experiments to characterize their process. It is important to obtain fundamental understanding of how to apply deposition knowledge acquired from small-scale systems (e.g., LENS equipped with a 500 W Nd:YAG laser, or other similarly sized lasers) to analogous large-scale systems (e.g., AeroMet, which manufactures components for the aerospace industry and uses an 18 kW CO_2 laser). Multiple process maps with various scales have been developed for predicting part features for the large-scale process via extrapolation by Birnbaum et al. (2013). The authors showed that the process maps can be applied over multiple process size scales by simply using dimensionless parameters, i.e., by changing the normalization temperature with changes in power range. Although the resulting prediction can be used to provide a possible range for the optimal process parameters in large-scale LBAM processes, the prediction may be inaccurate due to the error caused by model extrapolation. Therefore, there exists a great need to fill the gap between industrial applications that demand the use of large-scale deposition processes and the process development that occurs on small-scale processes in the laboratory conditions.

5.2.6 Limitation of Process Maps

Process maps are advantageous in that they provide a fundamental way to predict and thus control melt-pool size, stress, and material properties by presenting results in a form that process engineers can readily use. Bontha et al. (2009) further generalized the thermal process maps to establish the relation between dimensionless process parameters and solidification cooling rates in laser deposition processes, namely, solidification maps. The simultaneous control of residual stress and melt-pool size has also been addressed by Vasinonta et al. (2007). However, there also exist two major limitations in the existing process map methods. First, the current process maps are for limited part geometries—developed for thin wall and bulk shapes only. In other words, these process maps do not hold for other common shapes, let alone parts with complex geometries. One promising application of AM is to fabricate parts whose geometry is so complex that they cannot be produced using traditional manufacturing methods. Therefore, in order to make process maps useful in real-world manufacturing applications, future work is needed to develop process maps that characterize the thermal behaviors and mechanical properties of parts with various geometries. Second, process maps do not consider temperature-dependent material properties. Current process maps are based on Rosenthal's analytical solution for temperature with moving-heat-source boundary, and material properties are assumed independent of temperature. This may not be realistic assumption in real-world applications. In fact, process maps are used to approximate the underlying fabrication process when the temperature remains in a certain range.

5.3 Data-Driven Optimization Methods

Learning Objectives

- Understand the need of data-driven process optimization methods and models
- Understand the advantages of classical design of experiments methods (factorial, fractional, Taguchi) and their limitation
- Understand the concepts of space-filling designs and D-optimal designs
- Understand the models of response surface and kriging for characterizing experimental data
- Understand the concept of artificial intelligence

Although physics-based models are essential for thoroughly understanding the underlying LBAM processes, their development is extremely challenging due to the complexity associated with LBAM. Some research efforts have circumvented this challenge by utilizing data-driven methods that directly model how the process parameters affect the quality of final parts. Data-driven approaches involve (1) choosing an experimental design for generating data and (2) choosing a model to fit the data. In particular, existing research for the optimization of LBAM parts primarily rely on a trial-and-error procedure to determine optimal process parameters and achieve targeted properties of the fabricated product. Statistical design of experiments (DOE) provides a systematic framework to utilize the previous experimental data and plan future experimental trials with the minimal cost. An experimental design represents a batch or sequence of experiments to be performed, expressed in terms of factors (design variables) set at specified levels. Based on the experimentation data, simplified/approximated relations between part features and process parameters can be learned in an empirical way. Properly designed experiments and subsequent data modeling are essential for effective experimental analysis and process optimization. A common feature shared by data-driven methods is that process parameters and part features (e.g., mechanical properties) are empirically related based on experimental data sets. In other words, the developed methods are not completely dependent on the domain knowledge of a specific process and thus may be applied to other processes. Below, we review DOE methods together with the corresponding data models that have been applied for LBAM studies, as well as those can be potentially applied.

5.3.1 Full Factorial Designs

The most basic experimental design is a full factorial design. The main idea is to replicate all possible combination of the levels of the factors in each run of experiments. The number of design points dictated by a full factorial design is the product of the number of levels for each factor. For instance, if there are 2 factors, each with ℓ levels, each experimental run would investigate ℓ^2 design points. More generally, when there are k factors, each with ℓ levels, the total number of design points is ℓ^k. The most common are 2^k (for evaluating main effects and interactions) and 3^k designs (for evaluating main and quadratic effects and interactions) for k factors at 2 and 3 levels, respectively. A 2^3 full factorial design is shown in Figure 5.2.

The main effect of a factor refers to the resulted change in the response variable by changing the level in factor of interest. Figure 5.3 illustrates an example of a 2^2 design. Two levels are denoted by "+" and "−" representing "high" and "low" levels for the factors of interest, respectively. The main effect of a factor of a factorial design is defined as the difference between the average response

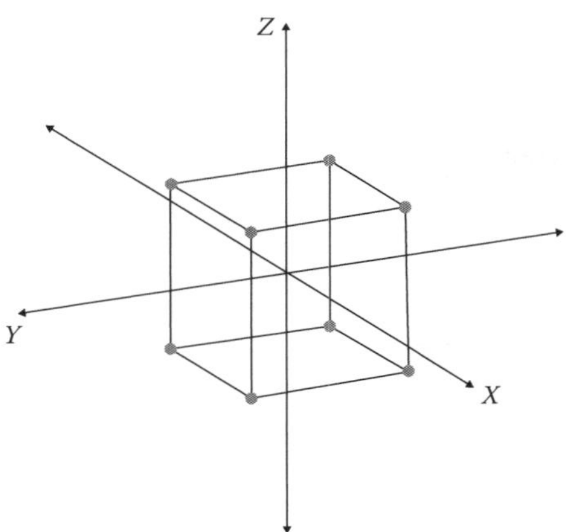

Figure 5.2 2^3 designs.

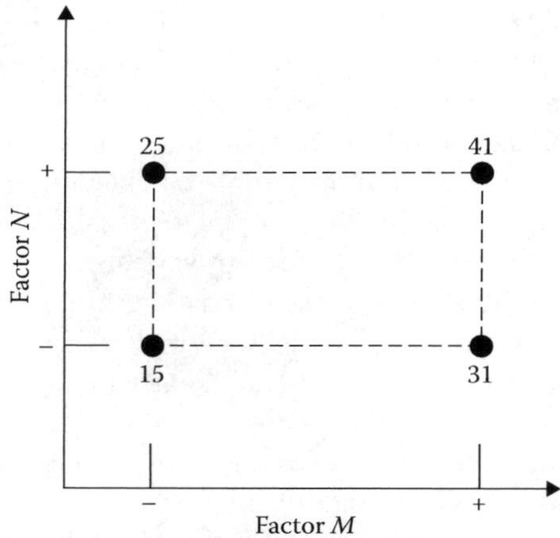

Figure 5.3 2² designs.

values at high and low levels of the factor of interest. In this example, the main effect of factor M is calculated as follows: Main Effect of Factor $M = (31+41)/2 - (15+25)/2 = 16$. The effect of a factor may depend on the level of others. In the case, there exists an interaction between these factors. Following the previous example, the magnitude of the interaction between factors M and N (i.e., MN effect) is calculated as $MN = ((39-21) - (11-27))/2 = 17$.

There are various ways to mathematically formulate factorial designs. For instance, a regression-type model can be developed to capture the main effects and interactions of two-factor as shown in Equation 5.1.

$$y = \alpha_0 + \alpha_1 x_1 + \alpha_2 x_2 + \alpha_{12} x_1 x_2 + \varepsilon,$$

where y represents the response value, α_i's are the coefficients which should be estimated, x_1 and x_2 are variables representing scaled values (–1 and +1 representing low and high levels, respectively) of factors M and N, respectively. Also, $x_1 x_2$ represents the interaction term between factors, and ε is the random error. After forming the model, statistical methods (such as ANOVA) can be employed to analysis and determine the statistical significance of each factor and interactions as well as conventional statistical tests and interpretations.

5.3.2 Fractional Factorial Design

The combinations of design points in full factorial design increase exponentially along with the number of factors. Hence, full factorial designs may not work well for cases in which a moderate-to-large number of factors are involved. Moreover, each design point should be replicated several times in order to make the experimental design capable of detecting the significant effects, leading to an unmanageable number of experiments. Fractional Factorial Design is a potential alternative when the experiments are expensive to run and a large number of factors and levels are involved. A fractional factorial design is a fraction of a full factorial design that aims to select a portion of all possibilities to reduce the number of required experiments. The general form of ℓ-level fractional factorial design is, ℓ^{k-p} where k–p is the fraction of original full factorial design which is to be run. For instance, a half fraction of 2^3 full factorial design includes $2^{3-1} = 4$ design points which

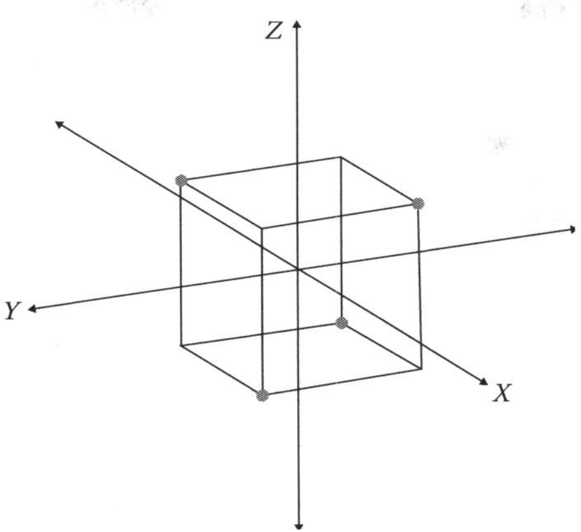

Figure 5.4 Fractional factorial designs.

are illustrated in Figure 5.4. Reducing the design space by shifting from full to fractional factorial design is also accompanied by an inevitable cost in terms of detecting the significant effects. For instance, applying 2^3 full factorial design not only considers all main effects (*x*, *y*, *z*) and all two factor interactions (*xy*, *xz*, *yz*) but also accounts for three factor interactions (*xyz*). On the other hand, half fraction of 2^3 full factorial design does not allow for estimation of three factor interactions. The interaction effects cannot be ignored unless they are known (or assumed) to be insignificant. An example of the application of 2^{l-p} design can be found in the following example:

Example 2: Two-Level Fractional Factor Design (Kummailil et al., 2005)

LENS is an AM process that can build complex, functional parts in metal, by slicing 3D objects into 2D layers of user-defined thickness and superimposing successive layers. Users must choose the layer thickness and match it with the deposition to keep the laser in focus throughout the build. If a mismatch exists between the layer thickness and the deposited thickness, the laser is no longer focused on the melt-pool surface, leading to geometric inaccuracy of parts, or even failures of build. The deposition thickness is typically governed by other process parameters, such as laser power, laser velocity, hatching spacing, and powder mass flow rate. To match the layer thickness with deposition, it is very important to understand and estimate the relationship between these process parameters and the deposition thickness so that the layer thickness can be optimized based on the chosen process parameters.

Kummailil et al. (2005) studied the effects of process parameters on deposition thickness and applied a two-level fractional factorial designs to develop empirical relationships between process parameters and deposition. Experiments were performed on a LENS 850 system using Ti–6Al–4V powder. The geometry of testing parts was a rectangular block with a width of 12.7 mm, length of 19.1 mm, and height dependent on the layer deposited (because the build may fail if the specified layer thickness mismatches the deposition). Two levels, high (+) and low (–), are chosen for each process parameter, as shown in Table 5.3. The minimum number of experiments needed is $2^4 = 16$. Experiments were performed according to the design matrix, as shown in Table 5.4, and the deposition thickness was measured for each testing sample.

The main effects of process parameters on deposition thickness are presented on Table 5.5. Laser power and mass flow rate have positive effects on deposition, whereas hatch spacing and laser velocity have negative effects on deposition. Mass flow rate and laser velocity have more significant impacts on deposition, whereas hatch spacing shows a less significant effect and laser power has the smallest impact on deposition thickness. Therefore, to achieve the targeted deposition thickness, powder mass flow rate and laser velocity may be the most effective parameters to adjust, based on the experimental data.

Table 5.3 Levels of Process Parameters

Process Parameter	–	+
Laser power (W)	250	350
Laser velocity (mm/s)	16.9	27.5
Hatch spacing (μm)	381	457
Powder mass flow rate (mg/s)	38.0	73.4

Table 5.4 Design Matrix of Factorial Designs

Design Point	*Laser Power*	*Laser Velocity*	*Hatch Spacing*	*Powder Mass Flow Rate*
1	–	–	–	–
2	–	–	–	+
3	+	–	–	–
4	+	–	–	+
5	–	+	–	–
6	–	+	–	+
7	+	+	–	–
8	+	+	–	+
9	–	–	+	–
10	–	–	+	+
11	+	–	+	–
12	+	–	+	+
13	–	+	+	–
14	–	+	+	+
15	+	+	+	–
16	+	+	+	+

Table 5.5 Main Effects of Parameters

Parameter	*Laser Power*	*Laser Velocity*	*Hatch Spacing*	*Powder Mass Flow Rate*
Main effect	6.97	–29.01	–12.36	30.56

An empirical model can also be developed based on the experimental data, since two major factors affecting the build process are material and energy. Specifically, process parameters affect the deposition by changing either the energy or amount of material available in the melt-pool area. The four process parameters can be grouped into two categories: (1) the energy density at the melt pool and (2) the powder flow rate. An empirical model was developed between deposition (μm) and the product of energy density (J/mm^2) and mass flow rate (g/s):

$$\text{Deposition} = \beta \times \text{energy density} \times \text{mass flow rate}.$$

The R^2 value of the fitted model is found to be 0.85, which indicates the goodness-of-fit of this linear model.

Fractional factorial designs are useful for screening factors to identify those with the greatest effects when a large number of factors are involved. In practice, it is usually assumed that the system is dominated by main effects and lower order interaction effects (e.g., quadratic effects). One specific family of fractional factorial designs vastly applied for screening are two-level Plackett-Burman (PB) designs. These are used to study $k = n-1$ factors, where n is an integer multiple of 4. By ignoring the interactions between factors, PB designs allow for using only one more design point than the number of factors to obtain unbiased estimation of main effects. To estimate quadratic effects, $3k$ or $3(k-p)$ designs can be used but often require an unmanageable number of design points. The most common second-order designs, configured to reduce the number of design points, are central composite and Box-Behnken designs. A central composite design (CCD) is a two-level factorial design, augmented by n_0 center points and two "star" points positioned at ± for each factor. Box-Behnken designs use the smallest number of factor levels in an experimental design. These are formed by combining 2^k factorials with incomplete block designs. They do not contain points at the vertices of the hypercube defined by the upper and lower limits for each factor. This is desirable if these extreme points are expensive or impossible to test. More information about central composite and Box-Behnken designs can be found in Montgomery (1984).

5.3.3 Taguchi Design

Taguchi design provides a balanced design of experiments that lays out the factors' levels in an equally weighted way. Taguchi design is an efficient method because it provides enough information by designing just a few design setups, and may be used a robust alternative to two- or three-level fractional factorial designs. The Taguchi design uses three sequential steps: (1) system design, which incorporate domain knowledge; (2) parameter design, which optimizes the settings of process parameters; and (3) tolerance design, which determines and analyzes tolerances around the optimal parameters. This subsection focuses on parameter design, which is a key step that incorporates statistical design of experimentation. Taguchi (parameter) design is developed based on the idea of orthogonal arrays. For a system with f factors, each with l levels, an orthogonal array is an N by k matrix denoted by L_N such that each possible combination of levels are repeated by the same number of times across the columns of this matrix.

A more rigorous definition of orthogonality is that the inter product of any two columns the design matrix is zero. Each row of an orthogonal array represents an experimental design setup, and the number included in the cells represents the level of each factor. The example of an L_4 design can be found in Table 5.6. Two levels, high and low represented by 1 and −1, respectively, are considered. There exist four possible combinations of factor levels: (1, 1), (1, −1), (−1, 1), and (−1, −1). For any two columns of the L_4 design matrix, each of four factor combinations appears exactly once. From another perspective, the inner product of any two columns is zero. Another example of 7-factor Taguchi design can be found in Table 5.7. These experiment designs used by

Table 5.6 *L4* Design

Experiment Number	Factors		
	f_1	f_2	f_3
1	−1	−1	−1
2	−1	1	1
3	1	−1	1
4	1	1	−1

Table 5.7 *L8* Design

Experiment Number	Factors						
	f_1	f_2	f_3	f_4	f_5	f_6	f_7
1	−1	−1	−1	−1	−1	−1	−1
2	−1	−1	−1	1	1	1	1
3	−1	1	1	−1	−1	1	1
4	−1	1	1	1	1	−1	−1
5	1	−1	1	−1	1	−1	1
6	1	−1	1	1	−1	1	−1
7	1	1	−1	−1	1	1	−1
8	1	1	−1	1	−1	−1	1

Taguchi orthogonal arrays can be considered as a fraction of full factorial designs. Specifically, orthogonal array L_N is a N/ℓ^k fraction of ℓ-level full factorial design with k factors. For example, orthogonal array L_4 can be considered a half fraction of a 2^3 full factorial design. By applying Taguchi designs, the number of design points decreases from 8 to 4. These arrays are constructed to reduce the number of design points necessary; two-level L_4, L_8, and L_{12} arrays, for example, allow 3, 7, and 11 factors/effects to be evaluated with 4, 8, and 12 design points, respectively.

Example 3: SLS Shrinkage Compensation (Raghunath and Pandey, 2007)

Selective Laser Sintering (SLS) is a powder-based AM process in which parts are built by selective sintering of layers of powder using a CO_2 laser. SLS can be used to produce functional parts for various applications, such as aerospace and rapid tooling. Shrinkage is a major issue that affects the accuracy of SLS parts. A common practice to resolve the issue of part shrinkage is to calculate or estimate the amount of shrinkage in each direction and apply the shrinkage compensation in the opposite direction in the digital model. Part shrinkage is found to be affected by various process parameters such as laser power, laser velocity, hatch spacing, powder bed temperature, and scanning length. To apply optimal shrinkage compensation to the digital file, it is important to identify the process parameters that govern part shrinkage in each direction, and understand the relation between process parameters and the amount of shrinkage. Raghunath and Pandey (2007) designed experiments using the Taguchi method and used polymer powder to fabricate cuboids of 30 mm × 30 mm cross section with different lengths along the laser scanning direction (i.e., scanning length).

The ranges of process parameters were chosen based on the minimum and maximum energy density E = Laser power/(Laser velocity × Hatch spacing). The energy density should be high enough so that sintering can occur. However, too high-energy density may cause the degradation of material properties. Raghunath and Pandey (2007) ascertained that the energy density should be at least 1 J/cm^2 for the sintering to occur; and that the polymer begins to degrade when the energy density is above 4.8 J/cm^2. Hence, the range of energy density was set in the range of (1–4.8) J/cm^2. The corresponding ranges of the laser power, laser velocity, and hatch spacing were selected to be 24–36 W, 3000–4500 mm/s, and 0.22–0.28 mm, respectively. In addition, the powder bed temperature and scan length were considered. Four levels of parameter values were considered for each process parameter. To select an appropriate orthogonal array, the total degrees of freedom need to be calculated. Four-level design results in three degrees of freedom for each of the five parameters. Plus one degree of freedom for the overall mean, the total degrees of freedom is $3\times5+1 = 16$. Hence, L_{16B} orthogonal array with 4 columns and 16 rows was used and is given in Table 5.8.

An analysis of variance (ANOVA) was used to analyze the shrinkage data in each direction to identify the parameters that have significant contributions to the total variance of shrinkage. If a factor

Table 5.8 L_{16B} Orthogonal Array

	Laser Power	*Laser Velocity*	*Hatch Spacing*	*Powder-Bed Temperature*	*Scan Length*
1	1	1	1	1	1
2	1	2	2	2	2
3	1	3	3	3	3
4	1	4	4	4	4
5	2	1	2	3	4
6	2	2	1	4	3
7	2	3	4	1	2
8	2	4	3	2	1
9	3	1	3	4	2
10	3	2	4	3	1
11	3	3	1	2	4
12	3	4	2	1	3
13	4	1	4	2	3
14	4	2	3	1	4
15	4	3	2	4	1
16	4	4	1	3	2

Table 5.9 ANOVA for Shrinkage in X-Direction

Factor	*Degrees of Freedom*	*Sum of Square (SS)*	*Mean Square (MS)*	*F-Statistics*
Laser power	3	26.39	8.80	2.51
Laser velocity	3	18.51	6.17	1.74
Hatch spacing	3	2.29	0.76	0.21
Part bed temperature	3	10.80	3.60	1.01
Scan length	3	91.61	30.54	8.60
Error	6	31.60	3.55	
Total	15	149.59		

is significantly influencing the process response (i.e., shrinkage in this example), the corresponding *F*-value would be large. For example, the ANOVA for the part shrinkage in the *X*-direction is presented in Table 5.9, which indicates that the scan length and laser power have the most significant impact on part shrinkage. Similar analysis was performed to shrinkage in *Y*- and *Z*-directions. It was reported that laser power and laser velocity have significant effects on the shrinkage in *Y*-direction, whereas, part bed temperature, laser velocity, and hatch spacing are more significant for the shrinkage in *Z*-direction.

Linear empirical models are developed to characterize the relation between process parameters and part shrinkage, and estimate the shrinkage compensation in each direction. Only significant process parameters identified using ANOVA are included to develop the empirical models. For instance, the developed models for shrinkage compensation in *X*-direction is

$$S_X = 1.611691 - 0.01615 \text{ laser power} - 0.009647 \text{ laser velocity}.$$

Similar empirical shrinkage models are also developed for the *Y*- and *Z*-directions. The developed models predict the shrinkage in percentage for any combination of process parameters to scale up the digital file for optimal accuracy.

5.3.4 More Advanced Designs and Modeling Methods

5.3.4.1 Space-Filling Designs

Besides these classical design-of-experiment methods, other designs exist, such as space-filling designs that treat all regions of the design space equally. Space-filling designs are useful for modeling systems that are deterministic or near-deterministic, such as a computer simulation, which involves many variables with complicated interrelationships. One criterion to fill the design space is to minimize the integrated mean squared error (IMSE) over the design space by using IMSE-optimal designs. Koehler and Owen (1996) describe several Bayesian and frequentist space-filling designs, including maximum entropy designs, mean squared-error designs, minimax and maximin designs, Latin hypercubes, randomized orthogonal arrays, and scrambled nets. A review of Bayesian experimental designs for meta-modeling is given in Kleijnen (2009).

5.3.4.2 D-Optimal Designs

D-optimal design is another type of design which is useful when classical designs (such as factorial and fractional designs) do not apply. In practice, standard factorial or fractional factorial designs may require too many runs for the amount of resources or time allowed for the experiment, or the design space may be constrained (the process space contains factor settings that are not feasible or are impossible to run). The design matrices generated by D-optimal designs are usually not orthogonal, and effect estimates are correlated. These types of designs are always an option regardless of the type of model the experimenter wishes to fit (e.g., first order, first order plus some interactions, full quadratic, cubic, etc.) or the objective specified for the experiment (e.g., screening, response surface, etc.). However, D-optimal designs are developed based on a chosen optimality criterion and the possible underlying model that will be used to fit the experiment data. Specifically, D-optimal designs maximize the determinant of the information matrix. This optimality criterion results in minimizing the generalized variance of the parameter estimates for a prespecified model. As a result, the "optimality" of a given D-optimal design is model dependent. In other words, an approximation model must be specified before the generation of design points.

5.3.4.3 Response Surface and Kriging

The response surface model (RSM) "is a collection of statistical and mathematical techniques useful for developing, improving, and optimizing process" (Myers et al., 2009). Since the true response surface function $f(x)$ is usually unknown, a response surface $g(x)$ is created to approximate $f(x)$. Low-order polynomial models are usually popular choices for the response surface $g(x)$. Depending on the needed curvature, various polynomial models can be developed. The more curvature needed to be incorporated, the higher order polynomial models are required.

After identifying the factors that have significant impacts on the response, the general RSM approach includes all or some of the following steps:

1. *First-order experimentation*: When the starting point is far from the optimum point or when knowledge about the space being investigated is sought, first-order models and an approach such as steepest ascent are used to "rapidly and economically move to the vicinity of the optimum" (Montgomery, 1984). The general form of first-order model is expressed as follows:

$$g(x) = \alpha_0 + \sum_{i=1}^{k} \alpha_i x_i.$$

The 3D response surface and 2D contour plot of a linear response surface model are demonstrated in Figures 5.5 and 5.6, respectively. The resulting response surface is a plane over the 2D design space. The first-order model is also called main effect model because it focuses on the main effects of the process parameters. Other than the main effects, interactions between the process parameters can be incorporated in the response surface model as follows:

$$g(x) = \alpha_0 + \sum_{i=1}^{k} \alpha_i x_i + \sum_{i<j} \sum_{j=2}^{k} \alpha_{ij} x_i x_j.$$

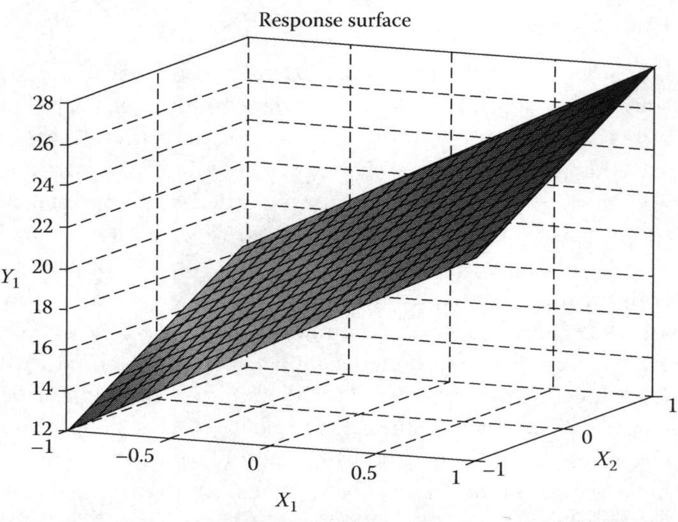

Figure 5.5 Response surface of $g(x) = 20 + 5x_1 + 3x_2$.

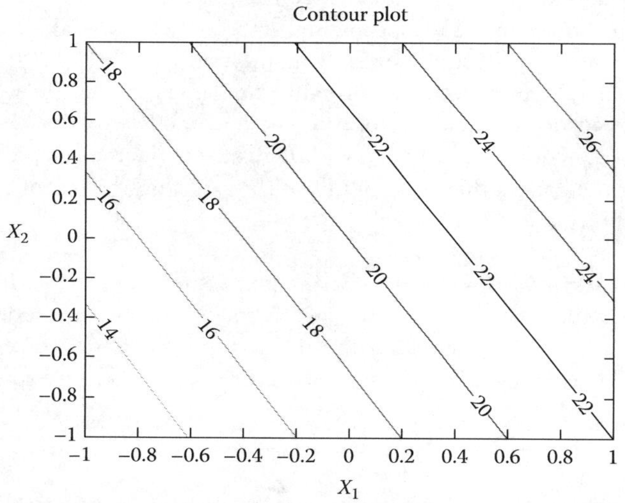

Figure 5.6 Contour plot of $g(x) = 20 + 5x_1 + 3x_2$.

Following the previous example, the 3D response surface and 2D contour plot of the first-order response surface model, with the consideration of parameter interaction, are demonstrated in Figures 5.7 and 5.8, respectively. As shown in the figures, the incorporation of the interaction term has imposed curvature to the response surface.

2. *Second-order experimentation*: After the best solution using first-order methods is obtained, a second-order model is fit in the region of the first-order solution to evaluate curvature effects and to attempt to improve the solution. The general form of second-order model is as follows:

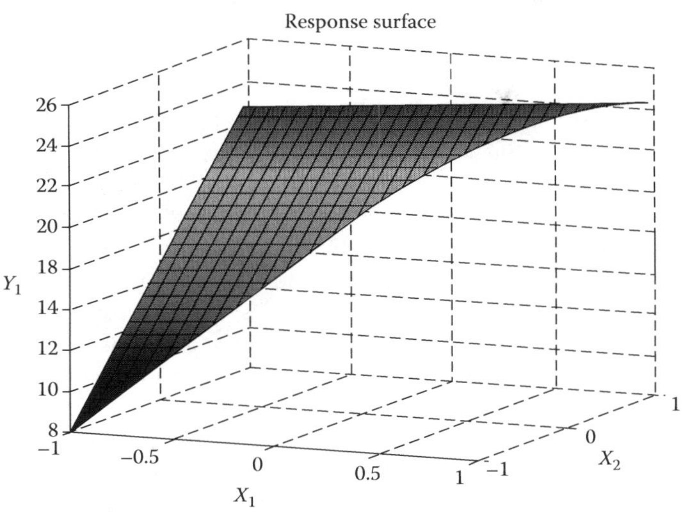

Figure 5.7 Response surface of $g(x) = 20+5x_1+3x_2-x_1x_2$.

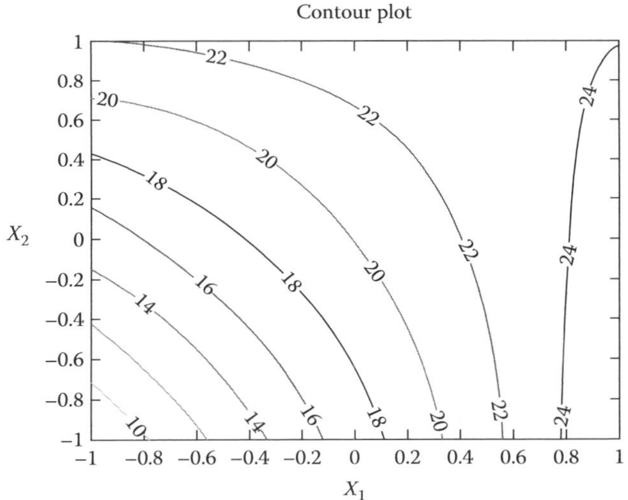

Figure 5.8 Contour plot of $g(x) = 20+5x_1+3x_2-x_1x_2$.

$$g(x) = \alpha_0 + \sum_{i=1}^{k} \alpha_i x_i + \sum_{i=1}^{k} \alpha_{ii} x_i^2 + \sum_{i<j} \sum_{j=2}^{k} \alpha_{ij} x_i x_j .$$

The parameters of the polynomials models are usually determined by least squares regression analysis by fitting the response surface approximations to existing data. These approximations are normally used for prediction within response surface methodology (RSM). A polynomial model may not seem a reliable and precise model for approximating the nonlinear functions of arbitrary shape over the whole design space; however, it is accurate enough for approximating relatively small portion of the true function, and it is difficult to take enough sample points in order to estimate all of the coefficients in the polynomial equation, particularly in high dimensions. A complete discussion of response surfaces and least squares fitting is presented in Myers et al. (2009). Similar to the response surface model, the method of kriging may also be used to characterize the responses as a combination of a polynomial model plus localized departures, characterized by a Gaussian process. Based on the observed experiment data and a covariance structure, the kriging method can predict the response value for any unobserved parameter setup. The kriging parameters can usually be estimated using the method of maximum likelihood estimates. Detailed discussion of parameter estimation can be found in Simpson et al. (1998). For small problems with relatively few sample points, fitting a kriging model is rather trivial. However, as the size of the problem increases and the number of sample points increases, the added effort needed to obtain the "best." Kriging may begin to outweigh the benefit of building the approximation. Besides, prediction with a kriging model requires the inversion and multiplication of several matrices which grow with the number of sample points. Hence, for large problems prediction with a kriging model may become computationally expensive as well.

5.3.4.4 Artificial Intelligence

Another category of data-driven methods that have been applied for AM process optimization is artificial intelligence (AI), which aims at "training" a black-box model based on a large training data set. More detail about the fundamentals of AI methods can be found in Russel and Norvig (1995). Theses algorithms include but are not limited to support vector machine, neural network, Bayesian network, and their extensions. With a large training data set, AI algorithms usually provide accurate estimations of parameter-feature relations. For instance, Lu et al. (2010) applied the method of least square support vector machine (LS-SVM) to investigate the relation between mechanical properties of parts and process parameters such as laser power, traverse speed, and the powder feed rate in LENS. Validated by using fabricated thin-walled parts, the method of LS-SVM is reported to accurately predict the deposition height when a large sample of parts is used to train the model. Similar successful applications of AI methods were performed by Casalino and Ludovico (2002), which uses feed forward neural network (FFNN) to model a laser-sintering process, and by Wang et al. (2009), which adopts Bayesian probability network to characterize a laser-bending process. DLD process optimization for controlled layer height can also be accomplished using advanced/intelligent computational methods such as the mutable smart bee algorithm (MSBA) and fuzzy interference system (FIS), and unsupervised machine learning approaches such as self-organizing maps (SOMs) (Fathi and Mozaffari, 2014). Despite these successful studies, the application of AI methods is rather rare in the literature of AM at large. This is because the key to a successful application of AI methods is enormous training data that can be used to estimate the process model, which usually results in extremely high experimental costs. Moreover, due to the proprietary nature of DLD experiment data, data sets are often hard to obtain.

5.4 Summary

The optimization of LBAM process is essential to building LBAM parts with enhanced mechanical properties and improved quality. This requires the combination of the optimization in powder material design/preparation and the corresponding optimal LBAM processing parameters. Thus, efficient process parameter optimization is essential for the establishment of a material process database, realizing a simplified, precise, and stable control of AM treatment of versatile powder materials for industrial applications.

In order to achieve enhanced or optimal quality for LBAM parts, it is essential to understand and characterize how LBAM process parameters affect thermal history, solidification, and eventually microstructural/mechanical properties. This remains an open research area due to the large number of process parameters (e.g., laser power, laser velocity, powder feed rate, layer thickness, hatching pitch, scanning pattern, etc.) involved during LBAM. Most of the existing studies seek only optimal process parameters via extensive experimental or simulation. A major limitation of this approach is that the resulting optimal process parameters may not be useful when experimental conditions (e.g., process or material) change, resulting in new experiments to-be-conducted from scratch. Further research is needed to (1) leverage the information from existing studies and (2) systematically characterize the relation between process parameters and part features so that the LBAM process can be optimized in a more efficient manner.

This challenge of process optimization is further compounded by the interactions among LBAM parameters. In reality, it may not be practical to incorporate all process parameters in either experimental or analytical studies. Ignorance of such higher order interaction effects, taking place during the LBAM process, causes systematic uncertainty in the resulting models and experimental results. Process uncertainty is associated with the initial (latent heat exchange) and evolutionary (dendritic) solicitation, and conductive, convective, and radiative heat transfer. For instance, the spatial/temporal scale for LBAM is relatively small for conduction heat transfer and thus thermal responses can be difficult to measure and model—especially for the material (which is not detectable). Such uncertainties will affect not only the microstructure but also the mechanical features of the fabricated parts. In addition, there lies considerable uncertainty in melt-pool depth (and other dimensions) due to uncertain heat transfer and fluid/part wetting behavior (contact angles unknown). Further research is needed to systematically incorporate uncertainty when optimizing the LBAM process.

5.5 Review Questions

1. What are the key process parameters that govern LBAM processes?
2. Provide an example of postmanufacturing processing procedure?
3. What is Buckingham's Π theorem?
4. What are the advantages and limitations of process maps?
5. What is the overall procedure of data-driven optimization methods?
6. What are the difference between full factorial design and fractional factorial designs?
7. What is the Taguchi parameter design?
8. Describe a scenario in which D-optimal design can be useful.

References

Birnbaum, A., P. Aggarangsi and J. Beuth (2003). Process scaling and transient melt pool size control in laser-based additive manufacturing processes, *Proceedings of the Annual International Solid Freeform Fabrication Symposium,* Laboratory for Freeform Fabrication, Austin, TX, pp. 328–339.

Bontha, S., N. W. Klingbeil, P. A. Kobryn and H. L. Fraser (2009). Effects of process variables and size-scale on solidification microstructure in beam-based fabrication of bulky 3D structures. *Materials Science and Engineering: A* 513–514: 311–318.

Casalino, G. and A. D. Ludovico (2002). Parameter selection by an artificial neural network for a laser bending process. *Proceedings of the Institution of Mechanical Engineers, Part B: Journal of Engineering Manufacture* 216(11): 1517–1520.

Fathi, A. and A. Mozaffari (2014). Vector optimization of laser solid freeform fabrication system using a hierarchical mutable smart bee-fuzzy inference system and hybrid NSGA-II/self-organizing map. *Journal of Intelligent Manufacturing* 25(4): 775–795.

Kahlen, F. J. and A. Kar (2001). Tensile strengths for laser-fabricated parts and similarity parameters for rapid manufacturing. *Journal of Manufacturing Science and Engineering* 123(1): 38–38.

Kleijnen, J. P. (2009). Keriging metamodeling in simulation: A review. *European Journal of Operational Research* 192(3): 707–716.

Koehler, J. and A. Owen (1996). 9 computer experiments. *Handbook of Statistics* 13: 261–308.

Kummailil, J., C. Sammarco, D. Skinner, C. A. Brown and K. Rong (2005). Effect of select LENS™ processing parameters on the deposition of Ti–6Al–4V. *Journal of manufacturing Processes* 7(1): 42–50.

Lu, Z. L., D. C. Li, B. H. Lu, A. F. Zhang, G. X. Zhu and G. Pi (2010). The prediction of the building precision in the laser engineered net shaping process using advanced networks. *Optics and Lasers in Engineering* 48(5): 519–525.

Montgomery, D. C. (1984). *Design and Analysis of Experiments*, John Wiley & Sons, New York.

Myers, R. H., D. C. Montgomery and C. M. Anderson-Cook (2009). *Response Surface Methodology: Process and Product Optimization Using Designed Experiments*, John Wiley & Sons, Hoboken, NJ.

Raghunath, N. and P. M. Pandey (2007). Improving accuracy through shrinkage modelling by using Taguchi method in selective laser sintering. *International Journal of Machine Tools and Manufacture* 47(6): 985–995.

Russel, S. J. and P. Norvig (1995). *Artificial Intelligence: A Modern Approach*, Prentice-Hall, Upper Saddle River, NJ.

Shamsaei, N., A. Yadollahi, L. Bian and S. M. Thompson (2015). An overview of direct laser deposition for additive manufacturing; Part II: Mechanical behavior, process parameter optimization and control. *Additive Manufacturing* 8: 12–35.

Simpson, T. W., T. M. Mauery, J. J. Korte and F. Mistree (1998). *Comparison of Response Surface and Keriging Models in the Multidisciplinary Design of an Aerospike Nozzle*. Institute for Computer Applications in Science and Engineering, NASA Langley Research Center, Hampton, VA.

Thompson, S. M., L. Bian, N. Shamsaei and A. Yadollahi (2015). An overview of direct laser deposition for additive manufacturing; Part I: Transport phenomena, modeling and diagnostics. *Additive Manufacturing* 8: 36–62.

Vasinonta, A., J. L. Beuth and M. Griffith (2007). Process maps for predicting residual stress and melt pool size in the laser-based fabrication of thin-walled structures. *Journal of Manufacturing Science and Engineering* 129(1): 101–109.

Wang, L., S. D. Felicelli and J. E. Craig (2009). Experimental and numerical study of the LENS rapid fabrication process. *Journal of Manufacturing Science and Engineering* 131(4): 041019–041019.

Yu, J., X. Lin, L. Ma, J. Wang, X. Fu, J. Chen and W. Huang (2011). Influence of laser deposition patterns on part distortion, interior quality and mechanical properties by laser solid forming (LSF). *Materials Science & Engineering A* 528(3): 1094–1104.

Chapter 6

Process Monitoring and Control

Prahalad K. Rao
University of Nebraska–Lincoln

Contents

CHAPTER OUTLINE

- Advantages of the additive manufacturing (AM) paradigm (Section 6.1)
- Examples of poor (defective) processing results from AM processes and need for online quality monitoring in AM (Section 6.1)
- Main challenges hindering wide application of AM (Section 6.2)
- Research gaps and challenges in quality control in AM (Section 6.2.1)
- Status quo of quality monitoring in AM (Section 6.2.2)
- Empirical studies for dimensional integrity assessment of AM parts (Section 6.3)
- Application of spectral graph theory (SGT) for dimensional integrity quantification of AM parts (Sections 6.3.3 through 6.3.6)
- Status quo of research in sensor-based process monitoring in metal AM (Section 6.4)

6.1 Introduction

Learning Objectives

- Advantages accrued by the additive manufacturing (AM) paradigm
- Understand the various quality-related challenges in AM

The goal of this chapter is to inform the reader of the pivotal research in process monitoring and control of laser-based additive manufacturing (LBAM) processes. In the context of process monitoring and control, this chapter approaches the topic from the following didactic perspectives:

1. How to quantify the dimensional integrity of AM parts? (Section 6.3)
2. Which sensing techniques can capture the relationship between process conditions and build quality in AM, especially directed energy deposition (DED) and powder bed-fusion (PBF)? (Section 6.4)
3. How some signal patterns from *in situ* sensors correlate with specific build defects? (Section 6.4). Before we proceed to answer these questions, we provide a brief motivation for the criticality of further research in AM quality monitoring and control via some practical examples, and subsequently, review the seminal literature in these areas in Sections 6.1 and 6.2, respectively.

The unique, layer-upon-layer method of AM allows creation of complex, freeform geometries that are difficult, if not impossible, to realize using conventional subtractive and formative manufacturing techniques. AM, thus, surmounts time-honored manufacturing constraints, allowing designers to explore unconventional functional facets and geometries.

The unprecedented ability to control the exact placement of material has led AM to be termed as a disruptive technological development, in that, AM can fundamentally change the existing paradigms in manufacturing and logistics [1]. A summary of various AM processes available for designers is delineated in Table 6.1 [2–5]; the possibility in terms of materials, details, and applications are potentially limitless. In the context of this chapter, viz., process monitoring and control, we focus predominantly on two AM processes, namely, PBF and DED.

Table 6.1 Nomenclature and Stratification of Various AM Process Categories (the Categorization as per ASTM 2792-12 a, F42 Standards Committee)

Category	*Powder Bed Fusion (PBF)*	*Directed Energy Deposition (DED)*	*Binder Jetting*	*Material Jetting*	*Vat Photopolymerization*	*Material Extrusion*	*Sheet Lamination*
Layer fusion mechanism	Thermal (laser or electron beam)	Thermal (laser or electron beam)	Binder/glue (epoxy)	Photopolymerization (UV curing)	Photopolymerization UV curing	Thermal	Thermal (ultrasonic) or binder
Material delivery mechanism	Roller or scraper	Motion of nozzle	Roller	Motion of jetting head or nozzle	Immersion into resin vat	Motion of nozzle	Placement of sheets
Material delivery element	Powder bed	Material hopper and nozzle	Powder bed	Jetting head or nozzle	Polymer Vat	Nozzle	NA
Layer shaping method	Scanning mirrors and f–θ lens for laser-based systems; magnetic coils for electron beam processes	Gantry-based motion of nozzle	Gantry-based motion of inkjet head	Gantry-based motion of jetting head or table	Motion of UV laser or projection of mask with a DLP projector	Gantry-based motion of nozzle	Ultrasonic energy, glue, and subtractive machining.
Layer deposition element	Laser/electron beam scans a line path	Nozzle deposits a volume of material along a line. Metal filament melted by an electron beam.	Inkjet head deposits discrete binder droplets	Jetting head deposits material droplets along a line	UV laser scans a line path, or a mask (area) is projected.	Nozzle deposits molten thermoplastic along a line	A sheet (volume) of material is deposited at a time

(Continued)

Table 6.1 (*Continued*) Nomenclature and Stratification of Various AM Process Categories (the Categorization as per ASTM 2792-12 a, F42 Standards Committee)

Category	*Powder Bed Fusion (PBF)*	*Directed Energy Deposition (DED)*	*Binder Jetting*	*Material Jetting*	*Vat Photopolymerization*	*Material Extrusion*	*Sheet Lamination*
Energy/layer binding material	Laser, infrared heaters, electron beam	Laser, electron beam.	Epoxy	UV light	UV laser, DLP projector, twin fiber lasers	Thermplastic polymerization	Ultrasonic horn, glue, friction.
Materials	Metals and polymers	Metals	Polymers, gypsum, metals	Photopolymers, wax	Photopolymers	Thermoplastic polymers	Metal, paper, plastic
Material form	Powder	Powder or wire	Powder	Liquid photopolymer resin	Liquid photopolymer resin	Filament (wires)	Sheets of paper, paper, metal, composites.
Process temperature	>800°C	>2000°C	None	None	None	230°C–250°C (ABS)	150°C–200°C (aluminum)
Trade names and other popular names	Laser sintering (LS) Direct metal laser sintering (DMLS; EOS) Selective laser melting (SLM; SLM Solutions) Laser Cusing (concept laser) Electron beam melting (EBM; Arcam)	Direct metal deposition (DMD; POM group) Laser cladding (LC) Laser engineering net shaping (LENS; Optomec) Electron beam free form fabrication (EBFFF)	3D printing (3DP; MIT, ZCorp now 3D Systems) Digital part materialization (ExOne) Color jet printing (CJP; 3D Systems)	Polymer jetting (PJ; Stratasys) Multi-jet printing (MJP; 3D Systems) Aerosol jet prining (AJP; Optomec)	Stereolithography (SL; 3D Systems) Direct light projection Mask projection Two photon micro-SLA	Fused filament fabrication (FFF) Fused deposition modeling (FDM; Sratasys)	Laminated objected manufacturing (LOM; Cubic Technologies) Ultrasonic consolidation (UC) Selective deposition lamination (SDL; MCor) Plastic sheet lamination (PSL; Solidimension/ Graphtec, Cubic Technologies)

(*Continued*)

Table 6.1 (*Continued*) Nomenclature and Stratification of Various AM Process Categories (the Categorization as per ASTM 2792-12 a, F42 Standards Committee)

Category	*Powder Bed Fusion (PBF)*	*Directed Energy Deposition (DED)*	*Binder Jetting*	*Material Jetting*	*Vat Photopolymerization*	*Material Extrusion*	*Sheet Lamination*
	Selective laser sintering (SLS; 3D Systems, UTexas,) Selective heat sintering (SH; uses IR heater)	Laser powder forming (LPF) Electron beam additive manufacturing (EBAM; Sciaky)					Paper lamination technology (PLT; Kira) Offset Fabbing (Ennex) Ultrasonic additive manufacturing (UAM; Fabrisonic) Computer-aided Manufacturing of Laminated Engineering Materials (CAM-LEM)
Most common materials used	Stainless steel 316L and 17–4PH; aluminum AlSi10Mg, titanium Ti6Al4V, cobalt-chrome, Inconel 718 and 625.	Titanium alloys, nickel-base super alloys, copper, aluminum, tungsten/ titanium carbide, 316 stainless steel	Gypsum, sand, nylon, polymers, Copper (ExOne)	Photopolymers	Photopolymers	ABS, PLA, nylon	Aluminum, copper, stainless steel, titanium, carbon fiber composite

(*Continued*)

Table 6.1 (*Continued*) Nomenclature and Stratification of Various AM Process Categories (the Categorization as per ASTM 2792-12 a, F42 Standards Committee)

Category	*Powder Bed Fusion (PBF)*	*Directed Energy Deposition (DED)*	*Binder Jetting*	*Material Jetting*	*Vat Photopolymerization*	*Material Extrusion*	*Sheet Lamination*
Distinctive application area	Automotive, aerospace, and biomedical	Aerospace, repair and remanufacturing	Multicolor mockups (architecture), engineering models (multicolor FEA models) Dental Implants, Detail Oriented (Jewelry)	Multimaterial builds, flexible elements, textiles	Ultrafine geometries (especially with 2 photon techniques)	Prototyping, popular at home solutions.	Embedded sensors and electronics
Popular machine tool manufacturers	EOS (Germany) 3D Systems (US) Concept Laser (Germany) Renishaw (UK) SLM Solutions (Germany) Phenix Systems (France, now 3D Systems) Arcam (EBM, Sweden)	Sciaky (EBAM, US) Optomec (US) POM (US)	Stratasys 3D Systems	Stratasys 3D Systems	Asiga Stratasys 3D Systems Envisiontec	Stratasys 3D Systems	Cubic Technologies MCor Fabrisonic

At the outset, we note the following salient advantages of AM:

1. Reduced prototyping and simplified design evaluation (what you see is what you build)—designs can be conceptualized in a few hours versus weeks; existing solutions can be easily reverse engineered and evaluated. Imagine disassembling a competitor's part, scanning the part using a 3D scanner, converting the scan data (3D point cloud) into a 3D model, printing the part with similar materials, and finally testing the performance of the part. This allows making improvements to an existing design instead of starting from a new concept.
2. Complexity and scalability are easy to achieve—changes the design for manufacturing paradigm to *build what we can design*. For instance, designers are no longer constrained by existing manufacturing thumb-rules; a square or oblong channel with an intricate internal geometry takes roughly the same effort to realize as a hole with a simple circular cross section. Furthermore, building a part in half scale versus full scale does not require any change in tooling or fixtures. For instance, Figure 6.1a is an x-ray image of a miniature Inconel turbine blade made using a direct metal laser sintering (DMLS) AM process. The turbine blade (Figure 6.1a) has two through channels for passage of cooling gases, which allow operation at much higher speeds compared to turbine blades bereft of such facets. Likewise, Figure 6.1b shows micro-computed tomography (micro-CT) scans of a titanium spinal implant having a complex lattice-like internal structure. The parts shown in Figure 6.1 were characterized using the micro-CT scanning facilities at Binghamton University. Thus, from a broader vista, AM can potentially shorten supply chains, reduce product development and design lead times, and facilitate sustainable green engineering [3].
3. Potential material variety, and the ability to mass produce graded and smart materials—the same machine tool can produce titanium or stainless steel parts without having to change over tooling. Moreover, it is possible to create an integral object from different materials; for example, in DED processes, certain sections of a part can be built using titanium and then switched over to stainless steel.
4. Reduced energy consumption and waste (low *buy-to-fly ratio*, reduced *chip tax*, and elimination of intermediate assembly steps)—instead of cutting away expensive materials in form of chips, the component is built from raw material, such as powder, wires, liquid, or sheets. This leads to savings in raw materials, cost of recycling, and energy. AM is largely friction free; no chips are produced or raw material shaped by forcing against a die and press; and contactless, as shaping of material is on accord of heat, chemical reactions, or flow.

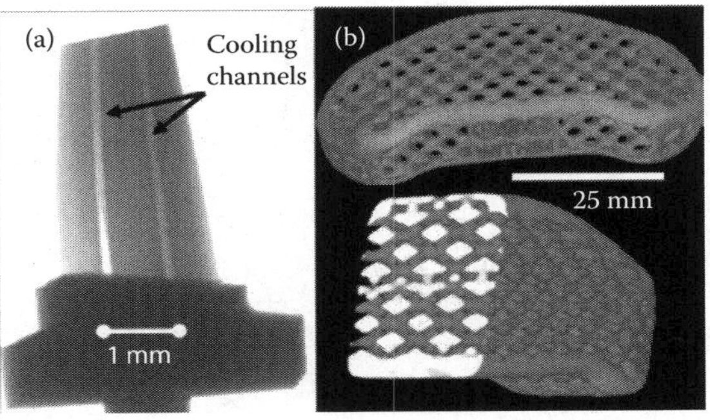

Figure 6.1 (a) X-ray image of a miniature turbine blade made using DMLS process. (b): A micro-CT scan of a titanium spinal implant crafted using DMLS.

5. Elimination of specialized tooling and setups—thus fixed costs are largely reduced as there is no need to make fixtures. In contrast, in machining, the cutting tool must be harder than the material to be cut; the tool invariably wears over time and has to be reground periodically. Similarly, with formative processes, such as die casting, injection molding, and forging, extrusion, the dies have to be replaced after certain number of cycles. With AM there is no need to build specialized tools and dies.
6. Single-point starting materials—which eliminate stocks of specialized materials and reduce inventory. With AM the size and geometry of raw material is no longer a constraint, hence inventory of high value, high cost items is no longer needed; for instance, visualize the logistics and planning for carrying parts on an aircraft carrier, submarine, or space shuttle. Instead of carrying engine nozzles, complex joints, critical parts, etc., raw materials such as metal powder, or filament can be stocked. This greatly simplifies logistical aspects of manufacturing.
7. Ability to repair broken parts—processes such as DED can be leveraged to repair damaged components, e.g., depositing material to repair a cracked turbine blade. Electron beam free form fabrication (EBFFF or EB3F) can be used to quickly replace broken features on dies and tooling.

Recently, a number of articles, workshop reports, and position papers, published by distinguished researchers and agencies, such as NSF, NIST, NASA, and DoD, articulate the need for research in the following areas in AM [6–15]:

1. Design guidelines
2. Materials
3. Quality control
4. Process physics, simulation and, modeling
5. Standardization of measurement and quantification

In these, the main challenges hindering the widespread adoption of AM processes are enumerated as follows:

1. Speed and efficiency of the process; builds can easily extend through several shifts.
2. Commercial unavailability of wide range of materials; pricing of materials can be prohibitive; and OEMs may restrict use of third-party materials.
3. Quality assurance challenges—e.g., lack of process repeatability and reliability, approaches for part qualification, and nondestructive testing hamper confidence in product quality [13].
4. Dearth of standards—materials testing, process performance capability databases, design guidelines, and process maps are not yet mature for AM technologies.

The following aspects are considered a priority in the context of quality control in AM:

- Dimensional integrity assessment of intricate geometry AM parts.
- *In situ* process monitoring and defect mitigation. Specifically, the need to address the dearth of understanding of existing correlations between materials, process conditions, sensor signals, and product quality.
- Process modeling and computational analytics to understand the physical root cause of poor quality. Specifically, the need for understanding failure mechanisms, and the effect of process conditions on part build quality from an *ab initio* perspective.
- Approaches for nondestructive testing and evaluation (NDE) of AM part quality.

The focus of this chapter is on repeatability and reliability of AM process, specifically monitoring of process status using sensor data.

Online quality monitoring in AM is a challenging research problem because minute drifts in process conditions can drastically affect build quality. The sensitivity of AM parts to evanescent process drifts is exemplified in Figure 6.2 [16]. The parts shown in Figure 6.2 are made using FFF, also called fused deposition modeling (FDM), a patented trade name of Stratasys, with (ABS) polymer material under identical processing conditions (layer height, feed, material flow, temperature, etc.) [16]. Yet, the part in Figure 6.2 (left) is replete with defects (overfilling, cracks, voids, etc.), whereas the part in Figure 6.2 (right) is closer to design specifications.

Keeping with the focus of this book, another example of quality-related challenges from the laser-based PBF process is exemplified in Figure 6.3. Figure 6.3 shows seven identical parts created on the same build platen, and under identical process conditions using an EOS 290M DMLS machine. The only difference is the manner in which the supports are designed. Of the seven parts, *all* barring two parts failed to build. Remarkably, the failure modes were distinctive for different parts:

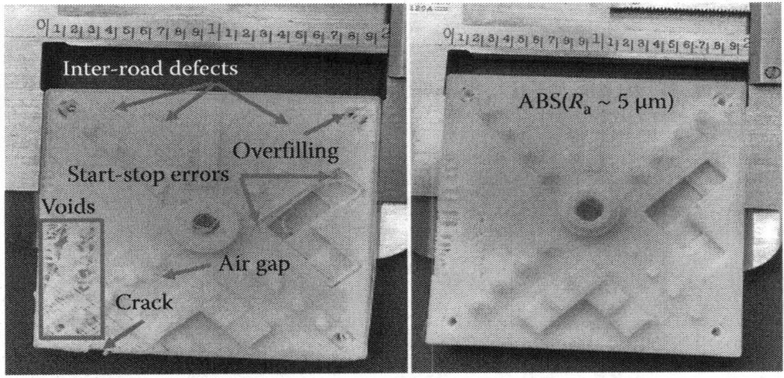

Figure 6.2 Two ABS polymer parts made under identical FFF process conditions yet showing markedly different quality.

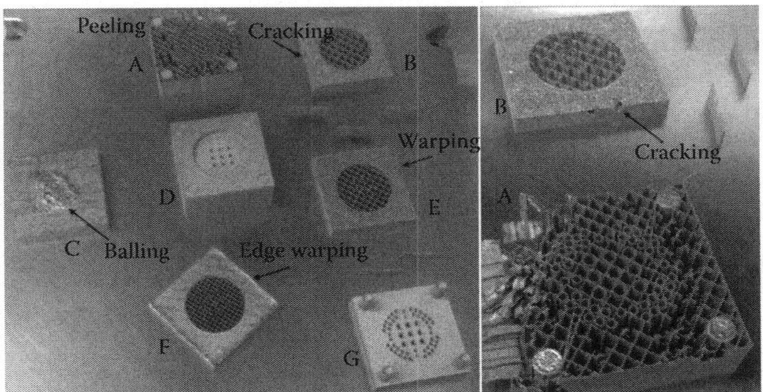

Figure 6.3 Seven identically designed 316 L stainless steel parts made under identical process conditions in an EOS M290 DMLS machine but with differing support structures on the same build platen.

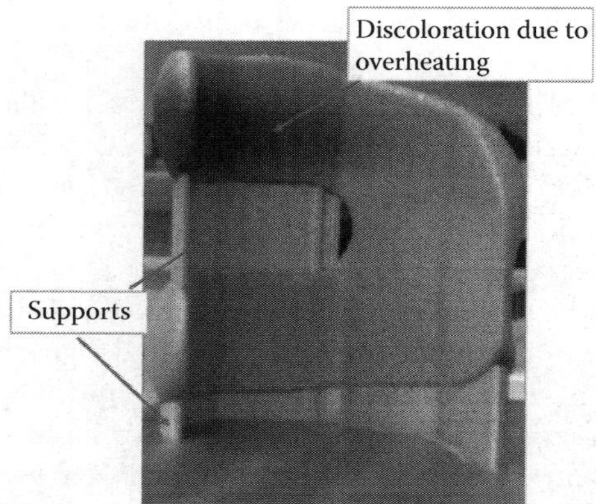

Figure 6.4 Stainless steel 316 L knee implant built on an EOS 290M machine showing significant discoloration of the surface due to overheating. The overheating is prominent in areas in contact with the supports, indicating the need for monitoring and adaptive control of the process. (Image courtesy of Dr. Ryan Willing, Binghamton University.)

- Part (A) shows catastrophic failure stemming from weak supports; the part peeled off from the platen, and the build had to be momentarily stopped.
- Part (B) shows cracking and voids along the edges.
- The center region of Part (C) shows evidence of poor surface finish and layer adhesion, probably due to melt-pool instabilities (*balling*).
- Parts (E) and (F) have warped along the edges (colloquially called *potato chipping*).
- Part F also shows severe surface defects due to the recoater making contact with the sintered part (super elevation).

Barring parts labeled (D) and (G), other parts failed to build; this highlights the importance of process monitoring in PBF. Given that a typical build in industry might last 20–30 h, and might consume several thousand dollars of material, it would be valuable if process defects were to be identified and corrected at an earlier stage.

Another motivating example attesting the compelling need for *in situ* sensing and monitoring in PBF is seen in Figure 6.4, which shows a 316 L stainless steel implant built on the EOS 290M machine at Binghamton University. Prominent discoloration of the workpiece is seen for areas, which are in contact with the support structures. This is owing to the poor thermal conductivity of the support structure. Although the microstructure of the discolored portion of the part was not examined, it is reasonable to assume that unchecked severe overheating may lead to cyclical remelting and resolidification of some layers and thus affect the physiomechanical properties of the build [17,18].

6.2 Need for Online Quality Monitoring in AM

Learning Objectives

- Research gaps and challenges in quality monitoring in AM
- Stratification of status quo of quality monitoring in AM

Lack of *repeatability and reliability* of key attributes, e.g., surface finish, dimensional fidelity, interlayer morphology, among others, which are key determinants of component functional integrity, is a major impediment for wider adoption of AM processes [6,14,15,19–22]. Despite enormous progress, quality assurance in AM remains an enduring research challenge, as shown in Figure 6.5. Because surface generation in AM involves material deposition, as opposed to material removal/forming, consequently, physical models and quality monitoring techniques developed for conventional manufacturing processes are incompatible for AM.

Figure 6.5 summarizes the connection between process–machine interactions, part quality, and functional fitness of AM parts. In this context, NIST in a recent road map report ascribed high priority for research in sensor-based measurement and physics-based modeling of AM processes and products [6]. Indeed, NSF, Air Force Research Laboratories, and the Institute for Defense Analyses (IDA) all recommend further research to overcome *the disconnect that exists between high-fidelity modeling research and real-time online process control efforts* [6,9,10,12,23].

Currently, quality assurance in AM is largely limited to *off-line data-driven techniques* and traditional empirical optimization approaches (e.g., design of experiments), leading to high scrap rates [24]. Moreover, because empirical studies require considerable time and resources, they contravene one of the main incentives of AM, i.e., going from design to realization without extensive retooling or process planning. Although sensor-based monitoring of AM has been introduced [22,25–51], these data-driven approaches are restricted for defect identification purposes; they do not have the physical basis to suggest the appropriate corrective action. The deficiency of existing physical models for capturing material–process–machine interactions, coupled with the dearth of customized *in situ* sensing strategies are critical limitations that must be overcome to accomplish effective quality control in AM [6,8,9,19,23,52–59].

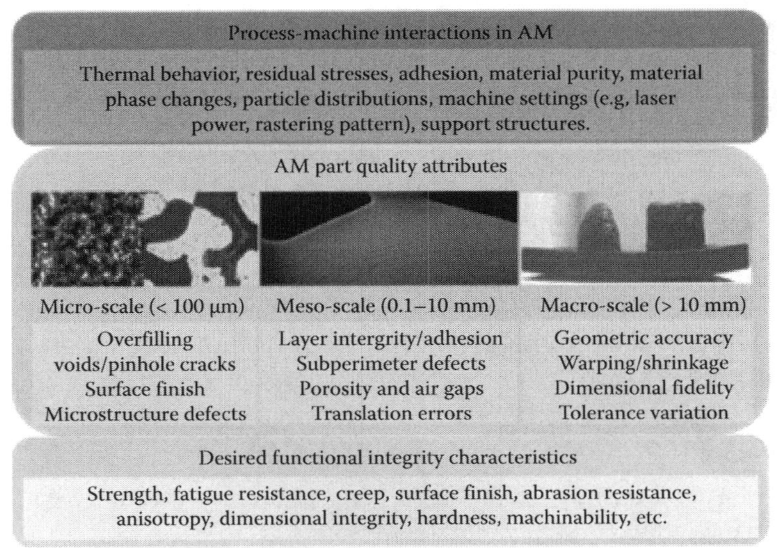

Figure 6.5 Some key process machine interactions, resulting part quality variables, and functional integrity in AM.

6.2.1 Research Gaps and Challenges in Quality Monitoring and Control in AM

The following unresolved challenges second the need for further fundamental research in the area.

1. *Need for novel metrology for benchmarking quality of AM components:* Surface morphology and dimensional fidelity of engineering components directly impacts their functional performance. Existing geometric dimensioning and tolerancing (GD&T) and surface metrology techniques, which are primarily intended for regular Euclidean features, are therefore not amenable for assessment of AM parts with complex freeform geometries [60–64]. In the absence of quantitative approaches for assessing surface morphology and dimensional fidelity, quality benchmarking of metal AM parts remains an unresolved impediment [6,62,65,66].
2. *Need for responsive sensing approaches:* Sensor-based monitoring in AM requires a different tack compared to traditional removal/formative processes, given the distinctive process mechanics (material deposition vs. material deformation). For instance, while accelerometers, force sensors, and acoustic emission sensors can detect variation in cutting regimes in machining, these are of limited utility in AM. Although sensing systems, including laser-based, acoustic, thermo-optical, infrared (IR) techniques, etc., have been implemented [21,22,25–49,67], further research in sensor technology, tailored to detect, as well as avert metal AM-specific anomalies, is necessary.
3. *Lack of physical models that capture process dynamics:* Current physical models suggested for AM do not have the capability to update results using information from sensor data. Consequently, these models have limited predictive capability, as they do not consider the evolving process–machine dynamics that govern functional integrity in AM. Hence, the possibility of compensating for process drifts based on real-time sensor information and physical models remains elusive [12,19]
4. *Dearth of physics-guided process adjustment:* In AM, quantitative physical models which invoke material science aspects, thermoelastic phenomena, and process–machine interactions are not well developed. Hence, current quality monitoring approaches in AM are either off-line, based on purely data-driven techniques (neural networks, mixture Gaussian modeling, statistical analysis), or lumped-mass formulations [36,68–72]. Therefore, their utility is largely relegated for detection of process anomalies. In the absence of physical models, the predictive value and process adjustment capability of data-driven models is limited. The lack of research toward combining physical models with sensor data is a drawback that should be surmounted in order to ensure closed-loop control in AM.

The Government Accountability Office (GAO) lists the following technical issues in the context of process monitoring and postprocess quality control (the stratification is the author's own) [7]. We note that there are several challenges in the areas of policy and technical education listed in Appendix IV of the GAO report, of which only a few are summarized in Table 6.2 (verbatim).

6.2.2 Status Quo of Quality Monitoring in AM

Approaches for quality assurance in AM are broadly demarcated in Figure 6.6. The four main research approaches are as follows:

1. *Statistical Modeling:* Refers to empirical process parameter optimization. Studies of this type involve, for instance, changing the scan speed and laser power given a representative

geometry and measuring a response variable, such as GD&T parameters, surface finish, and mechanical aspects [73]. The main disadvantages of such an approach is:

a. The time and cost of conducting experiments;
b. The limited region or space filled by experiments, in that it is difficult to extrapolate responses in untested regimes;
c. Inability to understand cause and effects layer by layer; and

Table 6.2 Some of the Materials, Quality Control, Process Modeling and Process Improvement, Education, and Policy Challenges in AM as Listed by the GAO [7]

Material and Material Testing
• Consensus methods and test data for qualification and certification are not sufficient
• Limited number of AM materials
• Unknown material properties, including material compatibility
Quality Control and Postprocess Inspection
• Need for improved suite of standards for AM
• Lack of inspection and quality standards for AM products
• Need for better process control to improve system performance and repeatability in AM
Process Modeling
• Fabrication speed too slow
• Need for appropriate design and mathematical analysis tools for AM products
• Need for appropriate design and physics-based analysis tools for AM processes
Process Improvement and Machine Design
• Need to print smaller-and larger-scale parts
• Improved part accuracy
• Improved surface finish
• Need to make more complex or integrated parts and products and not just shapes using AM
• Cyber security risks to AM equipment and design files
• Insufficient sharing of common data on AM process parameters, material, material properties, and testing
• Effect of AM equipment or software revisions on process qualification

(Continued)

Table 6.2 (*Continued*) Some of the Materials, Quality Control, Process Modeling and Process Improvement, Education, and Policy Challenges in AM as Listed by the GAO [7]

Technical Education and Training
• Need for improved collaboration and coordination among researchers
• Need for more pragmatic output from university research and development
• Need for more AM research and development infrastructure (labs, lab equipments)
• Hype and unrealistic expectations (e.g., disconnect between expectations and reality)
• Insufficiently skilled workforce and training
Policy and Legislation
• Potential safety concerns and uncertainty of product liability protections when AM is used by novices
• Need to avoid export controls/licensing that are impediments to US AM industry growth while still protecting US national security
• Need for greater consideration of US industrial base and national security concerns for AM
• Need for improved coordination among federal agencies on AM issues
• Lack of national strategy to enhance US competitiveness and engagement in the global marketplace
• Technology's reputation may be tarnished by bad experiences or by association with criminal activity

   d. Assumption of relatively stable process conditions and results from experiments carried out under lab conditions. For instance, often a representative standard test artifact is designed and experimental results are reported for this artifact. In reality, most AM applications rarely involve production over large batches. Indeed, AM is more viable for custom-designed, high value, low volume components.

2. *Sensing and Feature Extraction*: Machines are instrumented with sensors, such as IR cameras, photodiodes, and pyrometers, for monitoring process variables. The sensor patterns are correlated with process anomalies. However, the data analytics employed thus far are largely based on traditional statistical process analysis; correlation of sensor signal patterns with process conditions, material behavior, and product quality remains to be substantially explored.
3. *Physio-mechanical Characterization (Material Testing):* Test parts made under different process conditions are subject to mechanical tests such as fracture, fatigue, and creep [4,73,74]. Though this approach is similar to statistical modeling, a crucial difference is in the form of the response, for instance, microstructure characteristics, porosity are difficult to parameterize. The resulting correlations are also based on one factor at a time experiments; process interactions have not been thoroughly investigated.
4. *Computational Modeling:* Process phenomena are explained using techniques such as finite element and thermoelastic computational models [75–77]. The emphasis, thus, has been to

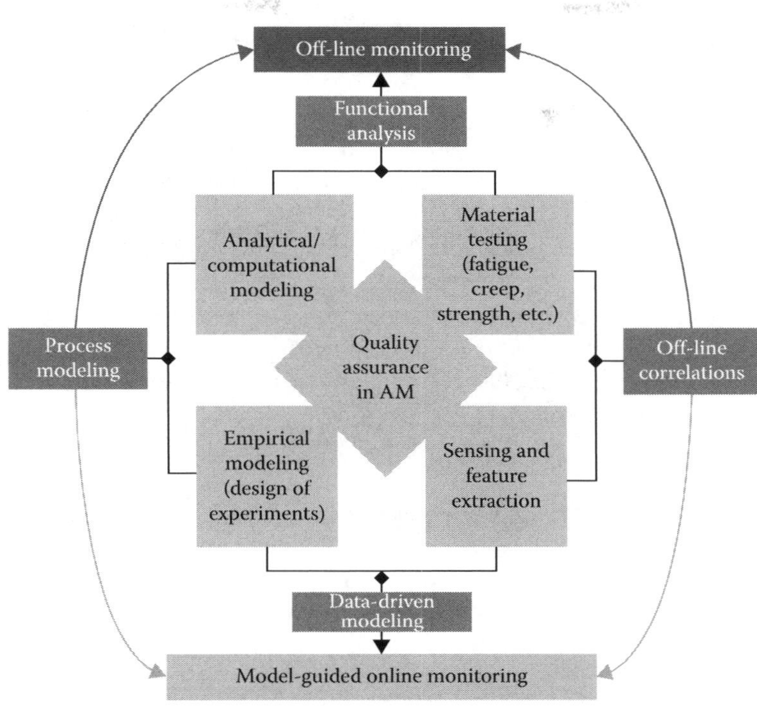

Figure 6.6 The fundamental approaches for quality monitoring in additive manufacturing.

explain residual thermal stresses as a function of scan speed, material flow rate, and temperature. Residual stresses develop due to repeated heating and cooling of the build (the laser heats underlying layers, besides the top layer), differential heat flow in the part as it is built (for instance, those portions on the part with supports in contact convey heat less amicably than portions without support), and temperature gradients in the part as the build height increases (e.g., the layers at the bottom are encased in powder).

From this stratum, researchers have sought to combine the advantages of each of these constituent parts.

- Materials testing and sensor-based modeling techniques are combined to obtain off-line correlations with part characteristics.
- Statistical modeling combined with sensor features for data-driven modeling.
- Computational models in conjunction with empirical models to obtain empirically verified process interactions.
- Computational models integrated with material testing data to obtain functional analytical models.

Finally, process models, off-line correlations, and data-driven models are essential for model-guided online monitoring. On the other hand, functional analysis, process models, and off-line correlations are ingredients for off-line monitoring.

6.3 Dimensional Integrity Assessment of AM Processes

Learning Objectives

- Empirical off-line studies for dimensional integrity in AM
- Understand how point cloud data can be useful for measurement of dimensional integrity of AM parts
- Understand the application of SGT for dimensional integrity quantification of AM parts

Empirical investigations in AM have tended toward off-line studies for optimizing three consequential quality variables, namely, dimensional/geometric precision, surface finish, and strength [78–95]. Recently, however, studies focused on eliminating build defects, such as voids, porosity, anisotropy, and delamination, are reported [84,96–102]. Key drawbacks evident in these efforts are:

1. Mostly invoke design of experiments-type studies for optimizing one response at a time
2. Rarely investigate temporal aspects, e.g., creep, fatigue, aging, and thermal cycling [95]
3. Largely off-line, with less focus on detection of evolving anomalies
4. Studies concerned with dimensional fidelity and surface morphology are restricted to external surfaces.

6.3.1 Off-Line Dimensional Integrity Studies in AM

Dimensional integrity studies in AM have, thus, largely relied on (1) categorical/visual evaluations [103]; (2) measurement of salient characteristics of specific facets/landmarks on the component [83]; and (3) GD&T measurements using coordinate measuring machines (CMM) [92]. These approaches provide only a partial perspective of the overall dimensional integrity of the component, and are noted to be fraught with limitations [61].

Accordingly, metrology of freeform geometry parts, particularly using 3D point cloud data, has recently garnered substantial research attention, primarily driven by emergence of AM techniques. In this context, a number of review articles have been published [60,104–106]. Point cloud data has been previously used in dimensional metrology of automotive parts, buildings, biomedical prosthetics, geospatial applications, etc. [60,106–108]. In this review, we will briefly summarize representative research reporting 3D point cloud data for quality assurance purposes.

From the broader manufacturing perspective, 3D point cloud data were hitherto predominantly used for reverse engineering purposes [109]. The emphasis in reverse engineering is to recreate the geometry of a component as an approximate CAD model from 3D point cloud data, and subsequently study its properties by either using analytical techniques (e.g., FEA) or physically testing a close replica. The conversion of point cloud data into a CAD surface involves several subtasks such as point matching, curve fitting, meshing, and segmentation, which by themselves are active research areas, particularly, in the machine learning, computer graphics, and functional modeling domains [110–116]. Coupled with the versatility of AM, and advent of inexpensive desktop/handheld 3D laser scanners, reverse engineering of complex geometries can now be accomplished within a few minutes [109].

Furthermore, with the considerable improvement in precision of laser scanning techniques and concomitant efficient data analysis algorithms in recent years, 3D point cloud acquisition is being increasingly considered for metrology purposes [106,117]. Nonetheless, a review of industry-based AM practitioners in Spain conducted by Manguia et al. [66] revealed that most users preferred conventional methods and implements, e.g., calipers (66%) and CMM (33%) for quality assurance of AM parts. No industry source reported using CAD or reverse engineering techniques for gaging dimensional integrity of AM parts.

For instance, Kainat et al. [118] developed a method for postproduction measurement of high strength steel pipes using a 3D laser scanner. Based on the acquired point cloud data, they compared measurements to an ideal best-fit cylinder estimated using point cloud rendering software (Geomagic by 3D systems). From their analyses, Kainat et al. subsequently detected and quantified the severity of various pipe imperfections, e.g., thickness variations, ovalization, seam weld errors, pipe diameter variations, etc. In a similar vein, Jiang et al. [105] used a function fitting and pattern tessellation approach for extracting statistical information from point cloud data in order to classify surfaces according to predefined functionally grouped classes.

Iuliano and Minetola [117] applied 3D point cloud data for CAD modeling of molds (used for plastic forming) after the molds were readied for use in production, i.e., after the punch and cavity have been precisely matched. Mold matching is a tedious process that requires considerable expertise, and is often a matter of conjecture. For instance, several intermediate finishing steps, e.g., grinding and polishing, are typically required after CNC machining of the mold parts from tool steel. Hence, the allowances on the as-machined mold are a matter of importance. Too much allowance and the time required for postprocessing will become prohibitive, whereas too less of an allowance may be lead to scrapping of the machined die. Most pertinently, because of material removal in intermediate processing steps, the net shape of the die set may differ significantly from the original design recorded in the blueprint. Iuliano and Minetola [117] captured the net shape of the finished die as a 3D point cloud. By doing so, geometry variations due to incremental changes made during mold matching were digitized as a CAD model. This is advantageous, because by knowing the final net shape of the die used in production, the number of intermediate processing steps and initial allowances on the as-machined mold can be minimized when the die set has to be repaired or remade.

Raja et al. [119] evaluated several rapid manufacturing methods, including AM processes and high speed machining, for assessing the competitive benefits offered by integration of emerging computer-based, intelligent data-driven techniques for the manufacture of complex aerospace parts. The studies by Raja et al. give an insight into the cost and time saving afforded by incorporation of data-driven quality monitoring approaches in manufacturing, e.g., shape point cloud analysis for quality monitoring. Thus, their work offers another contextual motivation of using point cloud data to assess part quality in AM.

Raja et al. investigated two specific scenarios, namely, (1) using AM in the manufacture of aerospace component castings and (2) the use of high speed machining to process complex components. Six different AM methods, viz., stereolithography (SLA), selective laser sintering (SLS), laser melting, 3D plotting with thermoform material, material jetting, and fused filament fabrication were used to construct two aerospace components. The parts were scanned with a laser-based profile scanner, and subsequently the component geometry was reconstructed from the acquired 3D point cloud data. The parts were quantified based upon morphological aspects, such as surface finish, geometric accuracy, and functional fitness, using the reconstructed geometry. Thereafter, the AM parts were processed into castings using the lost wax process, and the functional performance of these AM-patterned castings was tested. Furthermore, inspection was again conducted with point cloud data obtained for each finished casting, and a comparison was made to the CAD model. The following inferences pertinent to our work can be drawn from the results reported by Raja et al.:

- Surface finish and dimensional integrity of AM-patterned castings were independent. For instance, while the multimaterial jetting process affords good surface finish, the produced castings had one of the worst geometric accuracies. Thus, a quality variable such as surface finish should not be used as surrogate indicator for dimensional integrity.
- The AM-patterned castings showed presence of location-dependent anomalies, i.e., specific areas, e.g., underside of casting were not accurately reproduced, whereas the other areas were satisfactory. Hence, it is important to not only quantify the overall dimensional integrity but also localize dimensional variations contingent on the spatial geometry of the part. In other words, it is relevant to indicate the faulty locations on the component.
- Difficulty in quantitatively gaging the dimensional integrity of AM components; indeed, the majority of results reported by Raja et al. are based on qualitative criteria.

The approaches described, thus, employ 3D surface scanning, typically using laser interferometry. A major drawback on these approaches is that they cannot probe the internal geometry and material defects of a component; they are limited to the surface. In order to overcome this constraint, Kruth et al. [104] tailored a CT scanner for metrology applications. Their system, which integrates measurement and postprocessing algorithms, is capable of single-digit micrometer-level accuracies (2–5 μm). Arrieta et al. [61] and Bouyssie, et al. [120] also reported a CT scan approach for evaluation of biomedical replicas produced using SLA, although it would be difficult to CT scan relatively large parts.

Two main limitations are evident in the current state-of-the-art from this review:

1. Dearth of research reporting application of point of cloud-based for assessment of geometric integrity in AM.
2. Considerable focus exists on the processing and conversion of 3D point cloud data to a CAD computer model; however, only minimal in-depth analysis of the data itself. For instance, there is a lack of research investigating the use of point cloud data as an analytical tool to assess geometric deviations, and subsequently relating the result to the process for diagnostic purposes.

The spectral graph theory (SGT) approach forwarded by Rao et al. attempts to alleviate the foregoing gaps; this approach is described in detail herewith (please see also Ref. [121]). Further, mathematical details of SGT are detailed in a recent publication by Rao et al [122].

6.3.2 Spectral Graph Theoretic Approach for Assessment of Dimensional Integrity

6.3.2.1 Experimental Procedure

The NIST standard AM test artifact was used as the geometric reference for this investigation. Test components were manufactured at Oak Ridge National Laboratories (ORNL) using three different combinations of deposition technique and materials [123]. The three components are labeled as ABS chamber, ABS platform, and CF-ABS platform [124,125]. These experimental conditions are reported in Table 6.3.

Table 6.3 Material and Processing Conditions for the Three Types of Test Components Studied in This Work [121]

Build Characteristics ↓	*Component Type*		
	ABS Chamber Figure 6.8a	*ABS Platform Figure 6.8b*	*CF-ABS Platform Figure 6.8c*
Material	ABS*plus*-P430 (Stratasys)	Generic 1.75 mm ABS filament (makergeek.com)	Tailored carbon fiber impregnated ABS (CF-ABS) (see Ref. [124])
Machine	Stratasys uPrint SE Plus FDM machine	Solidoodle 3 desktop FFF machine	Modified Solidoodle 3 desktop FFF machine
Thermal distortion control	Heated build *chamber* (77°C)	Heated build *platform* (85°C)	Heated build *platform* (85°C)
Extrusion temperature (°C)	310	205	215
Layer height (mm)	0.330	0.20	0.40
Tool path and direction	Two perimeter contours, 45° sparse raster fill. Slicing algorithm was similar for each system.		

Subsequently, the component geometry was assessed with a linear laser scanning probe (FaroArm Platinum) and stored as a 3D point cloud. The test component coordinate measurements $(\mathcal{T}^{N\times d})$ were compared to the corresponding design reference coordinates $(\mathcal{R}^{N\times d})$ from the .stl CAD file, where N is the number of measurement points, and d the dimension of the coordinate axes (Cartesian). In our case, N is approximately 500,000 and $d = 3$. The difference between the test component and design reference coordinates, $\mathcal{T} - \mathcal{R} = \mathcal{X}^{N\times d}$, is termed the point cloud *deviation matrix*. Our aim is to classify the quality of AM test parts based on their dimensional integrity, as measured in terms of the 3D point cloud coordinate deviations contained in $\mathcal{X}$. The part measures approximately 100 mm × 100 mm × 8 mm (4 in. × 4 in. × 0.3 in.). Each feature is designed to evaluate a specific capability of the AM process based on the limits of dimensional accuracy [123]. The point cloud data, thus, obtained is shown in Figure 6.7.

The laser scanner records reflected light from the surface of a component as a point in 3D space, with a maximum volumetric deviation of ±43 μm. The point cloud for each part was then imported into a commercial software package (Geomagic by 3D Systems) for analysis. Standard functions within the software are used to remove outlier points and disconnected components. The error-corrected 3D point cloud data is subsequently converted to a polygon mesh for comparison against the reference CAD model in order to assess the geometric accuracy of the component.

The polygon mesh and CAD model were numerically compared by measuring the normal distance from a point on the mesh to the closest surface of the CAD model. The geometric 3D point cloud data, thus, obtained is structured as a matrix consisting of nine columns, which are: (1–3) Cartesian coordinates of the reference point on the CAD model $\mathcal{R}^{N\times 3} = \{\text{Xref}, \text{Yref}, \text{Zref}\}$; (4–6) the coordinates of the reference point from the measured polygon mesh $\mathcal{T}^{N\times 3} = \{\text{Xmesh}, \text{Ymesh}, \text{Zmesh}\}$; and (7–9) the deviation $\mathcal{T} - \mathcal{R} = \mathcal{X}^{N\times 3}$ between the measured and CAD coordinate.

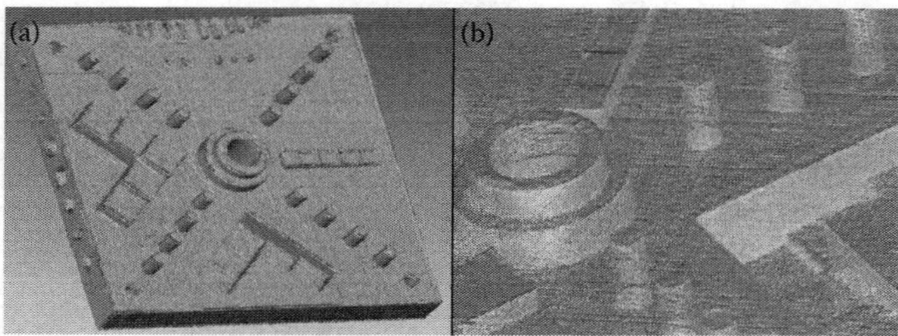

Figure 6.7 **(a) 3D point cloud of NIST part (100 mm × 100 mm × 8 mm; 4 in. × 4 in. × 0.3 in.) obtained using the FaroArm laser scanning probe. (b) Zoomed in section showing fine features (4× zoom). (From Rao, P., et al, *J. Manuf. Sci. Eng.*, 138(5), 051007, 2015.)**

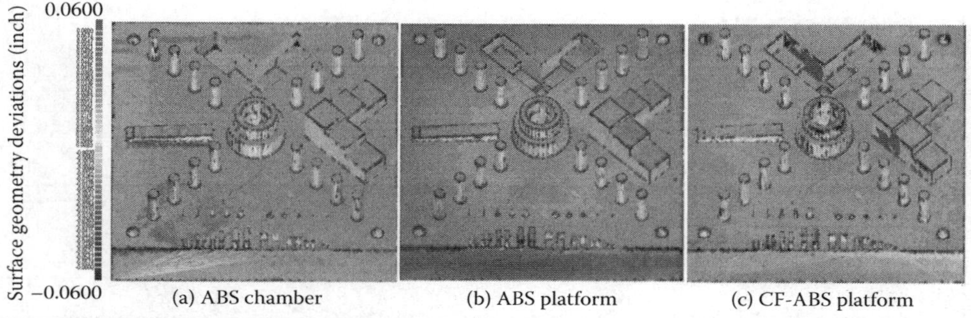

Figure 6.8 **Plots showing surface geometry deviations (inches) obtained from 3D point cloud data for three different AM samples (for further details, see Table 6.3). The AM samples shown here measure 100 mm × 100 mm × 8 mm (4 in. × 4 in. × 0.3 in.). Areas with a reddish hue have a positive deviation from the ideal design specifications (area has larger than specified dimension), while bluish-colored areas indicate a negative deviation from the specified dimensions (dimensions are smaller than specifications), and yellow color areas have part dimensions closer to specifications [121]. (a, b) Components produced using acrylonitrile butadiene styrene (ABS) thermoplastic but different processing conditions (Table 6.3). (c) Component produced using CF impregnated thermoplastic composite (CF-ABS).**

The deviations from the reference geometry can be visualized as flooded contour plots, as shown in Figure 6.8. Although the flooded contours of the geometric variations are an intuitive way to visualize the integrity of the part, quantification of these variations is challenging. In the forthcoming sections, a novel spectral graph approach devised by Rao et al. [121] for quantifying geometric integrity of complex AM parts from 3D point cloud data measurements is demonstrated.

6.3.3 Graph Theoretic Approach for Representing Point Cloud Data as a Network Graph

The objective of this section is to represent a sequence $\mathcal{X}$ of 3D point cloud data as a network graph G, i.e., achieve the mapping $\mathcal{X} \mapsto G$. Consider a sequence, $\mathcal{X} = x_1, x_2, \ldots, x_N$, where each $\mathcal{X}_i$ is a $1 \times d$ vector. Essentially, $\mathcal{X}$ can be recast in matrix form with its rows indexed by x as follows:

$$\chi = \begin{bmatrix} x_1^1 & \cdots & x_1^d \\ \vdots & \ddots & \vdots \\ x_N^1 & \cdots & x_N^d \end{bmatrix} \tag{6.1}$$

From a 3D point cloud perspective, in Equation 6.1 each row of $\mathcal{X}$ corresponds to a coordinate location along the Cartesian plane $\{x, y, z\}$, i.e., $d = 3$. At this juncture, no sampling conditions have been imposed on $\mathcal{X}$, and it is assumed that $\mathcal{X}$ is an arbitrary sample of size N. Different sampling conditions will be progressively imposed on $\mathcal{X}$ in Section 6.3.5.

As a further note, in this work, unless otherwise stated, the matrix $\mathcal{X}$ contains deviations of test parts from the ideal designed dimensions. The individual deviations x_i are obtained by subtracting the reference geometry $\mathcal{R}$ of the component from the originating CAD .stl file from the geometry $\mathcal{T}$ of test part (measured using a laser scanner), $\mathcal{T} - \mathcal{R} = \mathcal{X}^{N\times 3}$. The coordinates of the test part are aligned with the .stl file using coordinate registration software (Geomagic by 3D Systems). For the N rows of $\mathcal{X}$, pairwise comparison metrics w_{ij} are computed using a kernel function Ω [126]. In this work, the following radial basis kernel is used; mathematical implications of using this kernel are elucidated in Ref. [122]. The constitutive equations of the SGT approach are as follows:

$$\mathrm{w}_{ij} = \Omega(x_i, x_j) = e - \left(\frac{k(x_i, x_j)}{\sigma_{\mathcal{E}}^2} \right) \forall\, i, j \in \{1, \ldots, N\}. \tag{6.2}$$

$$k(x_i, x_j) = \|x_i - x_j\|^2; \; \mathcal{E}^{N\times N} = \left[k(x_i, x_j) \right]$$

$$\Theta(\mathrm{w}_{ij}) = \mathrm{w}_{ij} = \begin{cases} 1, & \mathrm{w}_{ij} \leq r \\ 0. & \mathrm{w}_{ij} > r \end{cases} \tag{6.3}$$

$$\mathbf{S}^{N\times N} = [\mathrm{w}_{ij}],$$

where x_i is essentially a row from the matrix $\mathcal{X}$; w_{ij} is a pairwise radial basis distances between two rows i and j; $r = \dfrac{\sum_{i=1}^{N} \sum_{j=1}^{N} \mathrm{w}_{ij}}{N^2}$; and $\sigma_{\mathcal{E}}^2$ is the overall statistical variation of the Euclidean distance matrix $\mathcal{E}$.

Equations 6.2 and 6.3 are the keystones of the SGT method. Equations 6.2 and 6.3 are particularly important because they convert a 3D point cloud data into an *unweighted undirected* graph.

The binary symmetric *similarity matrix* $\mathbf{S}^{N\times N} = \left[\mathrm{w}_{ij} \right]$ essentially contains an undirected network graph, G. Each row and column of S may be treated to be the vertex (or node) of the *unweighted undirected* graph, $G \equiv (V, E)$. with nodes (vertices) V and edges E [127]. The weight of an edge connecting a node i with another node j is w_{ij}. Thus, from Equations 6.2 and 6.3 we have represented a 3D point cloud data ($\mathcal{X}$) as an unweighted and undirected graph, $G \equiv (V, E)$.

6.3.4 Quantification of Graph Network Topology

Once the point cloud data $\mathcal{X}$ is represented as a graph G, relevant topological information is extracted from G, which is subsequently used for quantifying $\mathcal{X}$. For this purpose, the *degree* d_i

of a node i is computed, which is a count of the number of edges that are incident upon the node, and the diagonal *degree matrix* $\mathcal{D}$ structured from d_i is obtained as follows,

$$d_i = \sum_{j=1}^{j=N} \mathrm{w}_{ij} \quad \forall i, j \in \{1, \ldots, N\}, \tag{6.4}$$

$$\mathcal{D}^{N \times N} \stackrel{\text{def}}{=} \mathrm{diag}(d_1, \ldots, d_N) \tag{6.5}$$

Next, the *volume* $\mathcal{V}$ and the *normalized Laplacian* $\mathcal{L}$ of the graph G is defined as:

$$v(G) \stackrel{\text{def}}{=} \sum_{i=1}^{i=N} d_i = \mathrm{tr}(\mathcal{D}), \tag{6.6}$$

$$\mathcal{L}^{N \times N} \stackrel{\text{def}}{=} \mathcal{D}^{-\frac{1}{2}} \times (\mathcal{D} - \mathbf{S}) \times \mathcal{D}^{-\frac{1}{2}},$$

$$\text{where } \mathcal{D}^{-\frac{1}{2}} = \mathrm{diag}\left(1/\sqrt{d_1}, \ldots, 1/\sqrt{d_N} \right). \tag{6.7}$$

$\mathcal{L}$ is analogous to Kirchhoff matrix encountered in electrical networks [128]. Thereafter, the eigen spectrum of $\mathcal{L}$ is computed as:

$$\mathcal{L}v = \lambda^* \mathbf{v}. \tag{6.8}$$

Note that $\mathcal{L}$ is symmetric positive semi-definite, i.e., $\mathcal{L} \geq 0$, its eigenvalues (λ^*) are non-negative, and bounded between 0 and 2, i.e., $0 \leq \lambda_i \leq 2$. The smallest nonzero eigenvalue (λ_2) is termed the *Fiedler number* and the corresponding eigenvector ($\mathbf{v}_2$) as the *Fiedler vector* [127,129]. Barring pathological scenarios that rarely occur in practical circumstances, the Fiedler number is strictly bounded between 0 and 1, i.e., $0 < \lambda_2 < 1$ [127,129]. The graph *topological invariant* Fiedler number (λ_2) is used as a quantifier for $\mathcal{X}$. For a more detailed discussion of the mathematical properties of the Fiedler number, we refer the reader to Refs. [122,130].

This section closes with a remark concerning the Laplacian eigen problem (Equation 6.8) that must be solved in order to obtain the Fiedler number (λ_2). It is observed, albeit seldom, that the Laplacian matrix $\mathcal{L}$ may become defective, i.e., it does not have a full complement of eigen vectors and values. This is typically observed when there is a seemingly large cluster of nodes with almost identical weights. This problem can be overcome by a straightforward modification to the degree matrix, as follows,

$$\mathcal{D}^{N \times N} \stackrel{\text{def}}{=} \mathrm{diag}(d_1, \cdots, d_N) + \mathbf{I} \tag{6.9}$$

where $\mathbf{I}$ is the identity matrix. This modification barely perturbs the Laplacian matrix. For instance, the difference in Fiedler number estimates was found to be in the order of 10^{-4} when tested against experimental data. Furthermore, the manner in which the graph is constructed, based on the threshold function in Equations 6.2 and 6.3, ensures that pathological scenarios such as star graphs and bipartite graphs, which have a Fiedler number (λ_2) = 1, are avoided [127]. This is because the Heaviside step function Θ minimizes the effect of outliers due to thresholding, i.e., $r = \frac{\sum_{i=1}^{N}\sum_{j=1}^{N} w_{ij}}{N^2}$. In the same token, the addition of the identity matrix in Equation 6.9 ensures that $\lambda_2 \neq 0$. Hence, in our application, it can be claimed that $0 < \lambda_2 < 1$.

In summary, a mapping $\mathcal{X} \mapsto G$ can be achieved whose properties are characterized using the Fielder number (λ_2). More pertinently, it can be inferred that further the dimensions of a component deviate from the reference, the higher the Fiedler number (λ_2). Hence, a comparatively larger Fielder number (λ_2) signifies parts with poor dimensional integrity. This will be demonstrated further using experimental data.

6.3.5 Methods for Practical Application of SGT

In this section, three different sampling schemas are developed for analyzing the 3D point cloud data $\mathcal{X}$. This is necessary for the following reasons:

1. It is computationally intractable to obtain the Fiedler number over the complete part. For instance, given 3D point measurements ($\mathcal{X}$) over 500,000 spatial locations in Equation 6.1, would entail 25×10^{10} pairwise comparisons to obtain **S** (Equations 6.2 and 6.3). The Laplacian matrix ($\mathcal{L}$) in Equation 6.7 and the resultant eigen decompositions (Equation 6.8) would then be of the order of (10^{10}), and, therefore, untenable. We will explore methods to make the approach computationally tractable using different techniques to sample the 3D point cloud data.
2. Forward diagnostic opportunities, i.e., localize which region on a given part has large deviations from the nominal.

Confirm that the reported results are statistically consistent irrespective of how the point cloud data is sampled. More pertinently, sampling helps overcome the effect of outliers in the data.

6.3.5.1 SGT-Method 1: Sequential Sampling of Point Cloud Measurements with Moving Windows

This method samples the acquired 3D point cloud deviations , where $\mathcal{X}^{N\times d}$ is the number of data points and d is the number of dimensions (x, y, and z Cartesian coordinate axes), with contiguous, nonoverlapping windows. This is because the experimental data used in this work N is ~500,000, with $d = 3$. For such a large data set, computing the similarity matrix S (Equations 6.2 and 6.3), and subsequently solving the eigenvalue problem for the Laplacian matrix $\mathcal{L}$ (Equation 6.8), is neither computationally efficient nor practically tenable. Computing the similarity matrix alone requires $\mathcal{O}(N^2)$ pairwise radial basis comparisons (Equation 6.2). In order to overcome this constraint, the following two steps are taken:

- *Step 1*: Split $\mathcal{X}^{N\times d}$ into n smaller matrices $\mathbf{X}_h^{k\times d}$, $\mathrm{h} = \{1,\ldots,n\}$, such that $k \times n = N$.

The procedure is essentially equivalent to operating a (non overlapping) sliding window of size k on a signal $\mathcal{X}$ along the time domain. In block matrix form, this can be written as:

$$\mathcal{X}^{N\times d} = \left[\left[\mathbf{X}_1^{k\times d}\right],\left[\mathbf{X}_2^{k\times d}\right],\ldots,\left[\mathbf{X}_n^{k\times d}\right]\right]_{n\times 1}^{T} \tag{6.10}$$

The key challenge is to select a sufficiently large window size k that does not impose computation constraints, unfortunately, this is a heuristic choice. Based on trials, an appropriate value of k was observed to be in the vicinity of 0.05%–0.1% of the length of total data N, which in our case is ~500,000 data points. Accordingly, k was set to 500 for all subsequent analysis.

- *Step 2*: Estimate the Fiedler number λ_2^h using Equations 6.2 through 6.8 for each sub sample X_h in Equation 6.10. This step can be distilled as:

$$\Lambda_2 = \left[\lambda_2^1 \quad \lambda_2^2 \quad \cdots \quad \lambda_2^n\right]^{\mathrm{T}} \tag{6.11}$$

These steps are summarized in pseudo-code form in Figure 6.9. This windowing procedure is used again in Methods 2 and 3. A key advantage of this method stems from its simplicity. However, the computational effort is not trivial because the Fiedler number must still be computed for a large number of windows ($h \sim 1000$). More pertinently, because the procedure somewhat preserves the sequence in which the point cloud data was obtained, the Λ_2 sequence will also reflect any inherent measurement/instrument related bias in the system. Therefore, another method was devised based on a random sampling procedure.

6.3.5.2 SGT-Method 2: Random Sampling of Point Cloud Measurements

In this method, instead of directly windowing the 3D point cloud data $\mathcal{X}$ in a sequential manner, an intermediate sampling step was imposed. The steps are as follows:

- *Step 1*: From the point cloud sequence $\mathcal{X}^{N\times d}$ random sample m points without replacement, i.e., obtain $\mathcal{x}^{m\times d}$, $\mathcal{x} \subseteq \chi$. Essentially, randomly pick m rows from the matrix $\mathcal{X}$.

```
procedure windowing (𝒳^{N×d}, k)         // k is length of window, e.g, k = 500
n = ⌈N/k⌉;                               // n is number of windows
w = k;
i = 1;
For (h =1; h ≤ n; h++)
Compute λ_2^h;                            (Equations 6.2 through 6.8)
𝒳_h^{k×d} = 𝒳^{k×d} (i:w, 1:d);          (Equation 6.10)
i = 1 + w;
w = k + w;
end;
Return Λ_2 = [λ_2^1 λ_2^2,....,λ_2^n]^T;   (Equation 6.11)
```

Figure 6.9 Pseudocode for the windowing procedure.

- *Step 2*: Apply the windowing procedure described in SGT-Method 1 (Figure 6.9) to the sub-sample space $\mathcal{x}$.
- *Step 3*: Repeat Steps 1 and 2 for different m (i.e., change m). Thus, for each m, a sequence of Fiedler numbers $\Lambda_2(m)$ is obtained after applying the windowing procedure. Alternatively, to ease computation the average Fiedler number over the complete sequence $\Lambda_2(m)$ can be saved. The process can be stopped once $\Lambda_2(m)$ statistically converges, i.e., average of the entries in $\Lambda_2(m)$ converge after a sample size m has been reached.

The key advantages of this method are that it eliminates inherent instrument and measurement bias, and is computationally efficient as it can be stopped once the results have converged.

6.3.5.3 SGT-Method 3: Spatial Localization of Geometric Deviation from Point Cloud Measurements

The previous two approaches did not consider the spatial sequence of the data. Hence, they cannot be used for identifying the specific location of a defect on the component. In contrast, the present method preserves the spatial information by following these steps:

- *Step 1*: Sort the point cloud data deviations $\mathcal{X}^{N\times d}$ in either the x or y direction based on the corresponding reference coordinates $\mathcal{R}$. It is implicitly assumed that the material layers are deposited in the z direction. Let $\overline{\mathcal{X}}$ be the sorted point cloud deviations matrix. $\overline{\mathcal{X}}$ therefore preserves spatial information.
- *Step 2*: Split the sorted deviation matrix $\overline{\mathcal{X}}$ into n equal windows by apply the windowing procedure described in SGT-Method 1 (Figure 6.9).

However, in this method, the windowing procedure is slightly modified. Instead of fixing the window size k as done in SGT-Method 1 and SGT-Method 2, in SGT-Method 3 the number of windows n is fixed and then compute the window size k. This has the effect of slicing the component into n strips of identical width (W/n, where W is the width of the part in the direction sorted in Step 1), which allows equitable comparison across parts. If n is set at 500, as is done for all instances in this work, then each sampling window is approximately 10 mils in width (250 μm), and 0.3 in. (8 mm) in height for the NIST sample (Figure 6.7). Because of such spatial sampling, the sequence $\Lambda_2(n)=\left[\lambda_2^1 \quad \lambda_2^2 \quad \cdots \quad \lambda_2^n\right]^{\mathrm{T}}$ is mapped to a particular area on the component. Consequently, it is possible to track which facet has deviated along with the magnitude of the deviations from the design blueprint dimensions.

6.3.6 Application to AM Surfaces

Results from application of the SGT approach to experimental 3D point cloud data using the different methods discussed in Section 6.3.5 are shown in Figure 6.10 and Table 6.4. The following inferences are made based on these results:

1. *Referring to Figure 10a1 through c1,* which shows the mean Fiedler number (λ_2) (from the Fiedler number sequence λ_2) for different test parts obtained using the three SGT-methods developed in Section 6.3.5, the Fiedler number (λ_2) demonstrates a consistent trend, namely, λ_2 (CF-ABS platform) < λ_2 (ABS chamber) < λ_2 (ABS platform). This trend is unambiguous

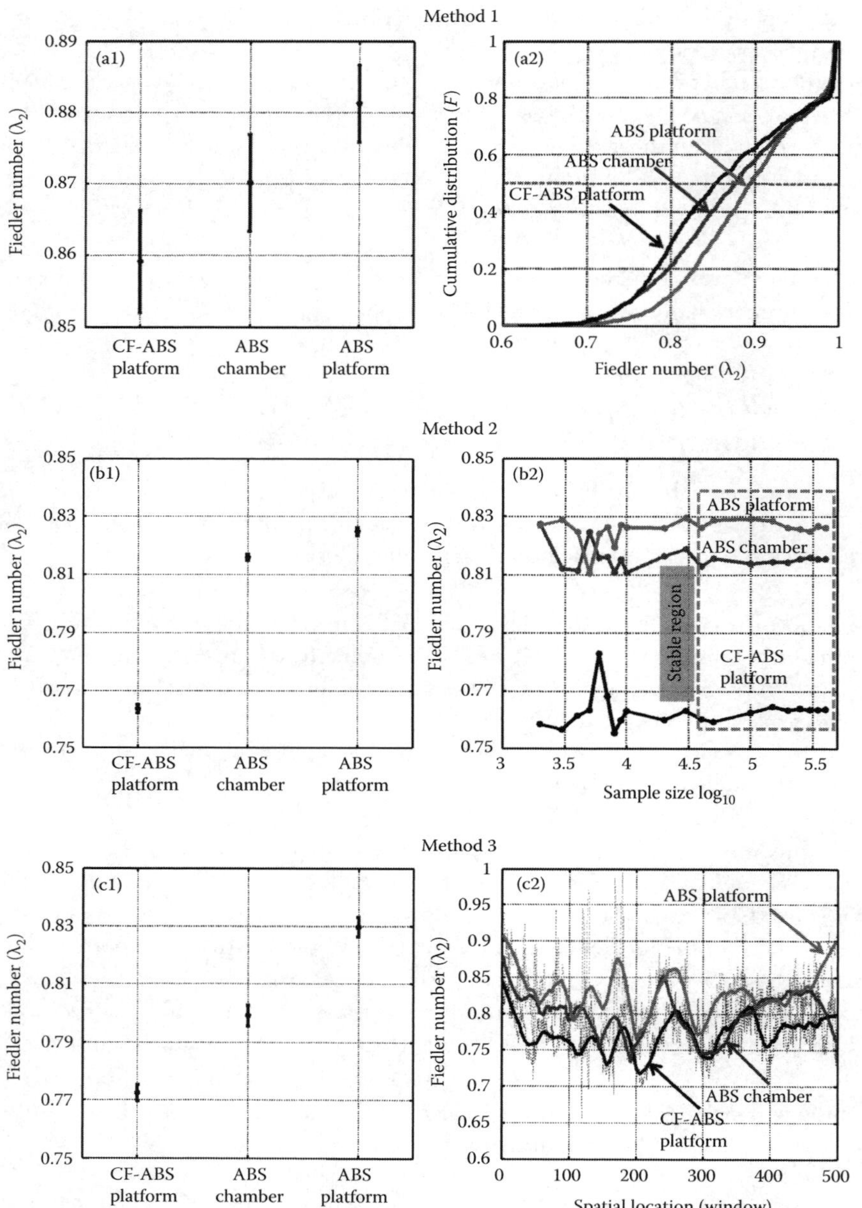

Figure 6.10 Results from applying the SGT approach to experimental test components [121]. (a1 through c1) The Fiedler number for the three different components using three different sampling methods described in Section 6.3.5. The difference in Fiedler number across test components is statistically significant ($p < .01$). The trends in Fiedler number are similar irrespective of the sampling method. The error bars represent the two-sided 95% CI on the mean. (a2) The cumulative distribution function of Fiedler number obtained from SGT-Method 1 is different across test components. (b2) The Fiedler number vs. sample size (m) for Method 2. The Fiedler number converges for $m > 50{,}000$. (c2) The Fiedler number vs. spatial location on component, notice the generally smaller magnitude of the CF component.

Table 6.4 Descriptive Statistics and Quantitative ANOVA Results for the Fiedler Number (λ_2) Obtained Using Different Methods

Analysis Method →	*SGT-Method 1, k = 500*			*SGT-Method 2, m = 300,000; k = 500*			*SGT-Method 3, n = 500*		
Test Component ↓	*Mean λ_2*	*Std. Error*	*±95% CI on Mean*	*Mean λ_2*	*Std. Error*	*±95% CI on Mean*	*Mean λ_2*	*Std. Error*	*±95% CI on Mean*
CF-ABS platform	0.8592	0.0036	0.0072	0.7638	0.00077	0.0015	0.7724	0.0015	0.0029
ABS chamber	0.8702	0.0035	0.0068	0.8160	0.00057	0.0011	0.7992	0.0018	0.0036
ABS platform	0.8813	0.0028	0.0054	0.8250	0.00070	0.0014	0.8297	0.0017	0.0034
SSE/dfe = MSE	11.8050/1797 = 0.0066			0.5123/1794 = 0.000285			2.1198/1491 = 0.0014		
SST/dft = MST	0.1471/2 = 0.0735			1.30353/2 = 0.6527			0.8186/2 = 0.4093		
Pooled Std. Dev	0.0811			0.0169			0.0377		
F-Statistic = MST/MSE	11.1933			2285.5			287.89		
F-Critical (0.95, dft, dfe)	3.0007			3.0007			3.0018		
Tukey pairwise distance	0.0110			0.0023			0.0056		

SSE, sum of squared errors; dfe, degrees of freedom for error; MSE, mean squared error; SST, treatment sum of squares; dft, treatment degrees of freedom; MST, mean of treatment sum of squares [121].

regardless of the sampling method applied. The error bars in Figure 6.10a1 through c1 indicate a two-sided 95% confidence interval (CI) on the mean Fiedler number (λ_2). In contrast, the statistical feature mining and facet examination techniques were not able to reliably capture differences between the test parts [121].

2. *Referring to Table 6.4,* which presents results from statistical analysis of the SGT method, such as I analysis of variance (ANOVA), it is evident that the difference in Fiedler number across different test components is statistically significant ($p < 1\%$) irrespective of the sampling method used for estimation. Tukey's comparison test revealed that the pairwise difference in Fiedler number for the various test components is also statistically significant ($p < 1\%$); the statistical significance was less than 1% for all pairwise comparisons between test components. Therefore, it can be statistically confirmed that λ_2 (CF-ABS platform) < λ_2 (ABS chamber) < λ_2 (ABS platform) with 99% confidence.
3. In light of the foregoing observations (1) and (2), and referring to the physical interpretation for the Fiedler number (λ_2), the dimensional integrity of the test components can be classified quantitatively in the following order: Carbon Fiber-ABS (CF-ABS) platform (best adherence to specifications), ABS chamber (midway), ABS platform (worst dimensional integrity).
4. *Referring to Figure 6.10a2,* which depicts the cumulative probability distribution function (cdf) of the Fiedler number sequence $\boldsymbol{\Lambda}_2 = \left[\lambda_2^1 \quad \lambda_2^2 \quad \cdots \quad \lambda_2^n\right]^{\mathrm{T}}$ (Equation 6.11) obtained using SGT-Method 1 (Section 6.3.5) for each of the test components; it is evident that the distributions have markedly different means. The CF (CF-ABS platform) component has the smallest mean Fiedler number, while the ABS platform component has the largest. This is an affirmation of the inference made in (1)–(3) above: λ_2 (CF-ABS platform) < λ_2 (ABS chamber) < λ_2 (ABS platform).
5. *Referring to Figure 6.10b2,* which depicts the mean of the Fiedler number sequence $\Lambda_2(m)$ for each of the three test components obtained from SGT-Method 2 (Section 6.3.5). The 3D point cloud sample size is varied from $m = 2000$ to $m = 400,000$. For small samples sizes, $m < 10,000$, the Fiedler number has an ambiguous trend for the three test components. Nonetheless, the Fiedler number estimates begin to show a clearer demarcation for sample sizes beyond $m > 10,000$, converging to a stable value for $m > 50,000$ (the Fiedler number estimates for $m = 300,000$ is shown in Figure 6.10b1). This result is in agreement with our previous inferences—the Fiedler number (λ_2) has a consistent trend, namely, λ_2 (CF-ABS platform) < λ_2 (ABS chamber) < λ_2 (ABS platform).
6. *Referring to Figure 6.10c2,* which shows the Fiedler number sequence $\Lambda_2(n) = \left[\lambda_2^1 \quad \lambda_2^2 \quad \cdots \quad \lambda_2^{n=500}\right]^{\mathrm{T}}$ across $n = 500$ spatial locations of equal area sorted along the x direction as described in SGT-Method 3 (Section 6.3.5). The bold lines in Figure 6.10c are smoothed approximations of the sequence $\boldsymbol{\Lambda}_2(n)$ obtained using a Savitsky-Golay filter. This is done in order to eliminate transient outliers in the data. It is evident that the Fiedler number sequence $\boldsymbol{\Lambda}_2(n)$ for the three test components is significantly different. In general, the CF-ABS platform component has the least Fiedler number (λ_2), while the ABS platform component has the largest. From a physical perspective, this again confirms that the CF component (CF-ABS platform) adheres closest to the specified dimensions.
7. *Referring to Figure 6.10c2,* a more pertinent pattern is evident at the edges (of Figure 6.10c2) where the Fiedler number tends to increase for all components, which indicates possible warping near the edges of the component, as often observed in polymer extrusion AM techniques. The trend is perceptibly most acute for the ABS platform component. This observation exemplifies the utility of SGT SGT-Method 3 for localizing defects.

These results are indicative of the effectiveness of the Fiedler number (λ_2) for tracking the dimensional integrity of AM components from 3D cloud data. We now chronicle some of the seminal literature in sensor-based process monitoring in AM, specially focused on DED and PBF.

6.4 Sensor-Based Process Monitoring in Additive Manufacturing

Learning Objectives

- Online process monitoring in DED AM process
- Online process monitoring in PBF AM process

6.4.1 Status Quo of Research

Repeatability and reliability is an acknowledged impediment that hinders broad acceptance of AM technology [6]. In order to detect evolving process anomalies, researchers have sought to incorporate sensing techniques (Refs. [36–41]) such as vibration, charge coupled device (CCD) video imaging, IR and ultraviolet (UV) imaging, pyrometers, photodiodes, and ultrasonic wave generators in AM machines. An early example (1994) was presented by Melwin et al. [44], who used a video-micrography apparatus bearing band pass and polarizing filters for observing the melt pool in SLS. DED and PBF AM systems are evidently most popular applications for incorporating sensors, perhaps, because of the high value of laser-engineered components, and also because these AM processes resemble well-known laser-based processes such as laser welding [45].

Elwany and Tapia [22] have conducted a comprehensive review of sensor-based process monitoring approaches, specifically focused on metal AM processes. More recently, Nassar et al. [131–133], Purtonen et al. [45], and Mani et al. [11] provide excellent reviews of the status quo of sensing and monitoring focused in metal AM, particularly, DED and PBS. From these review articles, it is evident that researchers are actively working on sensor-based monitoring of AM processes. In Sections 6.4.4 and 6.4.5, our aim is toward developing an understating of the approaches taken in some of the highly cited work in the area. The patent literature may also serve as a valuable resource for the interested reader; we note that several patents have been granted for embodiments that incorporate sensing and analytics in AM machines (see, e.g., Refs. [33,35,134,135].

6.4.2 Representative Examples of Sensor-Based Monitoring in Polymer AM

From a sensor-based monitoring perspective in polymer AM, Bukkapatnam et al. [71] investigated vibration that occurs in FFF, by comparing mechanistic lumped-mass models with experimentally obtained sensor data and demonstrated the ability to distinguish process abnormalities from sensor data. Fang et al. [50] used machine vision techniques to detect defects in FFF of ceramics based on optical imaging of each layer during the build. The optical images were used to visually identify various defects. Additionally, the authors were able to evaluate the geometrical integrity of the build by comparing the optical image of a layer to its original CAD design.

In a related research, Fang et al. [51] augmented their monitoring approach using online signature analysis. They mathematically represented (in the form of a function) the ideal build morphology for each layer from the CAD file, which is termed the *process signature*. Thereafter, the process signature was compared with the physical layer-by-layer build pattern after image analysis to detect process anomalies. Furthermore, Cheng and Jafari [136] examined the build surface using image intensity information. Build defects were classified into two types, namely, randomly occurring defects and anomalies due to assignable causes, e.g., improper extruder tool path. The randomly occurring defects were detected by 2D texture analysis, while the assignable defects were identified using the process signature technique developed previously by Fang et al. [51].

Cooke and Moylan [137] used a CCD camera (1.23 megapixel resolution) and a variable focus optical lens to acquire images of each build layer in binder jetting process. The CCD camera trigger is coupled to the carriage of the powder layering mechanism so that an image of a layer is acquired after it is finished, as opposed to continuously. This allows for comparison between successive layers. A total of six images are acquired after completion of every layer and averaged. Using edge detection image processing techniques, geometrical aspects, e.g., wall thickness, outer diameter, inner diameter, as well as positioning errors during manufacture of a cylindrical, were identified.

6.4.3 Representative Examples of Sensor-Based Monitoring in Metal AM

In a series of related works, Craeghs et al. [29–32] described optical-based approaches for monitoring build quality in SLM by imaging the thermal behavior at the melt pool. Craeghs et al. [30] were able to detect process defects such as deformation and overheating using their optical system [30]. Krauss et al. [38,39] designed an apparatus that uses long wave infrared (LWIR) thermal camera to record the melt-pool temperature profile. By comparing the irradiance of the melt pool with baseline (normal) conditions, Krauss et al. were able to detect formation of voids. A similar study, using a different design and a mid wave infrared (MWIR) thermal camera is reported by Wegner et al. [49]. Bartkowiak [25] described a DMLS apparatus integrated with a spectrometer for *in situ* measurement of the layer melt characteristics such as emissivity. Other researchers, e.g., Chivel et al. [27] and Jacobsmuhlen et al. [37], have also developed similar optical imaging systems for process monitoring in AM [27]. In a recent work, Rieder et al. [41] described a novel ultrasonic sensing system for detecting build defects in DMLS. A broadband ultrasonic sensor mounted on the underside of the build plated is used to detect voids, akin to acoustic microscopy. Du and Kovacevic incorporated a high frame rate camera with an IR filter to monitor layer morphology and laser photodiodes to track the powder flow rate in LENS. In a similar setup, Barua et al. [40] monitored the build quality in LENS by imaging the UV reflection in the UV spectrum. We will now detail some of these works in further detail in the forthcoming two sections, focusing on DED and PBF apparatus.

6.4.4 Studies in Directed Energy Deposition

DED, also alternatively referred to as direct metal deposition (DMD), laser cladding, LENS (trade name of Optomec), laser powder forming (LPF) is a processes descended from laser beam welding. A schematic of powder-based laser DED is shown in Figure 6.11 along with some of the consequential factors influencing build quality. In DED-based processes, a stream of metal powder (20–150 μm particle size) suspended in an inert carrier gas medium (typically Argon or Helium) is delivered onto a substrate via a nozzle (typical standoff 1–3 mm). A laser beam, introduced coaxially with the powder feed nozzle, melts the powder as it is delivered.

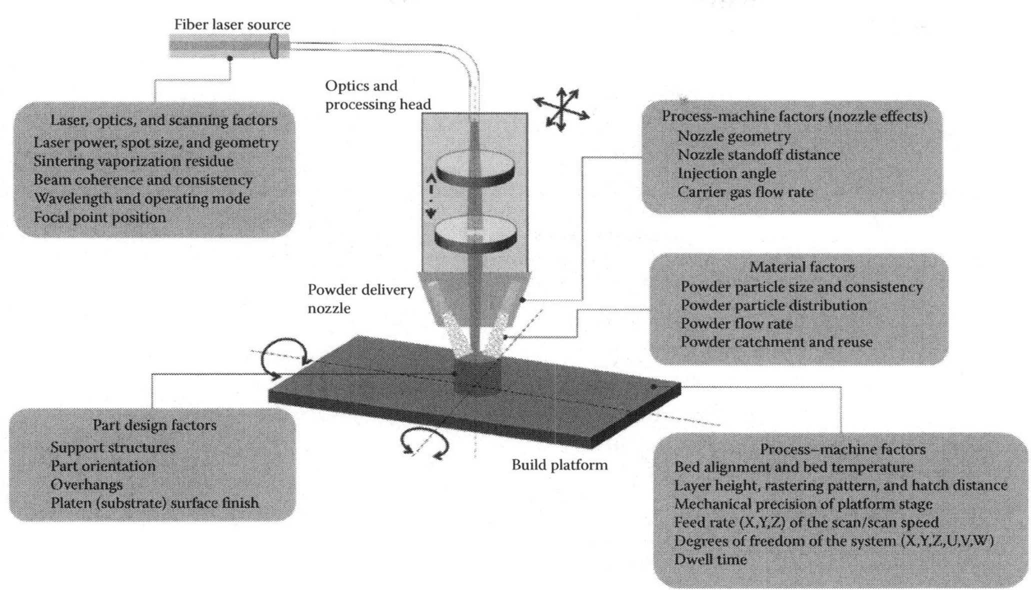

Figure 6.11 The schematic diagram of the powder-based DED along with key factors influencing build quality.

The carrier gas also acts as a shield, to avoid oxidation of a freshly deposited layer. In some embodiments, a separate orifice for introduction of a shield gas is reported. By translating the laser beam and substrate, a net shape can be formed in a layer-by-layer manner. The laser source is typically a 300–500 W Ytterbium fiber laser (larger systems are capable of 4 kW); gas (CO_2) and solid-state (Nd-YAG) diode laser sources are also prevalent. The active wavelength is typically in the Near Infrared (NIR) approximately 1000 nm region. In early works in the literature, researchers often coupled custom-built nozzles to the spindle of a five-axis CNC machining center. Other users have coupled commercial nozzles to robotic arms (e.g., Commonwealth Center for Advanced Manufacturing, CCAM in Disputanta VA, http://www.ccam-va.com).

Another derivative of the process uses wire-fed material and an electron beam for the energy source. The latter is often referred to as EBFFF or EB3F or electron beam additive manufacturing (EBAM; a trade name registered to Sciaky) to distinguish it from powder-fed directed energy deposition. Unless otherwise stated, the acronym DED is used in this chapter to refer to the powder-based variant. DED processes are capable of processing virtually any metal/metal alloys that can be delivered in powder form, e.g., stainless steel, aluminum, copper, rhenium, titanium and titanium alloys, nickel-based super alloys (Inconel, Wasp alloy, Hast alloy), Cobalt-based super alloys (Stellite), stainless steel, molybdenum, ceramics (Tungsten Carbide), and metal matrix composites. The distinguishing features of DED are:

1. Can be used for repair applications. For instance, a crack in a turbine blade or tooling can be filled.
2. Graded materials can be deposited, as multiple orifices in the powder delivery nozzle can be used to deposit different materials.

3. Not tightly constrained by build size, since the build chamber is not limited by a build platen size, powder reservoir, and scanning area of the laser as in PBF.
4. Higher material deposition rates (typically 0.2 lb/h) in laser-based PBF can be achieved. The layer thickness (250 μm) is also much higher than commercial laser PBF systems (20–40 μm).

The main processing disadvantages of DED stems from the somewhat degraded accuracy of the builds, as the kinematics of translation influences the precision of material deposition. Consequently, the part accuracy (~0.25 mm) and surface finish of DED parts is comparatively inferior (~25 μm, Ra) compared to PBF. Typical process defects include cracking/distortion due to the high cooling rates; improper fusion/bonding of layers; porosity due to powder contamination and gas entrapment; microstructure defects due to suboptimal melting; and interroad (hatch line) defects resulting from poor process planning [32]. Vetter et al. [138] listed the various material–process–machine interactions governing the DED process. We enumerate some of these consequential factors in accordance with the schema described by Craeghs et al. [32] albeit for the different process (PBF). A review of different control strategies for these different strategies is provided by Boddu et al. [27].

Pioneering works in DED process monitoring in the United States are by Dr. Jyoti Mazumder et al. [139–142] at the Center for Laser Aided Intelligent Manufacturing, University of Michigan (Ann Arbor); Dr. Edward Reutzel and group at the Center for Innovative Material Processing through Direct Digital Deposition (CIMP-3D), Penn State University [131–133]; and Dr. Fuewen Liou at Missouri University of Science and Technology (MUST, formerly University of Missouri–Rolla) [74,143]. The reader is encouraged to visit their websites (Table 6.5).

In the context of monitoring and control of DED processes, researchers are focused on three aspects:

1. *Melt-pool monitoring*: chiefly devolves into measuring the thermal aspects of the deposition process. The shape, intensity, and temporal aspects of the melt pool are measured, typically, using IR pyrometers and complementary metal-oxide-semiconductor (CMOS) or CCD camera fitted with an NIR filter. IR sensors are also often used to get a reading of the temperature of the melt pool.
2. *Powder delivery rate monitoring*: using laser photodiodes and imaging techniques to assess the adequacy of the powder flow rates. The sensing system is typically coupled to the delivery tube [36] or the nozzle side [143].
3. *Layer morphology monitoring*: involves assessing the shape and physical aspects of the deposited layer. Traditionally, a line laser is projected on a deposited layer, the reflection is captured using a camera, and subsequently analyzed using image processing techniques (to account for distortion). The output is the layer height (or clad height). Recently, laser spectroscopy has been investigated as a means to go beyond simple height measurement, toward measurement of the characteristics of a deposited road (hatch spacing) [132].

Boddu et al. [143] described a custom-built DED system combined with a five-axis CNC machining center; the energy source is a 2.5 kW Nd-YAG laser source active in the 1060 nm range. The powder is delivered via four orifices, which are coupled to two hoppers. The concept is to devise a hybrid system where the laser DED nozzle deposits material, which is finish machined without having to declamp the workpiece. The noteworthy aspect of this system stems from the integrated

Table 6.5 Boundary Conditions and Controllable Input Parameters in DED Processes

Boundary Condition Factors		*Controllable Input Parameters*		
Part Design Factors	*Material Factors*	*Environmental Factors*	*Process–Machine Factors*	*Laser Optics and Scanning Factors*
• Location of support structures • Contact area and type of supports • Part orientation • Part overhang • Platen (substrate) type	• Material type and purity • Powder particle size and distribution • Powder capture and reuse • Foreign residue as a result of processing • Powder flowability • Powder mixing in the hopper	• Oxygen concentration • Chamber temperature • Substrate temperature • Cleanliness of the lens and exhaust efficiency • Number of degrees of freedom of table and laser • Integrity/accuracy of machine elements	• Powder flow rate • Layer height • Carrier and shielding flow rate • Wavelength and operating mode • Nozzle standoff • Injection angle • Nozzle geometry	• Laser power, spot size, and geometry • Rastering pattern, scan speed, hatch distance • Beam coherence • Focus integrity and shape • Dwell time

Note: Information modified from Vetter et al. [138].

monitoring system. This sensor suite comprises of a melt-pool monitoring, layer height measurement, temperature measurement, and powder delivery monitoring. The intent is to use the outputs from these sensors in the context of data-driven closed-loop control.

A CMOS camera capable of frame rates approaching 300 frames per second (fps) is mounted coaxially with the nozzle, and is equipped with a beam bender to focus on the melt pool. The width and length (shape) of the melt pool were estimated by image processing techniques. In addition, the surface temperature of the melt pool was measured using a noncontact IR sensor operating in the short wave infrared (SWIR) region (1.5 μm). The height of the deposited layer is measured using a laser displacement sensor. The principle is that as the layer height increases, the output voltage of the sensor increases, thus a correlation can be drawn of process variables with deposited layer height. The powder deposition is estimated using an imaging system. The powder stream is imaged using strobe lights, and several images are averaged. This is because the powder being composed of discrete particles, it is difficult to estimate the powder flow characteristics, e.g., width of the powder flow at a certain distance from the nozzle with an instantaneous image. The averaged image sequence, thus, obtained is converted into black and white by thresholding. Tests are conducted under predetermined optimal conditions, and the width of the powder flow is manually measured by the operator and used as a baseline for evaluating possible drifts in the future.

In one of the earlier works, Vetter et al. [138] discussed the various factors and interactions in DED and explain how there are two physical interaction zones that significantly influence the quality of the build; these two interaction zones are as follows:

- The first interaction zone occurs in the region where the powder leaves the nozzle, and involves three entities, namely, the laser, gas, and powder.
- The second interaction zone occurs at the region where the powder contacts the substrate, and involves the substrate in addition to the three delineated in the first interaction zone.

To systematically characterize the effect of these interactions on the build quality, Vetter et al. [138] integrated multiple sensors into the machine. The material flow properties are ascertained using an optical sensor coupled to a spectrometer. The shape of the stream is captured using a CCD camera, while the temporal aspects of the flow are recorded with four silicon photodiodes, and finally, a pyrometer measures the melt-pool temperature. These sensors enabled the authors to map important aspects, such as powder flow rate and temperature gradients, at various locations between the nozzle and substrate. However, although this study elucidates the material–process parameters–sensor signal interactions, the tests were largely carried out in conditions where no parts are actually built.

Nassar et al. [132] measured the quality of the build by examining the elemental optical emission spectra of the workpiece during the build process. They captured the optical emission using a spectrometer. The central premise of this approach is that if there is incomplete fusion of roads (hatch lines) during a layer, then the optical emission spectrum would be markedly different from a well-fused line. The spectrum measurement is in the 200–1100 nm range, which spans the UV to NIR spectrum. The sampling rate of the spectrometer is maintained close to 8 Hz. Nassar et al. [132] demonstrated the effectiveness of this approach by depositing Ti–6Al–4V layers with varying hatch spacing; the hatch spacing was progressively increased during the deposition of each layer [132]. At the outset, the optical emission spectra is observed to be devoid of any sharp peaks for a well-fused portion of the layer, i.e., where the hatch spacing is smaller. The absence of sharp peaks in the optical emission spectrum means that distinctive elemental forms were not detected. Conversely, if clear spikes corresponding to the powder elements are detected, then it implies that the material is not well fused.

The experimental results reported by Nassar et al. showed spikes corresponding to Vanadium and Titanium in those portions of the workpiece where the hatch spacing was excessive. Taking this rationale forward, the authors quantified the optical emission spectrum. This quantification is done by using an approach that computes the ratio of the area under a particular wavelength band to the lower envelope of the area of the entire spectrum. This ratio is used as a monitoring statistic.

Spectroscopy has been investigated earlier (2010) as a means for online monitoring by Bartkowiak [25]. A spectrometer was used by Bartkowiak to observe the optical emission in the 247–472 μm, i.e., UV to the violet-blue region of the visible spectrum. Discernable difference in the emission spectra is reported for different processing conditions. However, quantification of the differences in spectra vis-à-vis processing conditions was not attempted. Song and Mazumder [141] have also used laser spectroscopy for monitoring the elemental composition of the deposit. They plotted calibration curves mapping the spectrum behavior versus material composition. Experiments were conducted with chromium-based tool steel powders with different chromium compositions. In a related work, Muzumder et al. [142] use laser spectrometry to identify elemental phase transformation in various powder compositions, including iron–nickel, iron–titanium, and iron–chromium binary powders.

In an alternative development, Mazumder et al. [139] described a system with three CCD cameras for measuring the height of the melt pool along with a dual color pyrometer for measuring the temperature of the melt pool. The CCD camera and pyrometer are connected to separate serially connected controllers. The CCD cameras are arranged at 120° intervals in the horizontal plane and at 45° in the vertical plane. The layer height can be estimated by triangulation with any pair of cameras. Thus, three estimates of the layer height are obtained with this system. The dual color pyrometer is set to operate in the NIR—SWIR range (1.3 and 1.4 μm). If the layer height deviates above the specified thickness, i.e., indicating excessive material is being deposited, the laser power is reduced so that material fusion abates. In contrast, if the layer height is lesser than the specified thickness, the temperature sensor reads a proportionally lower temperature. This in turn activates a separate controller that increases the laser power, thus, increasing the amount of material deposited to compensate for the drop in layer height. Mazumder et al. [139] demonstrated the effectiveness of this approach by depositing turbine blade sections. Using the controller devised by the authors, distortion in shape due to inordinate heating of narrow, thin wall sections was mitigated. In an earlier work, Song and Mazumder [140] described a state-space predictive controller-based on readings from the dual color pyrometer alone.

Bi et al. developed an approach using a single color IR pyrometer to attain closed-loop control of the laser power in DED process. The signals obtained from the IR pyrometer sensor are experimentally correlated with the laser power and quality of the deposited surface. The control strategy is to maintain the laser power (which was also separately measured) at a set point by correlation with the IR pyrometer. Essentially, the pyrometer signal is coupled to the laser power, in the pyrometer signal increases (decreases) from an *a priori* set threshold then the laser power is increased (decreased).

Bi et al. [144] also studied the effect of laser power on deposited layer quality in terms of presence of surface defects and oxidation. The abovementioned control strategy is tested on two powder material combinations, namely, stainless steel and Ni-base alloys. For instance, they measured the effect of laser power on stainless steel clad surface quality. They found that insufficient laser power to melt the stainless steel powder results in poor surface quality. An IR sensor signature trace reacts to the change in laser power; in near-optimal conditions, the IR signal mean is close to 0.8 V. The output from the IR sensor acts as a surrogate for adjusting the laser power to the correct value, and used as a means to control the clad quality.

As another example, Bi et al. [145] investigated the effect of insufficient shielding gas (Argon) supply on clad quality. Firstly, nickel-base powders are deposited under conditions where the shielding gas flow was unable to protect the melt pool, which was subsequently exposed to ambient air as the build progressed. Accordingly, the clad surface had a wavy morphology. The IR pyrometer signal acquired during the process shows abrupt disturbances matching with the wavy pattern of the clad. On correction of the problem by changing the direction of flow of the shielding gas, the deposition defect was mitigated; the defect-free deposition of the clad is reflected in the IR signal.

In a related work, Bi et al. [144] described an experimental deposition nozzle integrated with multiple sensors to detect the quality of the build, as well as the condition of optical components. During DED, some part of the powder vaporizes, also smoke and carbon residue often results during the process. In normal operating conditions, the carrier gas is able to deflect these by-products so that they do not coalesce on the colder parts of the machine, e.g., the optical components. Often, due to insufficient flow of the carrier gas, the residue might indeed settle on the lens, thus, affecting the focusing mechanism and laser power incident upon the substrate, which causes poor melting and suboptimal build geometry. In extreme instances, because the residue deposited on the lens absorbs a significant portion of the incident energy, damage to the lens and optical train can occur due to the high temperature.

Bi et al. [144] incorporated a Nd:YAG photodiode to measure the temperature of the lens via a system of dichroic mirrors mounted coaxially with the nozzle head. A dichroic beam splitter divides the incident laser beam, and the Nd:YAG photodetector measures the output power of the laser beam. In case the optical lens becomes coated with melt residue, the intensity of laser light will increase proportionally, which can serve as a monitoring statistic for detecting deterioration in machine condition. Additionally, Bi et al. attached three thermocouples to the nozzle head and optical train, presumably to also monitor the condition of the machine. The melt-pool temperature was monitored by a coaxial germanium IR photodiode (pyrometer), and the morphology of the melt pool was imaged using a CCD camera. The images acquired from the CCD camera are useful for controlling the standoff distance of the nozzle, since the area of the melt pool increases as the standoff distance increases. Using precalibrated settings, the melt-pool diameter can correlated with the standoff distance. In turn, the output from the germanium photodiode is used to control the laser power output, akin to the work described previously.

Davis and Shin [146] described a clad height monitoring system consisting of a line laser and CCD camera. The line laser (532 nm, 300 mW), which is mounted coaxially with the cladding head, projects a plane of light which is captured using a CCD camera. Using a custom-tailored triangulation-based approach, the actual height of the clad can be estimated. They verified the clad height estimated from their system with CMM measurements; a mean error in the range of 0.015–0.038 mm was reported.

Hu and Kovacevic [36] described a system with two sensors for simultaneously monitoring powder delivery and melt-pool information in DED process. The powder delivery rate is monitored using a laser diode–photodiode combination, while a high speed CCD camera is used for imaging the melt pool in the NIR (~700–1000 nm).

A photodiode–laser diode system is used for measuring material flow rate, the laser diode has an operating region of 600–710 nm (red light visible region), and has a power less than 500 mW. The laser diode is mounted on one side of the feed tube as it passes through a glass chamber. The photodiode sensor signals measures the light passing through the glass chamber; as the material flow increases, the lesser is the intensity of light passing through the chamber. Accordingly, an inverse relationship between material flow and the output (in volts) of the photodiode is reported by the authors.

The melt-pool sensing system consists of a CCD camera imaging at 800 fms mounted coaxially with the delivery head. The camera is fitted with filters to allow radiation in the 700–1000 nm range, the IR image is converted into a grayscale isotherm using an indirect calibration approach. This calibration step is necessary because the IR image is exceedingly bright to be used directly as a means to monitor the shape and morphology of the melt pool. An ultrahigh speed shutter camera with a pulsing nitrogen laser (337 nm) is used synchronously with the CCD camera to acquire images of the melt pool. The clear images, thus, obtained allows determination of the gray threshold value to detect the boundary of the melt pool using image processing (Canny filtering). Once the boundary is detected, the area of the melt pool can be estimated. The area of the melt pool is the monitoring statistic; this feature is used toward implementing forward control schema.

The authors demonstrate the utility of this sensing system in the context of relatively simple geometries built using H13 tool steel. A feed forward PID controller is used to monitor and correct for disturbances in the melt-pool morphology. In further development of the work, the authors modeled the thermal aspects of DED process using finite element analysis (FEA).

In a recent (2014) work, Qiling et al. [147] used a camera (the type of camera is not mentioned) to study the effect of process parameters on melt-pool morphology, namely, length, width, and area. For this purpose, as in other studies, the authors shield the camera with quartz glass, and affix an NIR filter to capture optical emissions in the 850 nm range. The length, width, and area of the melt pool were estimated using image analysis techniques; these are correlated with process parameters such as laser power, scan velocity, and powder feed rate. The melt-pool area is found to be directly proportional to the laser power, and inversely proportional to scan velocity. The trend of melt-pool area and powder feed has a more nonlinear behavior; the melt-pool area initially reduces with increasing laser power, and then increases. This is because during the initial increase in powder feed rate, the images are obscured, as the feed rate increases further for a given scan speed, the height of the melt pool increases. Due to the inclined nature of the camera setup, the increase in height is registered, erroneously, as an increase in area.

Ocylok et al. [148] studied the effect of process parameters on melt-pool area; however, instead of mounting the camera at an angle as in Qiling et al.'s setup [147], a CMOS camera is mounted coaxially with the nozzle. The advantage of using a coaxial camera setup is that the cross section of the melt pool can be estimated more accurately as the effect of distortion is mitigated. Ocylok et al. [148] also correlated the effect of process parameters such as laser power, and scan velocity (feed rate). The melt-pool area is found to be directly proportional to the laser power, and inversely proportional to the scan velocity.

In contrast to measuring the morphology of the melt pool in the IR spectrum as most researchers have described, Barua et al. [40] instead used *reflected* UV to gage the melt-pool shape. Their setup consists of a UV light source in the 320–400 nm (long-range UV) and UV LEDs in the narrow band of 390–395 nm. These are focused upon the melt pool, and the UV light reflected from the melt pool is captured using a CCD camera. Noting that the melt-pool radiation and process temperatures is primarily in the visible and IR region (~700 nm), there is negligible amount of activity in the UV spectrum that will interfere with the imaging process. Once the image of the melt pool is captured, the authors use image processing techniques to correct image distortion and estimate the melt-pool area.

6.4.5 Direct Metal Laser Sintering (DMLS)

PBF refers to a family of process in which progressive layers of material are raked across a build platen, and subsequently sintered/melted using a laser or electron beam. Depending on the manner in which the powder is fused, there is a difference between melting and sintering, as delineated in earlier chapters of this book [2,3]. The raking mechanism may also differ depending on the manufacturer, for instance, the 3D systems ProX series uses a roller that rotates in a direction opposite to the linear motion of the roller (visualize the roller has a "backspin"). The EOS series machines instead use a scraper blade with a rake angle, akin to a cutting tool. The blade material can be changed contingent on the material to be processed, typically, high speed steel is recommended (by EOS) for most materials, such as stainless steel, whereas with titanium powder material ceramic blades are used. The nature of the rake angle (positive or negative rake angle) of the blade is also a factor that may be adjusted. The materials that can be processed using PBF are vast; the earlier embodiments by Dr. Carl Deckard at University of Texas-Austin was restricted to polymer material, with modern PBF machines a range of metals can be processed. A caveat with using an electron beam as an energy source is in the processing of aluminum, because of the high reflectivity of aluminum, higher beam power is required to fuse the material. However, at higher electron beam power, aluminum vaporizes and coats the colder areas of the machine, akin to physical vapor deposition.

The family of PBF processes using laser as an energy source is referred to SLS (a trade name of 3D Systems, now often used for polymer materials), SLM, DMLS (a registered trade name of EOS), and laser cusing (LC; a trade name of Concept Laser). The heart of the process is a fiber laser source, typically, rated at 400 W with an active region in the vicinity of 1000 nm (NIR). A scanning mirror determines the rastering pattern of the laser, the focusing lens is a *f*-lens. The spot size is close to 50–100 µm in diameter. The linear scan speed can be varied in the 5–15 m/s range. Ytterbium (Yb) fiber lasers are ubiquitous in laser-based PBF systems, the laser system is sourced from two manufacturers SPI of Southampton, the United Kingdom (now a subsidiary of Trumf, Germany), and IPG Photonics of Oxford, MA [6]. The low cost, high efficiency, and ability to finely focus the beam are cited as the main reasons for Yb lasers replacing traditional CO_2 gas lasers [6].

A schematic of the PBF process is shown in Figure 6.12, along with consequential factors that influence build quality. Craeghs et al. [32] asserted that more than 50 parameters are at play in the process. We slightly modify the schema used by Craeghs et al. [32] to classify the process parameters. These are stratified as shown in Table 6.6.

The following methods are predominately used in the literature toward monitoring the PBF process:

1. *Melt-pool monitoring*: CCD cameras, IR cameras, and pyrometers are used to gage thermal, intensity, and morphological aspects of the melt pool. The visual systems may either be embedded coaxially with the scanning galvanometers using an appropriate optical train to capture the melt-pool behavior; or mounted outside the powder bed. The challenge with the latter, especially if an IR camera is used, is that the measured temperature profile is not directly representative of the actual temperature. This is because as the camera is mounted at an angle to the powder bed, the incident thermal radiation is not perpendicular to the

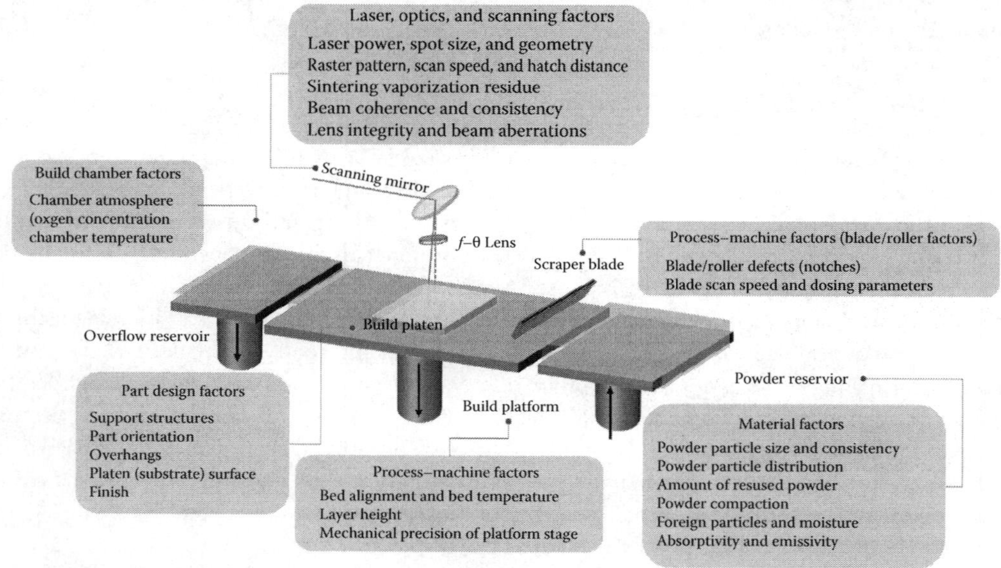

Figure 6.12 The schematic diagram of the laser-based powder-bed fusion (also called SLM, and DMLS) along with key factors influencing build quality.

Table 6.6 Boundary Conditions and Controllable Input Parameters in PBF Processes

Boundary Condition Factors		*Controllable Input Parameters*		
Part Design Factors	*Material Factors*	*Environmental Factors*	*Process–Machine Factors*	*Scanning Factors*
• Location of support structures • Contact area and type of supports • Part orientation • Overhang • Platen (substrate) finish	• Material type and purity • Powder particle size and distribution • Amount of powder reused from previous builds • Powder compaction • Foreign residue as a result of processing • Absorptivity and emissivity characteristics	• Oxygen concentration • Chamber temperature • Chamber evacuation gas (nitrogen, argon) • Cleanliness of the lens and exhaust efficiency • Presences of residue from previous builds	• Bed alignment (gap and skew) • Bed temperature • Layer height • Precision of machine elements • Blade type and rake angle • Blade/roller defects • Blade scan speed and dosing parameters	• Laser power, spot size, and geometry • Rastering pattern, scan speed, and hatch distance • Beam coherence • Lens integrity

sensing elements in the camera, which in turn causes aberrant readings. This problem can be overcome using a reference heat source in the powder bed.

2. *Powder bed monitoring:* Acoustic (ultrasonic) sensors, vibration (accelerometers), along with IR thermal cameras, have been proposed to monitor the powder bed conditions. Ultrasonic sensors mounted underneath the build platen can detect voids in the build. Vibration sensors located on the powder bed are used to identify faulty deposition of powder layers resulting from a damaged recoater blade, particularly in EOS series DMLS machines (the ProX series laser PBF machines by 3D systems instead uses a roller mechanism for layering). Instances of super elevations and poor surface finish are also localized using both visual and vibration sensors.

Craeghs et al. [32] elucidated the need for a melt-pool imaging system, which is coupled with sensors capable of monitoring status of process inputs such as distribution of the powder. This is because the melt-pool information, albeit valuable for monitoring the local thermal aspects, cannot be translated quickly into a corrective action, i.e., a delay is inevitable in mitigating process drifts. In other words, Craeghs et al. [32] recommended that a heterogeneous sensor suite be used for process monitoring PBF processes. This team at the Catholic University of Leuven, Belgium led by Dr. Kruth has published several influential articles in the area of quality monitoring and control, as well as in the general area of AM, a select few of these are cited herewith [24,26,29–32,104,108,149–153].

Accordingly, Craeghs et al. [32] incorporated two types of sensors into a custom-built laser-PBF machine, which allows flexibility in choosing process–machine conditions, such as laser power, hatch spacing, and scan speed which are not typically adjustable in a commercial machine. In the context of process monitoring, the authors have incorporated three sensors, namely, a visual camera to ascertain the characteristics of the powder raked by the blade across the build platen, a photodiode, and CMOS camera, all are used to monitor the melt pool.

In the context of monitoring the powder bed raked across the platen, Craeghs et al. [32] made two observations: (1) relatively gradual wear of the recoater blade (the blade material is high speed steel or ceramic material) that leads to small streaks uniformly across the platen and (2) severe damage to the recoater that causes deep scratches on certain parts of the powder bed. The effect of using a damaged recoater blade is shown in Figure 6.13; as discernable streaks are evident on the surface, as shown in Figure 6.13a, these are replicated on the final build, as shown in Figure 6.13b.

The damage to the recoater blade was detected by Craeghs et al. using a simple control chart-type strategy. The grayscale values from the visual camera are the used as the monitoring statistic. An example of the outcome is shown in Figure 6.14; the grayscale values for image of a layer deposited with a damaged blade is markedly different (clear spikes and higher standard deviation, as shown in Figure 6.14b) compared with a defect-free deposition, as shown in Figure 6.14a)

For monitoring the melt pool, the photodiode and CMOS camera were filters were used, which constraints the wavelength of light in the region of 780–950 nm. The upper limit is at around 1000 nm because beyond this wavelength range the semi-reflective mirrors used by the authors become 100% reflective. The bottom threshold is governed by the visible spectrum (400–700 nm). The sampling rate is 10 kHz; this translates to a sample every 100 μm, considering 1000 mm/s scan speed.

Using image processing techniques, the authors ascertain the melt-pool area and the length to width ratio of the melt pool. These quantities (area, and length and width ratio of the melt pool) are essentially the monitoring statistics, and are used to detect process drifts such as *balling* [17,18]. According to Craeghs et al., balling phenomena occurred when the melt-pool size progressively increases and eventually disassociates (or *balls*) into smaller parts. On account of splitting into smaller parts, the total surface area of the melt pool increases and hence it cools faster. This leads the separated melt-pool sections to coalesce again. The poor surface roughness at corners and overhangs is attributed to balling phenomena because the laser beam makes a so-called U-turn. The area where the U-turn is made is exposed to laser power for a longer time, and therefore, overheats in comparison to the rest of the part, which in turn is an initiator for balling phenomena.

As an illustrative example of the utility of monitoring process drifts in laser-based PBF process, the authors cite an example where the porosity of a part abruptly increased due to faults in the build platform stage motor. At certain instances, powder thickness equivalent to multiple layers were raked across the build platen. This unusually high layer thickness led to increase in porosity,

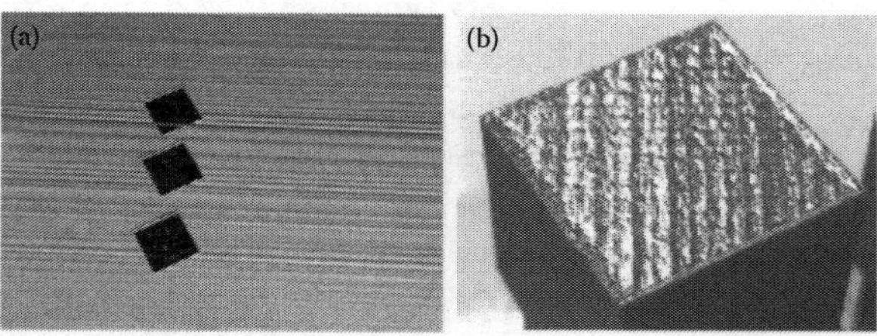

Figure 6.13 Effect of using a damaged recoater blade. (a) Streaks are seen as the powder is raked across the platen. (b) The poor surface finish of the build on account of the streaks [32].

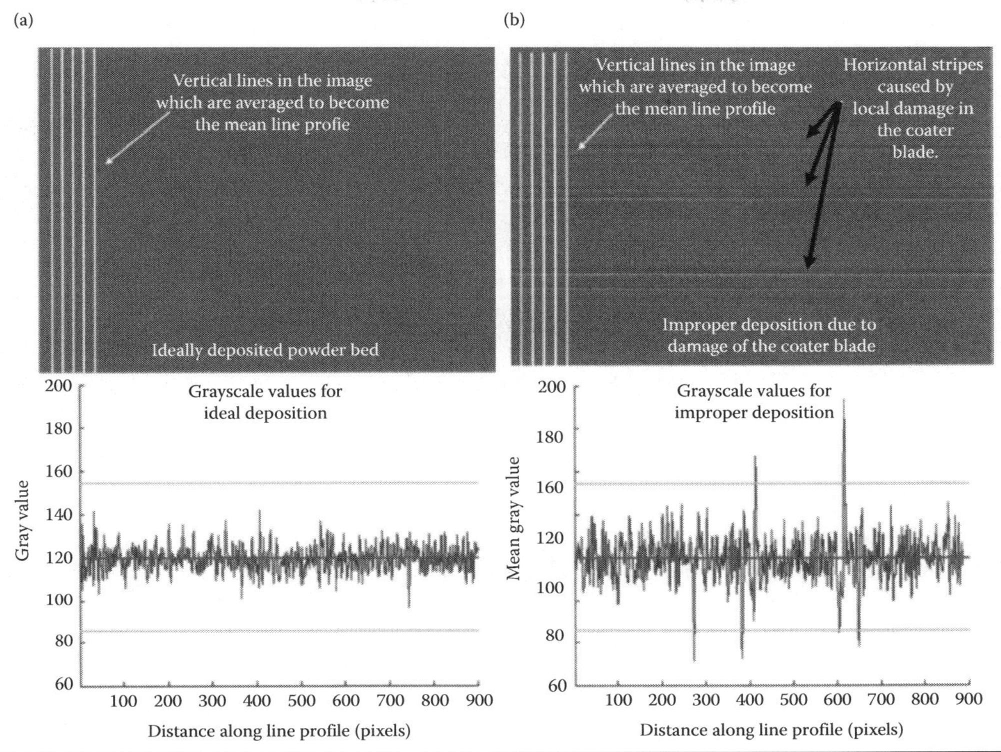

Figure 6.14 (a) The ideal deposited powder bed and corresponding grayscale values. (b) A powder bed with streaks on account of a worn/damaged blade and corresponding grayscale values; note the abrupt change in standard deviation in grayscale values. (From Craeghs, T., Clijsters, S., Yasa, E., and Kruth, J.-P., 2011, *Proceedings of Solid Freeform Fabrication Symposium*, Austin, TX.)

because the laser power was insufficient to melt the powder. The authors report that the photodiode signal depicts an inordinate increase in mean and standard deviation corresponding to layers with faulty deposition.

Jacobsmuhlen et al. [37] implemented an image-based monitoring approach specifically for detecting *super elevations* during PBF. Builds are said to be super elevated if the prior solidified layers protrude out of a freshly deposited powder. Thermal stresses are the main cause for builds to distort and become exposed; this is colloquially also called *potato chipping* on account of the resemblance to the edge "bowing up." Super elevation causes the recoated blade to contact the build plate, resulting in a "crash", and damage to the blade. The central theme of Jacobsmuhlen et al. work is to visually detect these super elevated regions and compare the results with a reference; eventually, this will allow adjustment of process parameters such as laser power and hatch spacing. The experimental results of Jacobsmuhlen et al. [37] indicated that super elevations can be reduced by decreasing laser power and increasing hatch distance. By detecting the occurrence of super elevation at an earlier stage, the layer height can be corrected, or the build can be cancelled.

We note that the presented work depends heavily on image processing algorithms such as Hugh transforms and areal operations on images (connectivity thresholding), which require *a priori* optimization of parameters. A CCD camera is coupled with a tilt shift lens and mounted on a geared head that gives the ability to traverse the camera in three directions. The tilt shift lens allows corrections of perspective distortions and enables the camera to maintain focus on the powder bed.

Using this setup, Jacobsmuhlen et al. [37] acquired images under varying conditions of laser power and hatch spacing. We note that the camera takes images of each layer initiated by a triggering mechanism embedded into the blade coater. The published images show that super elevation of parts due to poor selection of process parameters can be captured from the image acquisition and analysis proposed system.

Krauss et al. [38,39] incorporated a microbolometer-type IR camera operating in the LWIR, specifically in the 8,000–14,000 nm. The IR camera is mounted on the outside of the build chamber, and looks down on the powder chamber at an angle of 45° via a germanium window. This setup allows measurement of larger area of the powder bed, as opposed to small local areas as in coaxial measurement systems. The central theme of the author's work is to obtain the area and morphology of the heat affected zone (HAZ). They correlate the change in process parameters such as laser power, scan velocity with the area, aspect ratio (length to breadth ratio), hatch distance, and layer thickness. These correlations serve as the bases on which build quality can be monitored.

For instance, the authors created artificial flaws (pores). The irradiance morphology of the (artificial) defective build was compared with an ideal state. A significant difference is observed in the irradiance profile recorded for the ideal build versus defective build.

Pavlov et al. [154] describe a two color pyrometer sensor suite instrumented on Phenix PM-100 PBF machine. The pyrometer is mounted coaxially with the scan head; the field of view of the Indium Gallium Arsenide (InGaAs) pyrometers was 560 μm, which is considerably larger than the laser spot size (70 μm). They correlated the raw pyrometer outputs with process factors such as hatch distance and powder (layer) height. Hatch distance refers to the distance between two adjacent straight lines (tracks).

At this juncture, we note that a pyrometer measures the thermal radiation from the surface of a body. If the melted (hatch spacing) tracks are close enough that they contact each other, then instead of the heat radiating away from a freshly formed track, there is a transfer of heat between the two. This effect was noted by Pavlov et al. [29] when the hatch distance was less than 105 μm, the absolute readings from the pyrometer were near their minimum, and peak as the hatch distance increases, and gradually reduces as the hatch distance increases further beyond 120 μm. This is because when the hatch distance increases beyond a threshold, there are fewer tracks for a given cross section (10 mm × 10 mm in case of Pavlov et al.), and consequently the specific energy input into the layer is lesser.

Pavlov et al. contend that since the pyrometer signal is sensitive to the build characteristics, such as hatch spacing, it can be used to monitor the build quality. Accordingly, they present a case where the layer thickness is maintained (purposely) low so that the build is starved of material. In this experiment, they note that when a defective track is deposited, the pyrometer signal drops. The track failed to build due to the low layer thickness, and consequently, most of the laser energy was transmitted into the build platen, as opposed to being radiated into the chamber.

Rieder et al. [41] incorporated an ultrasonic sensor mounted on the underside of the build platen in PBF process. The probe is a 10 MHz unfocussed normal incidence ultrasonic sensor; the data is collected at 240 MHz sampling rate. The data acquisition process is synchronized with each build layer; a fresh layer triggers a sample collection cycle. A part with an artificially generated internal void is created, and the ultrasonic sensor signals reflected back are analyzed.

Reider et al. [41] reported that the ultrasonic sensor signal gets progressively contaminated with noise as the layer height builds, and interface echo (IE) and back wall scatter are observed. The signal reflected from the void is discernable, albeit barely, as a disturbance in the ultrasonic signal. The authors contend that the frequency spectrum of the ultrasonic signals as a discriminant of build quality.

Price et al. [155] measured the temperature evolutions in an electron beam melting (EBM) PBF process using an NIR camera. A LumaSense (model MCS640) NIR camera (spectral range 780–1080 nm wavelength) is mounted on the outside of a Arcam S12 EBM machine and is focused on the powder bed through the view port, and images are acquired at 60 Hz frame rate. Two different lenses are used the first has a larger field of view of 31 mm × 23 mm, while the second has a narrower 5.3 mm × 4.0 mm. A metal heat shield is placed over the powder bed. Measurements are taken during three phases of the EBM process, namely, preheating, contour melting, and hatch melting.

- Preheating is the phase where the beam is scanned at high speed across a freshly raked layer to raise its temperature uniformly.
- Contour melting is the process of creating the cross section boundaries, and has two distinct parts. The two parts only differ in the scan speed; in the first part the scan speed is relatively higher. Furthermore, during the contour melting phase the EBM beam is split into several parts to trace multiple boundaries at the same time.
- During the hatching phase, a single beam completes (hatches) the inside of the cross section; a straight raster pattern is employed.

The thermal evolutions of the EBM process captured using this system, can be useful for comparing defective versus ideal build conditions.

Schwerdtfeger et al. [46] incorporated an IR thermal camera (mode: FLIR A320) coaxially with an Arcam A2 electron beam AM machine; the camera has a spectral range of 7,500–13,000 nm. Thermal images of the build are recorded and visually analyzed. No effort was made to calibrate the thermal images with a reference radiation source. The authors induced artificial defects in the build by varying the focal distance of the beam. A qualitative comparison of the thermal image is made to discern instances of porosity and other internal defects.

Chivel and Smurov [27] implemented a coaxial CCD camera and two-color pyrometer (900–1700 nm) setup to monitor melt-pool morphology (100 μm, local focal diameter) and temperature in PBF process. The temperature distribution and intensity of the melt pool (from processing the CCD camera data) are correlated with the laser power. A linear trend in laser power at three levels (50 W, 100 W, 150 W) and surface trend is observed (viz., between c.1800°C and 2000°C). The pyrometer also enables mapping the surface temperature distribution.

In a work predating Chivel and Smurov [27], Bayle and Doubenskaia [47] used a similar setup with a coaxial IR camera (model FLIR Phoenix RDAS) along with a pyrometer with active wavelength of 1–1.5 μm mounted on a Phenix-PM100 laser PBF machine. Pyrometer readings are obtained over time for different layer thickness and hatch spacing settings. The IR camera is used to monitor the dynamics of particles, as they interact with the laser beam. Moreover, during the melting process, liquefied powder material is ejected from the melt-pool area. They report that a vast majority of powder particles displaced by the beam are in the same direction (~2.4–4.7 m/s) as that of the laser scan, a few are deflected backwards; the latter have much lesser velocity (~0.5 m/s).

Lott et al. [43] described a coaxial high speed camera setup illuminated by a photodiode laser. They imaged the laser scanning path, and collated the laser spot characteristics (as a scatter diagram) in different sections of the powder bed. This information is translated into quantitative information using spatial Fourier transforms. By monitoring the melt-pool dynamics, deviations from set points can be tracked.

6.5 Summary

At the outset of this chapter, we motivated the critical need for quality assurance in AM via some practical examples; we also noted the pitfalls and challenges that need to be addressed toward realizing repeatability and reliability in AM at a level commensurate with conventional machining and formative manufacturing techniques. This chapter reviewed some of the seminal literature in the pivotal area of process monitoring and control in AM. Two main areas of focus of this chapter were dimensional integrity quantification and sensor-based process monitoring.

For dimensional integrity measurement, we detailed a novel spectral graph theoretic approach developed by the author [121]. This approach relies on 3D point cloud data obtained from a laser scanner as opposed to tedious landmark and GD&T measurements obtained using a CMM. It can, thus, significantly reduce the measurement burden in AM.

In the context of process monitoring, this chapter reviewed sensing techniques used for assessing build quality in PBF and DED metal AM processes. Three key shortcomings are evident from the sensor-based monitoring literature, namely,

1. Lack of real-time process monitoring in order to detect the status of the build, most studies involved off-line analysis
2. Most efforts dedicated to detection, as opposed to prevention and compensation
3. No existing approach to integrate sensor data with physical models.

With the recent evolving research thrust in quality control and monitoring in AM, there is little doubt that in the coming decades the status quo of AM builds will extend beyond the current *print and pray*. The ultimate aim is to realize *reliable and repeatable* build quality in AM with closed-loop control, akin to some of the mature manufacturing processes.

6.6 Chapter Review Questions

1. By what other names has additive manufacturing (AM) been referred to historically?
2. State four advantages accrued by the additive manufacturing (AM) paradigm.
3. List three quality-related challenges in AM.
4. Find the materials that can be processed with DMLS machines. Specifically, do a web search on the materials that can be processed by the EOS machine. What are the typical particle sizes? Why can't the particles be made any smaller?
5. Find three applications where functional AM parts have been deployed in the field.
6. Find out what material the blade on the DMLS machine is made of; find out the reasons why the particular material is chosen.
7. Create a table listing the differences between PBF processes with respect to laser- and electron beam-based processes.

8. List five factors each from PBF and DED processes that are important to part quality.
9. Read the article by Elwani and Tapia [22] and understand the importance of emissivity in thermal measurement.
10. Find out the fiber laser manufacturers whose assemblies are used by the PBF and DED manufacturers. What is the typical spot size of the laser, and what is the spot distribution?
11. How is the debris during laser sintering in PBF prevented from contaminating the build?
12. What is the balling effect? When does it occur, and how does it affect the build quality?
13. Find out from the literature, the effect of the rastering pattern on part quality.
14. What different types of standardized test pieces have been proposed to assess geometric integrity in AM?

References

1. Lipson, H., and Kurman, M., 2013, *Fabricated: The New World of 3D Printing*, John Wiley & Sons, Indianapolis, IN.
2. Chua, C. K., and Leong, K. F., 2015, *3D Printing and Additive Manufacturing*, World Scientific, Singapore.
3. Gibson, I., Rosen, D. W., and Stucker, B., 2010, *Additive Manufacturing Technologies: Rapid Prototyping to Direct Digital Manufacturing*, Springer, New York.
4. Frazier, W. E., 2014, Metal additive manufacturing: A review, *Journal of Materials Engineering and Performance*, 23(6), pp. 1917–1928. doi:10.1007/s11665-014-0958-z.
5. Scheck, C., Jones, N., Farina, S., George, C., and Melendez, M., 2014, Technical overview of additive manufacturing, Naval Surface Warfare Center, Carderock Division, West Bethesda, MD.
6. NIST, 2013, Measurement science roadmap for metal-based additive manufacturing—Report prepared by Energetics Corporation, National Institute of Standards and Technology, Gaithersburg, MD.
7. GAO, 2015, GAO-15-505SP Additive manufacturing forum: 3D printing: Opportunities, challenges, and policy implications of additive manufacturing, United States Government Accountability Office, Washington, DC.
8. Bourell, D., Beaman, J., Leu, M., and Rosen, D., 2009, A brief history of additive manufacturing and the 2009 roadmap for additive manufacturing: Looking back and looking ahead, *Proceedings of US—Turkey Workshop on Rapid Technologies*.
9. Huang, Y., and Leu, M. C., 2014, Frontiers of additive manufacturing research and education—Report of NSF additive manufacturing workshop. Center for Manufacturing Innovation, University of Florida, USA, March, 1-35. http://plaza.ufl.edu/yongh/2013NSFAMWorkshopReport.pdf.
10. Kinsella, M. E., 2009, Additive manufacturing workshop: Results and plans, Airforce Research Laboratory, Dayton, OH.
11. Mani, M., Lane, B., Donmez, A., Feng, S., Moylan, S., and Fesperman, R., 2015, NISTIR 8036: Measurement science needs for real-time control of additive manufacturing powder bed fusion processes, NIST, Gaithersburg, MD.
12. Scott, J., Gupta, N., Weber, C., Newsome, S., Wohlers, T., and Caffrey, T., 2012, *Additive Manufacturing: Status and Opportunities*, Institute for Defense Analysis, Science and Technology Policy Institute, Washington, DC, pp. 1–29.
13. Waller, J. M., Parker, B. H., Hodges, K. L., Burke, E. R., Walker, J. L., and Generazio, E. R., 2014, NASA/TM-2014-218560: Nondestructive evaluation of additive manufacturing, NASA, Langley Research Center, Hamton, VA.
14. Petrova, V., 2013, *Advances in Engineering Research*, Nova Research, New York.
15. Cooper, K. P., 2003, Layered manufacturing: Challenges and opportunities, *Proceedings of Materials Research Society Symposia*, Cambridge University Press, New York, pp. 23–34.
16. Rao, P. K., Liu, J., Roberson, D., and Kong, Z., 2015, Sensor-based online process fault detection in additive manufacturing, *Journal of Manufacturing Science and Engineering* (Accepted).

17. Gu, D., and Shen, Y., 2009, Balling phenomena in direct laser sintering of stainless steel powder: Metallurgical mechanisms and control methods, *Materials & Design*, 30(8), pp. 2903–2910. doi:10.1016/j.matdes.2009.01.013.
18. Gusarov, A. V., Yadroitsev, I., Bertrand, P., and Smurov, I., 2007, Heat transfer modelling and stability analysis of selective laser melting, *Applied Surface Science*, 254(4), pp. 975–979. doi:10.1016/j.apsusc.2007.08.074.
19. Bourell, D. L., Leu, M. C., and Rosen, D. W., 2009, Roadmap for additive manufacturing: Identifying the future of freeform processing, The University of Texas, Austin, TX.
20. Wohlers, T., 2012, Wohlers Rapid report. Prototyping, Tooling & Manufacturing State of the Industry, Fort Collins, CO.
21. Rengier, F., Mehndiratta, A., von Tengg-Kobligk, H., Zechmann, C. M., Unterhinninghofen, R., Kauczor, H. U., and Giesel, F. L., 2010, 3D printing based on imaging data: Review of medical applications, *International Journal of Computer Assisted Radiology and Surgery*, 5(4), pp. 335–341. doi:10.1007/s11548-010-0476-x.
22. Tapia, G., and Elwany, A., 2014, A review on process monitoring and control in metal-based additive manufacturing, *Journal of Manufacturing Science and Engineering*, 136(6), p. 060801.
23. Huang, Y., Leu, M. C., Mazumder, J., and Donmez, A., 2015, Additive manufacturing: Current state, future potential, gaps and needs, and recommendations, *Journal of Manufacturing Science and Engineering*, 137(1), p. 014001.doi:10.1115/1.4028725.
24. Kruth, J.-P., Leu, M., and Nakagawa, T., 1998, Progress in additive manufacturing and rapid prototyping, *CIRP Annals—Manufacturing Technology*, 47(2), pp. 525–540.
25. Bartkowiak, K., 2010, Direct laser deposition process within spectrographic analysis in situ, *Physics Procedia*, 5, pp. 623–629. doi:10.1016/j.phpro.2010.08.090.
26. Berumen, S., Bechmann, F., Lindner, S., Kruth, J.-P., and Craeghs, T., 2010, Quality control of laser- and powder bed-based additive manufacturing (AM) technologies, *Physics Procedia*, 5, pp. 617–622. doi:10.1016/j.phpro.2010.08.089.
27. Chivel, Y., and Smurov, I., 2010, On-line temperature monitoring in selective laser sintering/melting, *Physics Procedia*, 5, pp. 515–521. doi:10.1016/j.phpro.2010.08.079.
28. Chivel, Y., and Smurov, I., 2011, Temperature monitoring and overhang layers problem, *Physics Procedia*, 12, pp. 691–696.
29. Craeghs, T., Bechmann, F., Berumen, S., and Kruth, J.-P., 2010, Feedback control of layerwise laser melting using optical sensors, *Physics Procedia*, 5, pp. 505–514. doi:10.1016/j.phpro.2010.08.078.
30. Craeghs, T., Clijsters, S., Kruth, J. P., Bechmann, F., and Ebert, M. C., 2012, Detection of Process failures in layerwise laser melting with optical process monitoring, *Physics Procedia*, 39, pp. 753–759. doi:10.1016/j.phpro.2012.10.097.
31. Craeghs, T., Clijsters, S., Yasa, E., Bechmann, F., Berumen, S., and Kruth, J.-P., 2011, Determination of geometrical factors in layerwise laser melting using optical process monitoring, *Optics and Lasers in Engineering*, 49(12), pp. 1440–1446. doi:10.1016/j.optlaseng.2011.06.016.
32. Craeghs, T., Clijsters, S., Yasa, E., and Kruth, J.-P., 2011, Online quality control of selective laser melting, *Proceedings of Solid Freeform Fabrication Symposium*, Austin, TX.
33. Griffith, M. L., Hofmeister, W. H., Knorovsky, G. A., MacCallum, D. O., Schlienger, M. E., and Smugeresky, J. E., 2002, Direct laser additive fabrication system with image feedback control, USPTO, Sandia Corporation.
34. Ilyas, I. P., 2013, 3D Machine vision and additive manufacturing: Concurrent product and process development, *IOP Conference Series: Materials Science and Engineering*, 46, p. 012029. doi:10.1088/1757-899x/46/1/012029.
35. Koch, J., and Mazumder, J., 2000, Apparatus and methods for monitoring and controlling multilayer laser cladding. United States patent US 6,122,564.
36. Hu, D., and Kovacevic, R., 2003, Sensing, modeling and control for laser-based additive manufacturing, *International Journal of Machine Tools and Manufacture*, 43(1), pp. 51–60. doi:10.1016/S0890-6955(02)00163-3.

37. Jacobsmuhlen, J. Z., Kleszczynski, S., Schneider, D., and Witt, G.,2013, High resolution imaging for inspection of laser beam melting systems, *Proceedings of 2013 IEEE International Instrumentation and Measurement Technology Conference*, IEEE, Minneapolis, MN, pp. 707–712.
38. Krauss, H., Eschey, C., and Zaeh, M.,2012, Thermography for monitoring the selective laser melting process, *Proceedings of the 23rd Annual International Solid Freeform Fabrication Symposium*, edited by D. Bourell, R. Crawford, C. Seepersad, J. Beaman, and H. Marcus, The University of Texas at Austin, Austin, TX, 2012.
39. Krauss, H., Zeugner, T., and Zaeh, M. F., 2014, Layerwise monitoring of the selective laser melting process by thermography, *Physics Procedia*, 56, pp. 64–71.
40. Barua, S., Sparks, T., and Liou, F., 2011, Development of low-cost imaging system for laser metal deposition processes, *Rapid Prototyping Journal*, 17(3), pp. 203–210. doi:10.1108/13552541111124789.
41. Rieder, H., Dillhöfer, A., Spies, M., Bamberg, J., and Hess, T.,2014, Online monitoring of additive manufacturing processes using ultrasound, *Proceedings of the 11th European Conference on Non-Destructive Testing*, Prague, Czech Republic October, pp. 6–10.
42. Li, X., 2001, Embedded sensors in layered manufacturing, PhD Thesis, Stanford University.
43. Lott, P., Schleifenbaum, H., Meiners, W., Wissenbach, K., Hinke, C., and Bültmann, J., 2011, Design of an optical system for the in situ process monitoring of selective laser melting (SLM), *Physics Procedia*, 12, pp. 683–690.doi:10.1016/j.phpro.2011.03.085.
44. Melvin III, L. S., Das, S., and Beaman Jr, S., 1994, Video microscopy of selective laser sintering, *Proceedings of the Solid Freeform Fabrication Symposium,* Austin, TX, pp. 34–41.
45. Purtonen, T., Kalliosaari, A., and Salminen, A., 2014, Monitoring and adaptive control of laser processes, *Physics Procedia*, 56, pp. 1218–1231.
46. Schwerdtfeger, J., Singer, R. F., and Körner, C., 2012, *In situ* flaw detection by IR-imaging during electron beam melting, *Rapid Prototyping Journal*, 18(4), pp. 259–263. doi:10.1108/13552541211231572.
47. Veiko, V. P., Bayle, F., and Doubenskaia, M., 2008, Selective laser melting process monitoring with high speed infra-red camera and pyrometer, *Proceedings of the SPIE*, 6985, pp. 1–8. doi:10.1117/12.786940.
48. Wang, Y., Zhao, W., Ding, Y., He, Z., and Lu, B., 2009, A detection and control method of resin liquid-level of stereolithography apparatus, *Rapid Prototyping Journal*, 15(5), pp. 333–338. doi:10.1108/13552540910993851.
49. Wegner, A., and Witt, G.,2011, Process monitoring in laser sintering using thermal imaging, *Proceedings of SFF Symposium*, Austin, TX, pp. 8–10.
50. Fang, T., Bakhadyrov, I., Jafari, M. A., and Alpan, G., 1998, Online detection of defects in layered manufacturing, *Proceedings of IEEE International Conference on Robotics and Automation, IEEE,* Leuven, Belgium, Vol. 1, pp. 254–259.
51. Fang, T., Jafari, M. A., Danforth, S. C., and Safari, A., 2003, Signature analysis and defect detection in layered manufacturing of ceramic sensors and actuators, *Machine Vision and Applications*, 15(2), pp. 63-75
52. Pal, D., Patil, N., Zeng, K., and Stucker, B., 2014, An integrated approach to additive manufacturing simulations using physics based, coupled multiscale process modeling, *Journal of Manufacturing Science and Engineering*, 136(6), p. 061022.
53. Pal, D., Patil, N., Nikoukar, M., Zeng, K., Kutty, K. H., and Stucker, B. E., 2013, An Integrated approach to cyber-enabled additive manufacturing using physics based, coupled multi-scale process modeling, *Proceedings of Solid Freeform Fabrication Symposium,* Austin, TX.
54. Patil, N., 2014, A novel numerical framework for simulation of multi-scale spatio temporally non-linear systems in additive manufacturing, PhD Dissertation, University of Louisville.
55. Patil, N., Pal, D., and Stucker, B., Predictive modeling capabilities for dimensional accuracy and surface finish in metal laser melting based additive manufacturing, *ASPE*, 7, pp. 65–79.
56. Witherell, P., Feng, S., Simpson, T. W., Saint John, D. B., Michaleris, P., Liu, Z.-K., Chen, L.-Q., and Martukanitz, R., 2014, Toward metamodels for composable and reusable additive manufacturing process models, *Journal of Manufacturing Science and Engineering*, 136(6), p. 061025.
57. King, W., Anderson, A., Ferencz, R., Hodge, N., Kamath, C., and Khairallah, S., 2014, Overview of modelling and simulation of metal powder-bed fusion process at Lawrence Livermore National Laboratory, *Materials Science and Technology*, 31(8), pp. 957–968.
58. King, W., Advancing metal AM at its most fundamental level, *JOM Magazine*, pp. 2202–2203.

59. Robinson, L., and Scott, J., 2014, Layers of complexity: Making the promises possible for additive manufacturing of metals, *JOM Magazine*, 66 (11), pp. 2194–2207.
60. Savio, E., De Chiffre, L., and Schmitt, R., 2007, Metrology of freeform shaped parts, *CIRP Annals—Manufacturing Technology*, 56(2), pp. 810–835. doi:10.1016/j.cirp.2007.10.008.
61. Arrieta, C., 2012, Quantitative assessments of geometric errors for rapid prototyping in medical applications, *Rapid Prototyping Journal*, 18(6), pp. 431–442. doi:10.1108/13552541211271974.
62. Brajlih, T., Valentan, B., Balic, J., and Drstvensek, I., 2011, Speed and accuracy evaluation of additive manufacturing machines, *Rapid Prototyping Journal*, 17(1), pp. 64–75. doi:10.1108/13552541111098644.
63. Dimitrov, D., Schreve, K., and De Beer, N., 2006, Advances in three dimensional printing–state of the art and future perspectives, *Rapid Prototyping Journal*, 12(3), pp. 136–147.
64. Seepersad, C. C., Govett, T., Kim, K., Lundin, M., and Pinero, D., 2012, A designer's guide for dimensioning and tolerancing SLS parts, *23rd Annual International Solid Freeform Fabrication Symposium*, Austin, TX, pp. 921–931.
65. Cooke, A., and Soons, J., 2010, Variability in the geometric accuracy of additively manufactured test parts, *Proceedings of 21st Annual International Solid Freeform Fabrication Symposium*, Austin, TX, pp. 1–12.
66. Munguía, J., de Ciurana, J., and Riba, C., 2008, Pursuing successful rapid manufacturing: A users' best-practices approach, *Rapid Prototyping Journal*, 14(3), pp. 173–179. doi:10.1108/13552540810878049.
67. Krauss, H., and Zaeh, M. F., 2013, Investigations on manufacturability and process reliability of selective laser melting, *Physics Procedia*, 41, pp. 808–815. doi:10.1016/j.phpro.2013.03.153.
68. Ziemian, C., and Crawn III, P., 2001, Computer aided decision support for fused deposition modeling, *Rapid Prototyping Journal*, 7(3), pp. 138–147.
69. Boschetto, A., Giordano, V., and Veniali, F., 2013, Surface roughness prediction in fused deposition modelling by neural networks, *The International Journal of Advanced Manufacturing Technology*, 67(9–12), pp. 2727–2742. doi:10.1007/s00170-012-4687-x.
70. Padhye, N., and Deb, K., 2011, Multi-objective optimisation and multi-criteria decision making in SLS using evolutionary approaches, *Rapid Prototyping Journal*, 17(6), pp. 458–478. doi:10.1108/13552541111184198.
71. Bukkapatnam, S., and Clark, B., 2007, Dynamic modeling and monitoring of contour crafting-: An extrusion-based layered manufacturing process, *Journal of Manufacturing Science and Engineering*, 129(1), pp. 135–142.
72. Bauereiß, A., Scharowsky, T., and Körner, C., 2014, Defect generation and propagation mechanism during additive manufacturing by selective beam melting, *Journal of Materials Processing Technology*, 214(11), pp. 2522–2528. doi:10.1016/j.jmatprotec.2014.05.002.
73. Vetter, P. A., Engel, T., and Fontaine, J.,1994, Laser cladding: The relevant parameters for process control, *Proceedings of Europto High Power Lasers and Laser Applications V*, International Society for Optics and Photonics, Vienna, Austria, pp. 452–462.
74. Boddu, M. R., Landers, R. G., and Liou, F. W., 2001, Control of laser cladding for rapid prototyping—A review, *Proceedings of the Solid Freeform Fabrication Symposium*, Austin, TX, pp. 6–8.
75. Bugeda Miguel Cervera, G., and Lombera, G., 1999, Numerical prediction of temperature and density distributions in selective laser sintering processes, *Rapid Prototyping Journal*, 5(1), pp. 21–26. doi:10.1108/13552549910251846.
76. Shiomi, M., Yoshidome, A., Abe, F., and Osakada, K., 1999, Finite element analysis of melting and solidifying processes in laser rapid prototyping of metallic powders, *International Journal of Machine Tools and Manufacture*, 39(2), pp. 237–252. doi:10.1016/S0890-6955(98)00036-4.
77. Michaleris, P., 2014, Modeling metal deposition in heat transfer analyses of additive manufacturing processes, *Finite Elements in Analysis and Design*, 86(0), pp. 51–60. doi:10.1016/j.finel.2014.04.003.
78. Agarwala, M. K., Jamalabad, V. R., Langrana, N. A., Safari, A., Whalen, P. J., and Danforth, S. C., 1996, Structural quality of parts processed by fused deposition, *Rapid Prototyping Journal*, 2(4), pp. 4–19.
79. Armillotta, A., 2006, Assessment of surface quality on textured FDM prototypes, *Rapid Prototyping Journal*, 12(1), pp. 35–41.

80. Bellini, A., and Güçeri, S., 2003, Mechanical characterization of parts fabricated using fused deposition modeling, *Rapid Prototyping Journal*, 9(4), pp. 252–264.
81. Sachs, E., Allen, S., Guo, H., Banos, B., Cima, M., Serdy, J., and Brancazio, D., 1997, Progress on tooling by 3D printing; conformal cooling, dimensional control, surface finish and hardness, *Proceedings of the Eighth Annual Solid Freeform Fabrication Symposium*, SME, Austin, TX, pp. 115–124.
82. Averyanova, M., Cicala, E., Bertrand, P., and Grevey, D., 2012, Experimental design approach to optimize selective laser melting of martensitic 17-4 PH powder: Part I—Single laser tracks and first layer, *Rapid Prototyping Journal*, 18(1), pp. 28–37. doi:10.1108/13552541211193 76.
83. El-Katatny, I., Masood, S. H., and Morsi, Y. S., 2010, Error analysis of FDM fabricated medical replicas, *Rapid Prototyping Journal*, 16(1), pp. 36–43. doi:10.1108/13552541011011695
84. Hopkinson, N., and Sercombe, T. B., 2008, Process repeatability and sources of error in indirect SLS of aluminium, *Rapid Prototyping Journal*, 14(2), pp. 108–113. doi:10.1108/13552540810862073.
85. Ippolito, R., Iuliano, L., and Gatto, A., 1995, Benchmarking of rapid prototyping techniques in terms of dimensional accuracy and surface finish, *CIRP Annals—Manufacturing Technology*, 44(1), pp. 157–160. doi:10.1016/s0007-8506(07)62296-3.
86. Kechagias, J., 2007, An experimental investigation of the surface roughness of parts produced by LOM process, *Rapid Prototyping Journal*, 13(1), pp. 17–22. doi:10.1108/13552540710719172.
87. Lanzetta, M., and Sachs, E., 2003, Improved surface finish in 3D printing using bimodal powder distribution, *Rapid Prototyping Journal*, 9(3), pp. 157–166. doi:10.1108/13552540310477463.
88. Liao, H.-T., and Shie, J.-R., 2007, Optimization on selective laser sintering of metallic powder via design of experiments method, *Rapid Prototyping Journal*, 13(3), pp. 156–162. doi:10.1108/13552540710750906.
89. Mallepree, T., and Bergers, D., 2009, Accuracy of medical RP models, *Rapid Prototyping Journal*, 15(5), pp. 325–332. doi:10.1108/13552540910993842.
90. Safdar, A., He, H. Z., Wei, L.-Y., Snis, A., and Paz, L. E. C. D., 2012, Effect of process parameters settings and thickness on surface roughness of EBM produced Ti-6Al-4V, *Rapid Prototyping Journal*, 18(5), pp. 401–408. doi:10.1108/13552541211250391.
91. Sager, B., and Rosen, D. W., 2008, Use of parameter estimation for stereolithography surface finish improvement, *Rapid Prototyping Journal*, 14(4), pp. 213–220. doi:10.1108/13552540810896166.
92. Weheba, G., and Sanchez-Marsa, A., 2006, Using response surface methodology to optimize the stereolithography process, *Rapid Prototyping Journal*, 12(2), pp. 72–77. doi:10.1108/13552540610652401.
93. Yadroitsev, I., Yadroitsava, I., Bertrand, P., and Smurov, I., 2012, Factor analysis of selective laser melting process parameters and geometrical characteristics of synthesized single tracks, *Rapid Prototyping Journal*, 18(3), pp. 201–208. doi:10.1108/13552541211218117.
94. Lin, F., Sun, W., and Yan, Y., 2001, Optimization with minimum process error for layered manufacturing fabrication, *Rapid Prototyping Journal*, 7(2), pp. 73–82.
95. Blattmeier, M., Witt, G., Wortberg, J., Eggert, J., and Toepker, J., 2012, Influence of surface characteristics on fatigue behaviour of laser sintered plastics, *Rapid Prototyping Journal*, 18(2), pp. 161–171. doi:10.1108/13552541211212140.
96. Ang, K. C., Leong, K. F., Chua, C. K., and Chandrasekaran, M., 2006, Investigation of the mechanical properties and porosity relationships in fused deposition modelling-fabricated porous structures, *Rapid Prototyping Journal*, 12(2), pp. 100–105.
97. Cooke, W., Tomlinson, R. A., Burguete, R., Johns, D., and Vanard, G., 2011, Anisotropy, homogeneity and ageing in an SLS polymer, *Rapid Prototyping Journal*, 17(4), pp. 269–279.
98. Storch, S., Nellessen, D., Schaefer, G., and Reiter, R., 2003, Selective laser sintering: qualifying analysis of metal based powder systems for automotive applications, *Rapid Prototyping Journal*, 9(4), pp. 240–251. doi:10.1108/13552540310489622.
99. King, W. E., Barth, H. D., Castillo, V. M., Gallegos, G. F., Gibbs, J. W., Hahn, D. E., Kamath, C., and Rubenchik, A. M., 2014, Observation of keyhole-mode laser melting in laser powder-bed fusion additive manufacturing, *Journal of Materials Processing Technology*, 214(12), pp. 2915–2925. doi:10.1016/j.jmatprotec.2014.06.005.
100. Qiu, D., and Langrana, N. A., 2002, Void eliminating toolpath for extrusion-based multi-material layered manufacturing, *Rapid Prototyping Journal*, 8(1), pp. 38–45.

101. Sun, Q., Rizvi, G., Bellehumeur, C., and Gu, P., 2008, Effect of processing conditions on the bonding quality of FDM polymer filaments, *Rapid Prototyping Journal*, 14(2), pp. 72–80.
102. Caulfield, B., McHugh, P., and Lohfeld, S., 2007, Dependence of mechanical properties of polyamide components on build parameters in the SLS process, *Journal of Materials Processing Technology*, 182(1), pp. 477–488.
103. Mahesh, M., Wong, Y., Fuh, J., and Loh, H., 2004, Benchmarking for comparative evaluation of RP systems and processes, *Rapid Prototyping Journal*, 10(2), pp. 123–135. doi:10.1108/13552540410526999.
104. Kruth, J. P., Bartscher, M., Carmignato, S., Schmitt, R., De Chiffre, L., and Weckenmann, A., 2011, Computed tomography for dimensional metrology, *CIRP Annals—Manufacturing Technology*, 60(2), pp. 821–842. doi:10.1016/j.cirp.2011.05.006.
105. Jiang, X., Scott, P., and Whitehouse, D., 2007, Freeform surface characterisation—A fresh strategy, *CIRP Annals—Manufacturing Technology*, 56(1), pp. 553–556. doi:10.1016/j.cirp.2007.05.132.
106. Bi, Z. M., and Wang, L., 2010, Advances in 3D data acquisition and processing for industrial applications, *Robotics and Computer-Integrated Manufacturing*, 26(5), pp. 403–413. doi:10.1016/j.rcim.2010.03.003.
107. Tang, P., Huber, D., Akinci, B., Lipman, R., and Lytle, A., 2010, Automatic reconstruction of as-built building information models from laser-scanned point clouds: A review of related techniques, *Automation in Construction*, 19(7), pp. 829–843. doi:10.1016/j.autcon.2010.06.007.
108. Cuypers, W., Van Gestel, N., Voet, A., Kruth, J. P., Mingneau, J., and Bleys, P., 2009, Optical measurement techniques for mobile and large-scale dimensional metrology, *Optics and Lasers in Engineering*, 47(3–4), pp. 292–300. doi:10.1016/j.optlaseng.2008.03.013.
109. Pal, P., 2001, An easy rapid prototyping technique with point cloud data, *Rapid Prototyping Journal*, 7(2), pp. 82–90.
110. Woo, H., Kang, E., Wang, S., and Lee, K. H., 2002, A new segmentation method for point cloud data, *International Journal of Machine Tools and Manufacture*, 42(2), pp. 167–178. doi:10.1016/S0890-6955(01)00120-1.
111. Schnabel, R., Wahl, R. and Klein, R. (2007), Efficient RANSAC for point-cloud shape detection. Computer Graphics Forum, 26: 214–226. doi:10.1111/j.1467-8659.2007.01016.x
112. Liu, G. H., Wong, Y. S., Zhang, Y. F., and Loh, H. T., 2003, Modelling cloud data for prototype manufacturing, *Journal of Materials Processing Technology*, 138(1–3), pp. 53–57.doi:10.1016/s0924-0136(03)00048-7.
113. Sun, W., Bradley, C., Zhang, Y., and Loh, H. T., 2001, Cloud data modelling employing a unified, non-redundant triangular mesh, *Computer-Aided Design*, 33(2), pp. 183–193.
114. Huang, J., and Menq, C.-H., 2002, Automatic CAD model reconstruction from multiple point clouds for reverse engineering, *Journal of Computing and Information Science in Engineering*, 2(3), p. 160. doi:10.1115/1.1529210.
115. Fabio, R., 2003, From point cloud to surface: The modeling and visualization problem, *International Archives of Photogrammetry, Remote Sensing and Spatial Information Sciences*, 34(5), p. W10.
116. Mémoli, F., and Sapiro, G., 2005, A theoretical and computational framework for isometry invariant recognition of point cloud data, *Foundations of Computational Mathematics*, 5(3), pp. 313–347. doi:10.1007/s10208-004-0145-y.
117. Iuliano, L., and Minetola, P., 2008, Enhancing moulds manufacturing by means of reverse engineering, *The International Journal of Advanced Manufacturing Technology*, 43(5–6), pp. 551–562. doi:10.1007/s00170-008-1739-3.
118. Kainat, M., Adeeb, S., Cheng, J. R., Ferguson, J., and Martens, M.,2012, Identifying initial imperfection patterns of energy pipes using a 3D laser scanner, *Proceedings of 2012 9th International Pipeline Conference*, American Society of Mechanical Engineers, Calgary, Alberta, Canada, September 24–28, pp. 57–63.
119. Raja, V., Zhang, S., Garside, J., Ryall, C., and Wimpenny, D., 2005, Rapid and cost-effective manufacturing of high-integrity aerospace components, *The International Journal of Advanced Manufacturing Technology*, 27(7–8), pp. 759–773. doi:10.1007/s00170-004-2251-z.

120. Bouyssie, J., Bouyssie, S., Sharrock, P., and Duran, D., 1997, Stereolithographic models derived from x-ray computed tomography reproduction accuracy, *Surgical and Radiologic Anatomy*, 19(3), pp. 193–199.
121. Rao, P., Kong, Z., Kunc, V., Smith, R., Love, L., and Duty, C., 2015, Assessment of dimensional integrity and spatial defect localization in additive manufacturing using spectral graph theory (SGT), *Journal of Manufacturing Science and Engineering*, 138(5), p. 051007. doi:10.1115/1.4031574.
122. Rao, P. K., Beyca, O. F., Kong, Z., Bukkaptanam, S. T., Case, K. E., and Komanduri, R., 2015, A graph theoretic approach for quantification of surface morphology and its application to chemical mechanical planarization (CMP) process, *IIE Transactions*, 47(10), pp. 1088–1111.
123. Moylan, S., Cooke, A., Jurrens, K., Slotwinski, J., and Donmez, M. A., 2012, A review of test artifacts for additive manufacturing (Report Number: NISTIR 7858), National Institute of Standards and Technology (NIST), Gaithersburg, MD.
124. Love, L., Kunc, V., Rios, O., Duty, C., Elliot, A., Post, B., Smith, R., and Blue, C., 2014, The importance of carbon fiber to polymer additive manufacturing, *Journal of Materials Research*, 29(17), pp. 1892–1898. doi:10.1557/jmr.2014.212.
125. Tekinalp, H. L., Kunc, V., Velez-Garcia, G. M., Duty, C. E., Love, L. J., Naskar, A. K., Blue, C. A., and Ozcan, C., 2014, Highly oriented carbon fiber–polymer composites via additive manufacturing, *Composites Science and Technology*, 105, pp. 144–150
126. Shi, J., and Malik, J., 2000, Normalized cuts and image segmentation, *IEEE Transactions on Pattern Analysis and Machine Intelligence*, 22(8), pp. 888–905.
127. Chung, F. R. K., 1997, *Spectral Graph Theory*, American Mathematical Society, Providence, RI.
128. Mohar, B., 1991, The Laplacian spectrum of graphs, *Graph Theory, Combinatorics, and Applications*, 2, pp. 871–898.
129. Fiedler, M., 1973, Algebraic connectivity of graphs, *Czechoslovak Mathematical Journal*, 23, pp. 298–305.
130. Rao, P. K., 2013, Sensor-based monitoring and inspection of surface morphology in ultraprecision manufacturing processes, PhD Thesis, Oklahoma State University.
131. Foster, B. K., Reutzel, E. W., Nassar, A. R., Dickman, C. J., and Hall, B. T.,2015, A brief survey of sensing for additive manufacturing, *Proc. SPIE 9489, Dimensional Optical Metrology and Inspection for Practical Applications IV,* 94890B (May 14, 2015), Baltimore, Maryland, United States, April 20 pp. 94890B-94890B-94810. doi:10.1117/12.2180654.
132. Nassar, A., Spurgeon, T., and Reutzel, E.,2014, Sensing defects during directed-energy additive manufacturing of metal parts using optical emissions spectroscopy, *Proceedings of Solid Freeform Fabrication Symposium*, University of Texas at Austin, Austin, TX.
133. Reutzel, E. W., and Nassar, A. R., 2015, A survey of sensing and control systems for machine and process monitoring of directed energy, metal-based additive manufacturing, *Rapid Prototyping Journal*, 21(2), pp. 159–167. doi:10.1108/RPJ-12-2014-0177.
134. Kovacevic, R., Hu, D., and Valant, M. E., 2006, System and method for controlling the size of the molten pool in laser-based additive manufacturing, Southern Methodist University. U.S. Patent No. 6,995,334.
135. Toyserkani, E., Khajepour, A., and Corbin, S. F., 2006, System and method for closed-loop control of laser cladding by powder injection. United States patent US 7,043,330.
136. Peyre, P., Aubry, P., Fabbro, R., Neveu, R., and Longuet, A., 2008, Analytical and numerical modelling of the direct metal deposition laser process, *Journal of Physics D: Applied Physics*, 41(2), p. 025403
137. Cooke, A. L., and Moylan, S. P., 2011, Process intermittent measurement for powder-bed based additive manufacturing, *Proceedings of the 22nd International SFF Symposium-An Additive Manufacturing Conference*, NIST, Austin, TX, pp. 8–10.
138. Vetter, P. A., Engel, T., and Fontaine, J., 1994, Laser cladding: The relevant parameters for process control, *Proc. SPIE 2207, Laser Materials Processing: Industrial and Microelectronics Applications*, Vienna, Austria, pp. 452–462. doi:10.1117/12.184751.

139. Song, L., Bagavath-Singh, V., Dutta, B., and Mazumder, J., 2012, Control of melt pool temperature and deposition height during direct metal deposition process, *The International Journal of Advanced Manufacturing Technology*, 58(1–4), pp. 247–256.
140. Song, L., and Mazumder, J., 2011, Feedback control of melt pool temperature during laser cladding process, *IEEE Transactions on Control Systems Technology*, 19(6), pp. 1349–1356.
141. Song, L., and Mazumder, J., 2012, Real time Cr measurement using optical emission spectroscopy during direct metal deposition process, *IEE Sensors Journal*, 12(5), pp. 958–964.
142. Song, L., Wang, C., and Mazumder, J.,2012, Identification of phase transformation using optical emission spectroscopy for direct metal deposition process, *Proceedings of SPIE LASE, International Society for Optics and Photonics*, San Francisco, CA, USA, January 21, 2012 pp. 82390–82399.
143. Boddu, M. R., Landers, R. G., Musti, S., Agarwal, S., Ruan, J., and Liou, F. W., 2003, System integration and real-time control architecture of a laser aided manufacturing process, *Solid Freeform Fabrication Conference,* Austin, TX.
144. Bi, G., Schürmann, B., Gasser, A., Wissenbach, K., and Poprawe, R., 2007, Development and qualification of a novel laser-cladding head with integrated sensors, *International Journal of Machine Tools and Manufacture*, 47(3), pp. 555–561.
145. Bi, G., Sun, C. N., and Gasser, A., 2013, Study on influential factors for process monitoring and control in laser aided additive manufacturing, *Journal of Materials Processing Technology*, 213(3), pp. 463–468. doi:10.1016/j.jmatprotec.2012.10.006.
146. Davis, T. A., and Shin, Y. C., 2011, Vision-based clad height measurement, *Machine Vision and Applications*, 22(1), pp. 129–136.
147. Qilin, D., Wei, F., Dianbing, C., and Peng, C., 2014, Measurement of the molten pool image during laser cladding process, *International Conference on Mechatronics, Electronic, Industrial and Control Engineering* (*MEIC 2014*), Atlantis Press, Shenyang, China.
148. Ocylok, S., Alexeev, E., Mann, S., Weisheit, A., Wissenbach, K., and Kelbassa, I.,2014, Correlations of melt pool geometry and process parameters during laser metal deposition by coaxial process monitoring, *Physics Procedia,* 56, pp. 228–238.
149. De Chiffre, L., Carmignato, S., Kruth, J. P., Schmitt, R., and Weckenmann, A., 2014, Industrial applications of computed tomography, *CIRP Annals—Manufacturing Technology*, 63(2), pp. 655–677. doi: 10.1016/j.cirp.2014.05.011
150. Elsen, M. v., Al-Bender, F., and Kruth, J.-P., 2008, Application of dimensional analysis to selective laser melting, *Rapid Prototyping Journal*, 14(1), pp. 15–22. doi:10.1108/13552540810841526
151. Kruth, J.-P., Levy, G., Klocke, F., and Childs, T., 2007, Consolidation phenomena in laser and powder-bed based layered manufacturing, *CIRP Annals—Manufacturing Technology*, 56(2), pp. 730–759.
152. Kruth, J.-P., Vandenbroucke, B., Vaerenbergh, V. J., and Mercelis, P.,2005, Benchmarking of different SLS/SLM processes as rapid manufacturing techniques, *Proceedings of International Conference on Polymer and Moulds Innovation (PMI)*, Gent, Belgium, April 20–23.
153. Kumar, S., and Kruth, J.-P., 2010, Composites by rapid prototyping technology, *Materials & Design*, 31(2), pp. 850–856.
154. Pavlov, M., Doubenskaia, M., and Smurov, I., 2010, Pyrometric analysis of thermal processes in SLM technology, *Physics Procedia*, 5, pp. 523–531.
155. Price, S., Cooper, K., and Chou, K., 2012, Evaluations of temperature measurements by near-infrared thermography in powder-based electron-beam additive manufacturing, *Proceedings of the Solid Freeform Fabrication Symposium*, University of Texas, Austin, TX, pp. 761–773.

Chapter 7

Functionally Graded Materials

Samuel Tammas-Williams and Iain Todd
The University of Sheffield

Contents

7.1 Introduction

In modern engineering terminology, functionally graded materials (FGMs) refer to an advanced composite consisting of two or more phases. What makes an FGM distinct from a standard composite material is the spatial variation of the phase composition of the FGM, which is deliberately modified with the intention of grading the material properties. Typically, this involves utilizing the inherent properties of each of the constituent phases in order to alter the material response at different locations. Unlike standard composite materials, they do not contain distinct boundaries between the phases. The first example of a modern FGM with a radically altering composition is often considered to be a metallic–ceramic composite developed for a rocket casing in the early 1980s in Japan (Sobczak and Drenchev, 2013). Engineers required a material that could withstand a maximum temperature of 2000 K and a temperature gradient of 1000 K over a 10 mm width, without developing high thermal stresses between interfaces. FGM, also sometimes referred to as compositionally graded material, can also refer to subtler changes in the composition and microstructure of a part. A simple kitchen knife requires only the cutting edge to be hard, while the rest of the material is strong and tough. Again, the defining feature of such material is the grading of the material properties with position. Other examples where FGMs have been used, or have potential to be exploited, are found in aerospace, biomaterials, defense, energy conversion, optoelectronics, and ultrasonic transducers (Jedamzik et al., 2000; Mahamood et al., 2012). Unfortunately, the improved performance is often offset by an increase in manufacturing costs.

The potential advantages of FGM have led them being regarded as the pinnacle of the modern material hierarchy (Jha et al., 2013). As pure metals have been superseded by alloys and composites in many high-value applications, in extreme environments, such as the rocket casing described above, standard engineering materials are unable to withstand the conditions and must be replaced by FGMs. Although it has only been relatively recently that engineers have started to utilize the potential of advanced composite FGMs, they have existed in nature for millions of years (Bruet et al., 2008). A "living fossil," a fish almost identical to one living around 96 million years ago, was found to have scales that exhibit a variation in hardness depending on the distance from the scale surface. As such, this helped to protect the fish from penetrating bite attacks by transferring load from the stiff outer layer to the softer inner layer and dissipating energy by plasticity without penetrating the scale. Such functional grading of material properties is typical in natural structures, where the discrete changes in material properties often observed in human-engineered structures are rare. Another example from the multitude available in the natural world is bamboo, where the fiber density of a stem varies with distance from the surface, with the highest density of fiber occurring at the outer edge of the bamboo (Nogata and Takahashi, 1995). It is no coincidence that this is where the highest stress is recorded during wind loading of the stem. Even our skin is functionally graded to provide different elastic, tactile, and toughness properties depending on the location of the body and depth (Jha et al., 2013). In nature, the use of raw materials tends to be very efficient, and biological structures have thus provided inspiration for various engineering FGMs (Oxman et al., 2011).

Before FGMs can be used to their full potential, there must be changes to not only the material manufacturing methods but also the engineering design process. Currently, when generating a computer-aided design (CAD) model with a commercial package, material is defined discretely; there is little scope for including smoothly varying material properties. CAD software must, therefore, be adapted to allow designs to be used to generate an FGM by assigning material properties/chemistry/phases before models can be directly used in manufacturing. Moreover, when trying to predict the response of FGM to environmental loading conditions, new mathematical techniques must be employed (Jha et al., 2013).

This chapter provides an overview of FGM and how it can be of benefit when utilized in engineering components. Possible manufacturing routes are discussed, in particular, how the rapid advances in the field of additive manufacturing (AM) have opened up new avenues for the manufacture and application of FGMs. However, simple FGMs, such as the blade mentioned earlier, had been produced long before AM was established as a viable manufacturing route.

7.2 Nonadditive Manufacturing Methods

7.2.1 Early Examples of FGM

While AM has allowed new ways to construct FGM, the manufacture of FGMs was established long before AM gained the recognition it now enjoys. The manufacture of FGM is often considered to have begun in Japan in the 1980s, but there are much older examples of less complex grading of material than the metal–ceramic composite that was developed by Japanese engineers. The hard cutting edge and tough center of a cutting blade mentioned in the introduction is a good example of a basic FGM. Once again, Japan provides an early example of this form of FGM. Japanese swordsmiths used carbonization and decarbonization of steel plates, which were then forged together to ensure that the outer surface of their blades had a much higher carbon content, and thus, higher hardness but lower toughness, than the internal steel. This effect was then amplified by the deliberate variation of the cooling rate experienced by the material. For many metals, where variation in the thermal treatment causes changes in the local microstructure, the material properties can be altered by changing the thermal history of materials. The same Japanese swordsmiths who changed the carbon content of their steel blade also used clay to selectively thermally insulate some areas of the sword blade prior to quenching the hot steel in water, and thus, superimposed mass and heat transport processes. This allowed the formation of very hard and brittle martensite at the cutting blade edge, while further from the surface remains a softer, at the same time tougher pearlite and ferrite structure (Inoue, 2010). A high-resolution image of a Japanese cutting blade is given in Figure 7.1a, showing clearly the different phases. This nonuniform distribution of phases and carbon content in the steel ensures that the blade remains sharp but with enough ductility to absorb the bending moment during a cutting operation.

Perhaps unsurprisingly, another example of early human use of different strength phases to create a more effective overall structure was also a weapon. In the 16th to 18th century, Damascus steel (named after the city in Syria) blades had many amazing and legendary properties ascribed to them. What makes Damascus steel blades instantly recognizable is the distinctive patterning on their surface (see Figure 7.1b), which consists of clustered iron carbide particles (cementite). They have also been found to contain carbon nanotubes and cementite nanowires. Their reputation for invincibility stems from their excellent hardness as well as toughness and strength. Unfortunately, the techniques used to manufacture this steel have been lost and the limited remaining examples make studying their properties difficult (Reibold et al., 2006; Wadsworth, 2015).

7.2.2 Thermal Processing

Thermal processing is still used in modern component production to grade the hardness of steels, e.g., flame hardening, induction hardening, and laser hardening. By heating and rapidly cooling

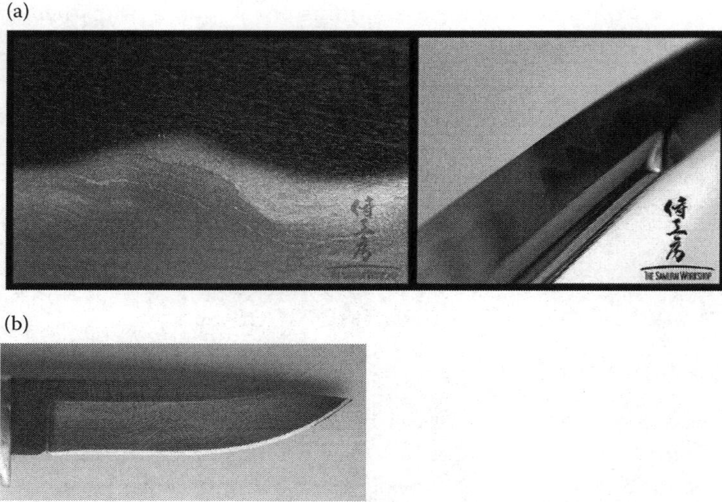

Figure 7.1 Functionally graded cutting blades: (a) Photographs of the high carbon content martensitic cutting blade (light) of a Japanese sword and the more ductile main body (dark). (Courtesy of Jeffrey Ching, The Samurai Workshop.) (b) Damascus steel blade with patterned soft and hard phases. (From Ralf Pfeifer, Wikipedia.)

the surface, a gradient in hardness is achieved. During an induction or flame hardening process, the cooling is often achieved by water spray. During laser hardening, where the temperature gradient is much higher and the heated region much thinner, the surface can be quenched by the rapid transfer of thermal energy from the surface to bulk material, negating the requirement for a water spray (Mortensen and Suresh, 1995).

7.2.3 Carburizing and Nitriding

The altering of steel properties via a change in carbon content is still used today. Nitrogen as well as carbon is used to increase the hardness of steel at the surface by the transport of atoms at the surface. These processes are referred to as steel nitriding and carburizing, respectively. Steel carburizing predates even the work of the Japanese swordsmiths, with axes produced ~3000 years ago in ancient Egypt, exhibiting a carbon content that varied from near zero in the center to around 0.9% at the cutting tip, and with a corresponding hardness of 70 to 444 HB. In modern manufacture of steels, carbon and nitrogen can be supplied by gas, solid compounds, or salt baths. The concentration profile, and thus, material properties can be controlled by the temperature and concentration profile at the surface. In a simple binary system, where diffusion is independent of concentration, the variation in concentration can be described by Equation 7.1 (Mortensen and Suresh, 1995).

$$\frac{C_x - C_s}{C_o - C_s} = \text{erf}\left[\frac{x}{2 \cdot \sqrt{D \cdot t}}\right], \tag{7.1}$$

where C_x, C_s, and C_o are the concentrations at position x, the surface, and the bulk; D is the diffusivity; and t is the time.

The profile predicted by Equation 7.1, with a peak at the surface followed by a rapid drop off, is often not what is desired by materials engineers. To flatten this peak and get a deeper layer of high concentration, a two-stage diffusion process is employed. The first stage is conducted by holding the component surface at a higher concentration than desired, the surface concentration is then lowered, but still above the bulk concentration, to allow the high concentration at the surface to diffuse deeper into the component. The result is an approximately uniform case of similar concentrations followed by a gradual decrease in concentration toward the bulk material.

Clearly, the Japanese method mentioned previously, of forging together steel plates with different carbon centration, is somewhat different from steel carburizing, where the diffusion of atoms from the surface creates an FGM. Instead, the interdiffusion of carbon atoms between the plates removed a sharp interface and lead to a smooth gradient in carbon concentration (Inoue, 2010).

In 2015, General Electric was awarded a patent using the nitriding of steel for a very different purpose than surface hardening (Dial et al., 2015). They used a stainless-steel alloy with a high Cr content that was positioned just below the austenite transition temperature on the steel phase diagram. The sheets of steel are laminated with an impervious polymer, and by removing this polymer only from the desired areas, the locations where nitrogen can diffuse into the material are controlled. The diffusion of nitrogen atoms into this alloy during high-temperature gas nitriding results in transformation into an austenite structure, but only where not protected by polymer. Thus, the sheets of steel are functionally graded with respect to their magnetic properties, i.e., only magnetic in the ferrite regions protected by the lamination. This could lead to improved performance in electrical motor systems where typically holes must be cut in a sheet to allow magnets to be inserted and there must be a trade-off between magnetic and mechanical properties.

7.2.4 Casting Techniques

On a much larger scale, the movement of entire particles of material in a melt can also be used to generate FGM. Metal matrix composites (MMCs) are a specific class of composite and can be manufactured by introducing particles of a second constituent into a melt pool of the first constituent to create a slurry. Particles can also be formed *in situ* during cooling of the melt. Gravity is present everywhere on Earth, and thus, sedimentation and floatation of the particles must be accounted for if a homogeneous distribution of particles is required. However, by exploiting the fact that particles of different size and density move differently in a liquid, the distribution of particles can be modified. By calculating how the different particles will move, the thermal conditions, and thus, solidification velocity can be altered to trap particles where desired (Sobczak and Drenchev, 2013). Sedimentation casting has been employed to grade the acoustic impedance of material in one direction. WC particles were dispersed in a Cu/Mn melt, where sedimentation led to a gradient in the particle concentration from 0% to 60% from top to bottom. By removing a discrete change in acoustic impedance, between low-impedance body tissue and a high-impedance ultrasound transducer, the FGM allows better transmission of sound waves into, and reflections out of, the body.

Centrifugal casting is considered one of the most effective methods for the production of FGM (Sobczak and Drenchev, 2013). Rather than allowing gravity to control the distribution of particles in a melt, a cylindrical mold is rotated to produce centrifugal and Coriolis force. These forces are so strong relative to the force of gravity that gravity can typically be neglected. Mathematical descriptions of the forces acting on the particles and the effect it has on the concentration gradients are available elsewhere (Gao and Wang, 2000; Drenchev et al., 2003). FGM aluminum MMC reinforced by SiC particles are well established where high surface wear resistance is required. For example, a functionally graded A356-SiC_p composite with an initial uniform concentration of 15% reinforcing phase particles in the melt was altered by the centrifugal force such that the concentration varied from ~45% at the edge to near zero away from the mold. As expected, this resulted in a large variation in hardness values recorded across the part, with a hardness of 100 BHN at the edge dropping to below 60 BHN (Rajan et al., 2010). Centrifugal casting can also be used with particles of a lower density than the melt. These particles will move to the center of the mold during rotation while if particles of a higher density are present, they will move to the edge. Thus, functionally graded MMC with two or more types of reinforcing particles can be produced. Further grading can be achieved when the particles are platelet in geometry, where the orientation can be graded with position (Watanabe et al., 2001).

Centrifugal and gravity casting has also been used with different alloys in the same mold. By pouring the two different compositions successively, an FGM can be generated where the first composition makes up the majority of the material at the edge or bottom of the mold for centrifugal or gravity casting, respectively, whereas the remaining volume consists of the second composition (Güntner and Sahm, 1999). The thickness can be controlled by altering the volume of the first material that is allowed to solidify before pouring the second.

Magnetic fields can also be used to grade the composition of a slurry in a mold. The concentration of ferromagnetic particles can be graded by the application of a magnetic field to the slurry. In addition, by the utilization of Lorentz forces, which act differently on different particles, and solidification rate adjustment, it is possible to form FGM with multiple layers of different compositions and gradual interfaces between the layers (Song et al., 2005).

Another alteration of the material properties can be achieved using only a single alloy feedstock while altering the relative density of the solidified part. Casting can be used to generate FGMs of varying densities by pouring the melt into a mold containing preforms. For example, preform spheres made from salt, flour, and water were added to a mold to generate an open pore pattern, before pure molten aluminum was infiltrated with high gas pressure (Zaragoza and Goodall, 2013). Following solidification of the aluminum, the preforms were removed by simply immersing the part in water. The size of the pores can be altered by changing the size of the preforms added to the mold. A somewhat crude method of generating FGM was demonstrated by Zaragoza and Goodall by first casting two separate samples with different pore sizes, then joining them together to form a material with a discrete change in pore size. However, they also showed that a more elegant solution is available where preforms are graded in the mold prior to casting, which allows a single infiltration and solidification step to produce a sample with graded density. An example of such a sample is given in Figure 7.2. The objective of this study was to optimize the heat transfer coefficient of the samples when fluid was passed through them. An interesting result of the experiment was that heat transfer was best when the pore sizes were graded, in particular, when the first pores encountered by the flow were largest. This was attributed to changes in the fluid flow behavior and illustrates how advantages can be gained by functionally grading material even in nonextreme environments. A thorough review of the production of cellular metals and metal foams is available elsewhere (Banhart, 2001).

The preform in the mold can also be retained to alter the properties. Tungsten/copper FGMs can be manufactured by infiltrating liquid copper into a preform of sintered tungsten powder. The material was functionally graded by layering the tungsten powder with different volumes of porosity prior to sintering. Tungsten/copper FGMs have been suggested suitable for use as heat sinks in fusion reactors due to their high thermal conductivity (Itoh et al., 1996). Electrochemical processing and infiltration has also been used to generate an FGM with a tungsten content varying from 80% to 0%, with the balance copper (Jedamzik et al., 2000).

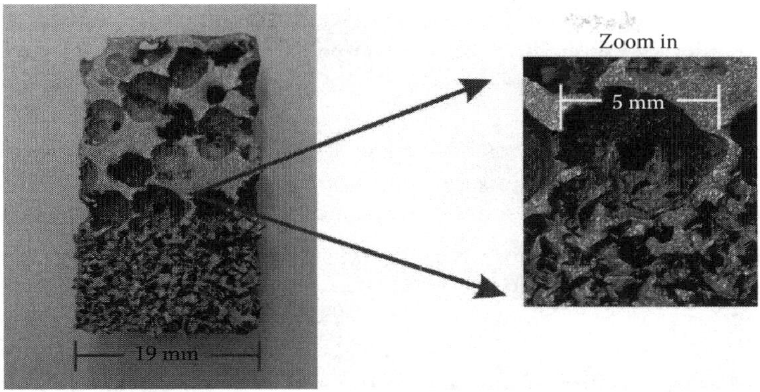

Figure 7.2 Example of an aluminum open pore foam sample produced by infiltration of aluminum melt into a mold with preforms. By using different sizes of preforms, the size of the pores can be graded to optimize the heat transfer properties of the part. (Courtesy of Russell Goodall, University of Sheffield.)

7.2.5 Powder Processing

Powder processing is perhaps the most common way of manufacturing FGM (Sobczak and Drenchev, 2013). Sintering or pressure-assisted hot consolidation can be used to generate a solid part from a powder preform in the desired geometry. Functional grading of the powder can be achieved using a number of methods (Kieback et al., 2003). Gradients can be achieved in porosity and pore sizes, chemical compositions, phase volume fractions, or grain sizes.

Simple sedimentation due to gravity can be used to form a continuous gradient of compositions and porosities within a powder mixture due to differences in particles' size and density. Centrifugal sedimentation can also be used to segregate different powder sizes/densities. To ensure the powder maintains the gradient formed during the sedimentation process, a binder can be used to ensure the green part has sufficient strength.

Die compaction of powder layers is a simple and well-established method to produce stepwise, rather than continuous, alterations in the powder mixture (Kieback et al., 2003). However, it can only be used to manufacture parts of limited size ($<100\,cm^3$) with relatively large layer thickness (≥ 1 mm). Thinner step variations ($<50\,\mu m$) in powder composition can be achieved using wet spraying of powder suspensions, which can be applied to a flat, curved, or rotating surface. By allowing the suspension to dry, multilayers of varying composition can be stacked one on top of the other.

Sintering is commonly used to create a solid part from the powder preforms generated using the methods mentioned earlier. During solid-state sintering of powder, the diffusion length is usually small enough to preserve any concentration gradient present in the powder compact prior to sintering. While the diffusion length can be much larger during liquid-state sintering, it is still possible to generate graded structures by sintering layers of different powder composition fractions (Colin et al., 1993).

A stepped gradient in powder feedstock has also been used to fabricate FGMs via a self-sustained thermal reaction (Zlotnikov et al., 2005). Blended Ti–C–Ni powder was arranged in steps of varying fractions of B_4C particles before a preheated ram was used to compact and ignite the mixture. The exothermic reaction produced enough thermal energy to fully consolidate the stacks of powder with various fractions of the nonreacting B_4C. Unfortunately, the nonreaction of the B_4C means that its addition reduces the exothermicity of the powder, such that the maximum volume fraction is limited. However, this research does highlight an interesting alternative to sintering in order to generate a dense part.

The stacking of sheets can also be used to reduce stepwise gradients in green parts. Powder rolling can be used to generate thin sheets of a given composition. Stamping or laser cutting can be used to achieve the required sheet size before lamination processes can be used to stack and bond the sheets (Zhang et al., 2001). Clearly, using different composition sheets allows the composition to be controlled, with a resolution limited by the thickness of the sheet. Bonding of the sheets has been achieved in a number of ways, including brazing, diffusion bonding, and cold rolling (Mortensen and Suresh, 1995), hot pressing (Kieback et al., 2003), combustion synthesis techniques or self-propagating high-temperature synthesis (Zhang et al., 2001), and spark plasma sintering (Guo et al., 2003).

While an overview of the manufacture of FGM via powder metallurgy has been provided above, the importance of powder-processing techniques in the nonadditive manufacture of FGM means that other sources are available that review the processes in more detail, e.g., Sobczak and Drenchev (2013), Kieback et al. (2003), and Mortensen and Suresh (1995).

7.3 Additive Manufacturing Methods

It has been demonstrated earlier that AM is only a subset of the manufacturing processes available to manufacture FGMs. Furthermore, it is important to understand processes that could be considered forerunners of AM-based FGM manufacture. For example, the coating processes such as plasma spray, electroforming, vapor deposition, and laser cladding, all involve the addition of material to a substrate, even though they are not usually considered as AM processes. Therefore, FGM coating processes will be briefly discussed first, before FGM manufacture using the techniques more generally accepted as AM are reviewed. However, it is important to realize the potential benefits of using AM to manufacture FGM. The very nature of AM, where material is added layer by layer, means that in principle any AM technique could be used to develop an FGM structure similar to those manufactured by the stepwise powder gradients discussed in the previous section (Hofmann et al., 2014a).

7.3.1 Coating Processes

The development of functionally graded coatings has had much research efforts focused on it, e.g., a German National Science Foundation (DFG) program on FGM (Schulz et al., 2003). This is due to many benefits that can be gained by having a gradual, rather than discrete change, from the substrate to the coating material.

Plasma spray is a well-established process used to add a coating to a substrate to improve the wear resistance, corrosion protection, high-temperature protection, or other properties of the substrate (Sampath et al., 1995). During plasma spray forming, powder particles are introduced to a plasma flame and very rapidly heated and accelerated. The particles melt prior to hitting a substrate, which causes them to flatten and allows the creation of low porosity coatings on the target material. It was noted in the 1990s that by changing the feedstock, layers of coating of different composition can be applied sequentially. If two powder supplies are used, then the volume fraction of each constituent can be controlled by altering the feed rate.

Electroforming has been used to generate coatings of both binary metal and metal ceramic types (Mortensen and Suresh, 1995). For example, graded Cu–Ni and Cu–Zn foils have been deposited on a conductive polymer. Depending on the conditions used to deposit the two components, postmanufacture annealing was used to smooth the composition gradient. Ceramic-particle-reinforced metal coatings have also been generated by adding the particles to the electrolytic solution. Two ways to grade the volume fraction of ceramic particles in the coating have been identified. Perhaps, most obvious is the variation in the particle concentration fraction in the electrolytic solution with time. However, testing this method required that the plating solution be replaced for each layer, which slowed the deposition process. Alternatively, when depositing graded layers of nickel and copper, by altering the current density, the volume fraction of particles was found to increase from about 5% to 22%.

Vapor deposition allows very thin layers to be deposited, permitting the achievement of very high-concentration gradients. However, such steep gradients can negate some of the advantage of FGM and can lead to reduced stability at elevated temperatures (Mortensen and Suresh, 1995).

Laser cladding involves adding of a small quantity of cladding material to the surface of a substrate and then using a laser to melt both the added material and a thin layer of the substrate to weld the material to the substrate (Jasim et al., 1993). This process is repeated to build up layers of new cladding material on the substrate (perhaps be considered an AM process). However, when trying to apply FGM coating of SiC particles to a nickel alloy substrate, it has been found that SiC particles are only present in a significant fraction in the final layer. Problems were also encountered with porosity and cracking of the coating. Laser cladding is very similar in principle to the blown powder AM systems that will be discussed next.

7.3.2 Blown Powder AM Systems

As stated earlier, any AM technique has the potential to be used to manufacture FGM. However, some AM processes are fundamentally more favorable for the manufacture of FGM. A key requirement is the ease with which the feedstock can be changed during a build. This makes systems where the feedstock is introduced directly into the melt pool, such as blown powder systems, more easily adaptable for the manufacture of FGM. In the direct laser deposition (DLD) system, the multiple powder feeders, arranged around the YAG laser, make dynamic alteration of the composition easier, with a near infinite number of combinations of materials compositions and gradients possible (Hofmann et al., 2014a). Schematic examples of the possible material composition changes that could be achieved using DLD, as well as some of the non-AM methods described previously, are shown in Figure 7.3. An ideal linear gradient transition from one alloy, or element, to another (A to B) is shown in Figure 7.3a. Of course, the actual gradient following manufacture will be dependent on both the layer thickness and the mechanism with which the composition can be controlled. The discrete change in an alloy, as shown in Figure 7.3b, will only work if there are no brittle phases formed at the interface and no diffusion between the phases, which will be discussed later. Layers of another material can be inserted only at certain regions, as shown in Figure 7.3c, to take advantage of differences in the thermal expansion or magnetism properties. More than two alloys can be used in the gradient, as shown in Figure 7.3d, which could be used to avoid unfavorable phases formed from the combination of alloys B and C. If one of the powder types has a much higher melting temperature than the other, then an MMC can be formed with graduation of the insoluble material in the matrix, as shown in Figure 7.3e. Alternately, MMCs can be created by utilizing chemical desegregation, as shown in Figure 7.3f, which is a technique often used in the production of bulk metallic glasses (Hofmann et al., 2008, 2014a).

An early example of AM being used to alter the material chemistry with position was conducted using the laser engineered net shaping (LENS) process, a variant of the DLD process (Banerjee et al.,

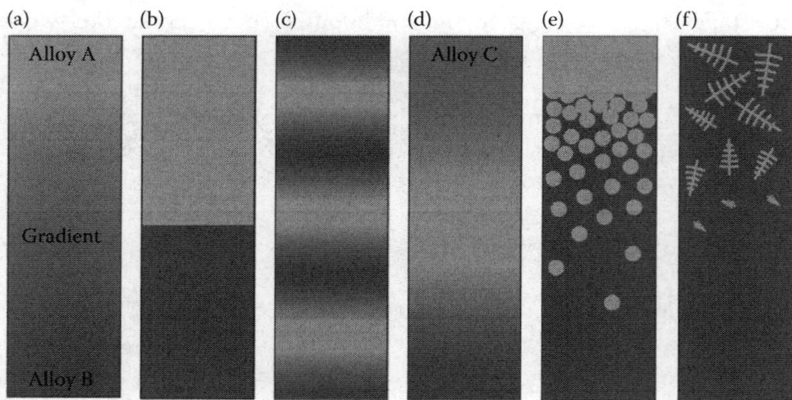

Figure 7.3 Schematics of possible composition gradients possible with DLD. (From Hofmann, D.C. et al., *J. Mater. Res.*, 29(17), 1899–1910, 2014a.)

2003). In this commercially available system, nozzles were used to inject metal powder with an inert carrier gas into a melt pool generated by a high-power laser. By using a double nozzle arrangement with separate powder hoppers for each nozzle, it was possible to control the chemistry by controlling the powder feed flow rate. Banerjee et al. used pure elemental titanium powder in one hopper and elemental titanium and vanadium powders mixed in the ratio of 70% and 30%, respectively, in the other. Previous experiments had been used to determine the optimum powder flow rate; by starting the build with the optimum flow rate composed of purely the titanium powder before gradually increasing proportion of the flow rate of the vanadium powder, it was possible to alter the chemistry of the deposited material during a build. In this example, the smooth one-dimensional grading of material composition from 0 to 25 wt% vanadium was used to systematically examine the effects of the vanadium on the microstructure. While the hardness of the material was measured and found to vary with the vanadium content, it was not the mechanical or thermal properties Banerjee et al. were looking to exploit; instead they were utilizing AM of FGM to gain understanding of phase transformations and concurrent microstructural evolution in $\alpha+\beta$ titanium alloys. Thus, an often-overlooked application of FGM is highlighted to better understand the effect of alloying elements on material microstructures.

A blown powder system was also used to develop a radial gradient in a part manufactured by Ti–6Al–4V powder mixed with pure vanadium powder (Hofmann et al., 2014b). It was noted that the multicomponent phase diagram (in this case Ti–Al–V) should be analyzed to examine the phases that will appear during the transition from one powder to another. It was observed that the gradient from Ti–6Al–4V to V was advantageous because it avoided the formation of brittle, ordered phases. This was in contrast to many of the other gradients that were tested, which cracked during manufacture, e.g., Ti to Invar and Ti to 304 L stainless steel. It was suggested that attention must be paid to the chemical composition of the gradient path to avoid the formation of these brittle phases in the region between the two extreme compositions.

A graphical representation of the solution suggested by Hofman et al. (2014) to this problem is shown in Figure 7.4. The aim is to smoothly grade the material from the alloy defined in (1) to that defined in (2). A simple linear transition, such as shown by the red line in Figure 7.4, on the phase diagram would result in a brittle phase forming and the build failing from cracking. However, by utilizing the knowledge of the three-component phase diagram, it is possible to define an arbitrary path from one alloy to another that avoids the formation of any brittle phases, shown in light

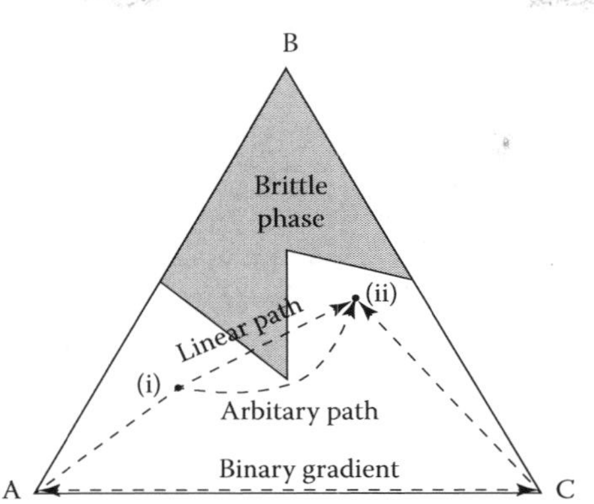

Figure 7.4 Schematic diagram of a hypothetical ternary phase diagram showing possible composition paths from one alloy (i) to the other (ii). Different paths may be chosen with the intention of avoiding brittle phases or simplifying material properties calculations. (From Hofmann, D.C. et al., *J. Mater. Res.*, 29(17), 1899–1910, 2014a.)

blue in Figure 7.4 (Hofmann et al., 2014a). Alternatively, to simplify the modeling of the material response, a binary transition first from (1) to A, then A to B, and finally from B to (2) can be adopted, shown in dark blue in Figure 7.4. If information is available about the crystal phases that form in such multicomponent systems, the gradient path can be defined to optimize the material properties of the gradient region.

Using an arbitrary path between the two desired compositions was shown to be of benefit when using the LENS process to manufacture Ti–6Al–4V structures onto an SS410 stainless-steel substrate (Sahasrabudhe et al., 2014). When the Ti–6Al–4V was deposited directly onto the substrate, brittle intermetallics and residual stresses lead to delamination and build failure within a few layers. However, by first depositing a thin (~750 μm) layer of NiCr, thermal and residual stresses were minimized. By such an approach, bimetallic structures were successfully fabricated.

A good example of a number of phases being generated when transitioning from one material to another was provided by Lin et al. (2006). They used laser rapid forming (LRF), another blown powder technique, to grade a solid structure with rectangular profile from pure titanium to 60% Rene88DT (a nickel-based super alloy). A series of phase evolutions were observed, namely: $\alpha \rightarrow \alpha+\beta \rightarrow \alpha+\beta+Ti2Ni \rightarrow \beta+Ti2Ni \rightarrow Ti2Ni+TiNi \rightarrow TiNi$. Microstructure selection models, based on the maximum interface temperature criterion, were used to construct a microstructural selection map for the Ti–Ni system that was experimentally investigated. It was found that the experimental results agree with the model prediction, and that the measured hardness of the sample could be correlated to the volume and morphologies of the phases present (Lin et al., 2006).

The examples discussed thus far have covered the smooth profile between two and three alloys shown schematically in Figure 7.3a and d, respectively. The LENS process was also used to provide an example of discrete changes in composition, as shown in Figure 7.3b, as well as a gradual transition between a stainless steel (SS316) and a nickel alloy (IN625) (Savitha et al., 2015). However, it was found that even when the powder feedstock was discretely changed from one alloy to another, layers of grading were present at the interface between the two alloys. This was due to the remelting of the previous layers, which is a feature of the LENS process and is used to ensure sufficient bonding between newly and previously deposited layers. This remelting results in some of the material from the previous layer, with the different composition, entering the melt pool. Since only a small fraction of the previous layer is melted, the melt pool, and hence newly deposited layer, is still primarily composed of the new composition. However, this transport of material up through the layers meant that, in the experimental conditions tested, it took three-layer depositions before the stainless steel was completely replaced by the nickel alloy. The transport of material through layers must be recognized when designing FGM parts, and in particular the tolerances with regard to their material gradient.

In addition to change of phases, the mixing of powder during melting can result in change in internal energy (Schwendner et al., 2001). Using the LENS process, Schwendner et al. (2001) mixed elemental titanium powder with elemental niobium and chromium powders to create two alloys, both with a 10 at% ratio. They found that the positive enthalpy of mixing associated with the Ti–10% Nb alloy resulted in a slower rate of solidification, an inhomogeneous alloy, and poor intermixing. In comparison, the negative enthalpy of mixing associated with the Ti–10% Cr alloy resulted in a more rapid rate of solidification, and better intermixing. This work demonstrates another factor that must be accounted for when planning the transition from one chemical composition to another.

7.3.3 Powder-Bed AM Systems

Powder-bed laser systems, known as selective laser melting (SLM) systems, have also been used to vary the chemistry of samples with position (Mumtaz and Hopkinson, 2007). The SLM process utilizes layers of powder spread by a wiper and, as such, layers are limited to a single composition. Therefore, the gradient in composition is restricted to one direction, i.e., the build direction. In addition, it is more difficult to dynamically change the powder composition as it is supplied from a prefilled hoper. However, the deposition of thin layers of powder allows the composition to be varied with good control over a small length scale. It has been shown that it is possible to functionally grade a nickel-based alloy (Wasp alloy) with ceramic zirconia by changing the powder used to spread the layers of 0.4 mm thick, during SLM. By simple modifications of the melt strategy, the relative density of the FGM was kept above 99.6%, while the volume fraction of zirconia was varied from 0% to 10%. However, while the average volume fraction zirconia was smoothly varied in the build direction, the material produced contained regions of higher and lower content of zirconia in the transverse direction. It was suggested that this was due to the different material properties of the two phases and buoyancy and Marangoni effects in the melt pool. The ultimate goal of this work is to provide a method of adding a Thermal Barrier Coating (TBC) to Wasp alloy without the need for a bond coat, but there are clearly problems to be overcome.

As stated previously, changing the powder composition during a build can be difficult when manufacturing samples with SLM. However, it is possible to alter the local microstructure by modifications to the melt strategy alone (Niendorf et al., 2014). When manufacturing samples with a stainless-steel alloy (AISI 316 L), the laser power was significantly increased, from 400 to 1000 W, in certain regions while keeping the overall energy density similar. The regions melted with the high power exhibited a much larger columnar grain size and stronger fiber texture aligned with the build direction. In contrast, the lower power regions were found to contain both smaller grains and a weaker texture. The boundary between these two regions was found to be clearly defined both by examination of electron backscatter diffraction (EBSD) maps and hardness profiles of the material. When samples were tensile tested using digital image correlation (DIC) to map the strain distribution, the lower Young's modulus and yield stress associated with the larger grains resulted in higher strain in the high-power regions. The overall Young's modulus was found to be reasonably well estimated by weighing the volume fraction of each phase (see Equation 7.3). Thus, it has been demonstrated that variations in mechanical properties can be achieved without alteration of the feedstock, allowing FGMs to be fabricated by AM with a single alloy.

The discrete change in microstructure, texture, and mechanical properties observed by Niendorf et al. in the stainless-steel sample described above is contrary to the observation of Savitha et al. (2015) who noted that discrete changes in feedstock composition resulted in a more gradual change in the as-built part composition. Clearly, using a single alloy removes the possibility of alloying elements being transported in the melt and altering the material composition. The temperature variations due to the different melt strategies must also be well contained within the target regions. This indicates that some methods of FGM manufacture will result in a better control of the variation in properties with spatial position. This is also another factor to be considered when designing an FGM part and defining the manufacturing route.

The ability of AM to manufacture "lattice" structures allows the relative density of structures to be altered with position. This allows the mechanical and thermal properties of the FGM to be altered with position. In addition to density, the layout of the lattices can have great effect on the properties of the material. An example of a graded density Ti–6Al–4V sample with diamond lattice morphology is shown in Figure 7.5. The various densities of the lattice layers allow the material properties to be altered while keeping the chemistry uniform. It was found that the yield stress of each layer was close to the yield stress of a uniform lattice of corresponding density (Van Grunsven et al., 2014). The elastic modulus was then found to be adequately described by the rule of mixtures for the isostress condition (see Equation 7.3).

One application of functionally graded lattice structures is in orthopedic implants, where a dense stiff core gradually transitions to a porous layer that allows bone ingrowth without a sudden change in properties. Bone itself is functionally graded, with stiff outer region ($E = 20\,\text{GPa}$) transitioning to a less dense spongy center ($E = 0.5\,\text{GPa}$). Titanium is often thought of as the most biocompatible material, but has both a much higher elastic modulus and density than the bone it replaces (Parthasarathy et al., 2011). This can cause problems due to bone's tendency to react and reconfigure itself to the load applied to it. The titanium implant shields the surrounding bone from mechanical stress, and as a result, the bone is reabsorbed, which can lead to the implant becoming loose and failing prematurely. By altering the density of a cubic lattice with position, Parthasarathy et al. (2011) showed with finite element modeling that it was possible to tailor the mechanical properties of implants to match a particular patient's requirements, which should restore better function and aesthetics than current methods allow.

Other work, which used finite element modeling to predict the mechanical response of FGM femur bone replacement, showed that by adjustment of Young's modulus with position, the amplitude of bone stress shielding could be adjusted (Ataollahi Oshkour et al., 2014). By better understanding the effect of material properties gradients, the FGM could be optimized for this

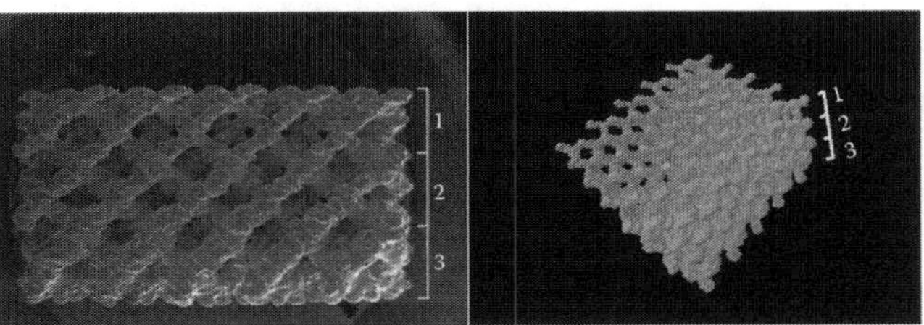

Figure 7.5 Images of graded lattices: (a) secondary electron scanning electron microscope image and (b) 3D reconstruction of a micro CT scan. Numbers 1–3 correspond to the smallest to the largest strut thickness, respectively. (Reprinted from van Grunsven, W. et al., *Metals*, 4(3), 401–409, 2014.)

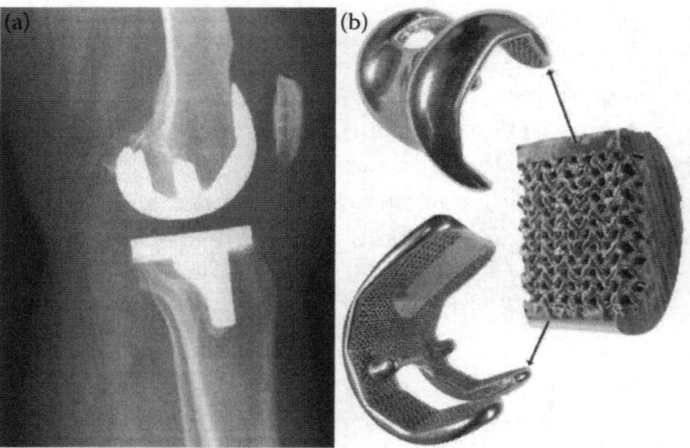

Figure 7.6 Knee implant: (a) x-ray image of total knee replacement and (b) example of a Co–29Cr–6Mo alloy femoral knee implant manufactured by EBM. Inset image shows the porous inner surface (*p*). (Reprinted from Murr, L.E. et al., *Int. J. Biomater.*, Article ID 245727, 14, 2012.)

application, e.g., it was found that a radial change in material properties was more effective than a longitudinal change in protecting the bone from reabsorption. This highlights that, in addition to perfecting the manufacture of advanced FGM by AM, the effect of the material properties gradient on the component *in situ* must be understood.

An example of the use of graded lattices structures in medical devices is provided in Figure 7.6. The electron beam melting (EBM) powder-bed process was used to manufacture a part that, despite a uniform Co–29Cr–6Mo composition, was optimized for the different requirements at each location. The porous inner surface makes bone cell ingrowth more effective, while the stiffness can be tuned to avoid stress shielding and bone decay. Single alloy graded implants are good examples of an FGM product that has been implemented into industrial production. European CE certification was granted in 2007, and thousands of FGM components fabricated by AM have been successfully implanted into human patients since then (Murr et al., 2012).

7.3.4 FGM Inspired by Nature

In addition to engineers trying to mimic natural material properties gradient to allow the production of better medical implants, some of the many natural examples of FGM have provided inspiration for FGM manufacture. The bamboo mentioned in the introduction was found to have the greatest density near the plant surface where the mechanical loading was highest. Palm trees and cancellous bone exhibit the same structure and were used to inspire the additive manufacture of variable density concrete foams. By varying the volume of aluminum powder and lime, which react to produce during hydrogen, the density of the concrete could be altered. Variable density concrete foam was manufactured using an extrusion head mounted on a robot arm. A maximum porosity volume fraction of 40% was achieved, and the compressive strength was found to be well predicted by the ceramic strength model found in the literature. In addition, theoretical calculations, using the experimentally found mechanical properties, showed that to support a given bending stress using an FGM concrete allowed a 9 wt% saving in comparison to a solid cylindrical column of the same dimensions (Oxman et al., 2011). While this is somewhat separated from the metal FGM that is the focus of this chapter, it does highlight how AM lattices, such as that shown in Figure 7.6, could be harnessed for beneficial effects by looking at how nature uses materials so efficiently.

7.4 Thermomechanical Behavior

7.4.1 Benefits of FGM

In the preceding sections, there has been mentioned of the mechanical and thermal benefits that can be derived from FGM. To summarize:

- The strength of bonding between dissimilar materials, such as ceramic and metals, can be greatly improved (Sampath et al., 1995).
- The magnitude of stress concentrations at free edges of materials can be reduced (Giannakopoulos et al., 1995).
- Thermal stresses due to differences in thermal expansion coefficients between different materials can be greatly reduced (Drake et al., 1993; Williamson et al., 1993).
- Plastic yielding, and subsequent component failure, can be delayed by controlling the stress distribution (Drake et al., 1993; Williamson et al., 1993).
- While the stress intensity factor, the driving force for crack growth, can be higher in FGM than homogeneous materials, the crack growth rate through different materials can be reduced by removal of the distinct boundary in material properties (Erdogan, 1995).
- Stress concentrations and the plastic strain around surface notches, critical for fatigue applications, can be reduced (Giannakopoulos and Suresh, 1997; Suresh et al., 1997).

Unfortunately, modeling the behavior of FGM structures is more complex than that of simple single component systems, or even composite materials with anisotropic properties, due to the variation of performance with position. Various ways to predict the material properties have been suggested and are described later.

7.4.2 Predicting the Material Properties of FGM

While the composition and mechanical properties of an FGM can vary with position, at a given location the material can be considered a composite of fixed phase volume fraction. The mechanically response of the FGM at the given location can then be estimated using models normally applied to composite materials. By altering the input used with these models to account for changes in phase composition with position, a simple method of estimating the change in mechanical properties with spatial displacement can be obtained.

Perhaps, the simplest model of the elastic properties of a composite of two materials with varying Young's modulus is provided by the rule of mixtures (Suresh and Mortensen, 1997). Two rules of mixture models are available depending on whether it is assumed that the stress or the strain is equal in the two composite phases. If the loading of the composite material results in an equal strain in both phases, such as when a unidirectional fiber reinforced composition is loaded in the fiber direction, then composite Young's modulus (E_c) can be predicted with the Voigt model, Equation 7.2.

$$E_c = E_1 \cdot f_1 + E_2 \cdot f_2, \tag{7.2}$$

where E and f are the isotropic Young's modulus and the volume fraction of Phases 1 and 2, respectively, and $f_1 + f_2 = 1$. If the Poisson ratios of the two constituent phases differ, then Equation 7.2 will produce a slightly inaccurate estimation of the elastic modulus.

Alternatively, if the strain in the two phases is different but the stress is the same, the composite Young's modulus can be estimated with the Reuss model, Equation 7.3.

$$E_c = \left(\frac{f_1}{E_1} + \frac{f_2}{E_2} \right)^{-1}, \tag{7.3}$$

where the symbols have the same definition as described in Equation 7.2. The Reuss model can be applied when unidirectional reinforced fiber is loaded transverse to the fiber direction.

In addition to the rule of mixtures, other, more complex methods of calculating the material properties have been proposed. Following is only a brief summary of the models available. In many cases, more details regarding the models are available in the references given. In a recent review of functionally graded plates, Jha et al. (2013) listed the models available and are given in the following sections.

7.4.2.1 Self-Consistent Scheme

The macroscopic elastic modulus of a two-phase composite can be estimated by the "self-consistent scheme" mechanics model proposed by Hill (1965). This method allows the properties of a composite to be estimated, for any concentration of two phases with either isotropic or anisotropic properties. However, the composite must be assumed to take the form of a matrix with ellipsoidal inclusions. The result is independent of which phase is used as the matrix phase (Jha et al., 2013). Thus, it is particularly suitable for determining the elastic properties of interconnected skeletal microstructures. However, when the method was applied to conductivity problems, it was found to be less useful (Hashin, 1968).

7.4.2.2 Mori–Tanaka Scheme

This scheme is applicable to two-phase composites with a discontinuous second phase (Mori and Tanaka, 1973; Benveniste, 1987). It can be used with an anisotropic matrix phase and aligned or randomly orientated small second phase particles.

7.4.2.3 Composite Sphere Assemblage Model

In this model, it is assumed that the composite is filled with a fractal assembly of spheres embedded in a matrix. Isotropic properties of composite materials can then be obtained analytically (Hashin, 1962; Hashin and Shtrikman, 1963).

7.4.2.4 Composite Cylindrical Assemblage Model

This model is applicable to fiber-reinforced composite and requires that both phases are isotropic (Hashin and Rosen, 1962; Hashin, 1979). In planes perpendicular to the fiber direction, microstructural representative volume elements must be transversely isotropic.

7.4.2.5 The Simplified Strength of Materials Method

This method can be used for a matrix phase reinforced with, and bonded to, fibers (Jha et al., 2013). It assumes that the reinforcement phase takes the form of a periodic array of square fibers, and can be used to estimate the orthotropic strength of fiber-reinforced composites from the properties of the two phases and the volume fraction. Its relative ease of use and computation efficiency makes it one of the most popular modeling methods.

7.4.2.6 The Method of Cells

It is similar to the simplified strength of materials method described earlier, but uses a larger representative volume element to estimate the material properties. As a result, this method is more computationally rigorous (Jha et al., 2013).

7.4.2.7 Micromechanical Models

Finite element simulations of representative volume elements can be used for either isotropic or orthotropic composite materials. Different volume fractions can be modeled to predict the mechanical or thermal response of the composites. Thus, curves can be fitted to the result of the FE models to estimate the response of an arbitrary volume fraction. This is considered the most accurate model since the microstructure is directly modeled with finite elements (Tarlochan, 2013), but one drawback of this method is that multiple models must be constructed in order to determine material properties for various volume fractions, which may require significant computation (Reiter et al., 1997).

7.4.3 Mathematically Idealized FGM

To allow the material performance of FGM to be modeled, it is useful to consider their material properties as smoothly graded with position in a structure. Without such assumptions, calculating the material response of FGM would become inefficient. In one-dimensionally graded materials, such as in plates, there are two commonly used estimations of the material properties. When engineers are concerning with modeling the fracture mechanics of FGM, the most common idealization is the exponential law (Jha et al., 2013), Equation 7.4.

$$P(z) = P_t \cdot \exp\left(-\lambda \cdot \left[1 - \frac{2 \cdot z}{h}\right]\right), \tag{7.4}$$

where $\lambda = 1/2 \cdot \ln(P_t / P_b)$.

Here, the local material property, $P(z)$, is a function of the position within the FGM, z; the thickness of the FGM, h; and the material property at the top (P_t at $z = h/2$) and bottom (P_b at $z = -h/2$). The material property estimated using Equation 7.4 could refer to the Young's modulus, shear modulus, Poisson's ratio, density, or other properties.

When conducting stress analysis, it is more common to apply the power law stated in Equation 7.5, to predict the behavior of FGMs.

$$P(z) = (P_t - P_b) \cdot \left(\frac{z}{h} + \frac{1}{2}\right)^k + P_b. \tag{7.5}$$

The same notation as Equation 7.4 is applied and the new term, k, is a material grading index. Among other applications, the power law relationship has been recently used successfully to model the large deflection of simply supported FGM thick plates (Khabbaz et al., 2009), the magnetic properties of FGM (Wu et al., 2010), and the natural vibrational frequency of FGM structures (Jha et al., 2012).

7.5 Challenges Facing FGM

As can be inferred from the examples given earlier, many FGM produced by AM are in fact not functional, rather they are small samples used to demonstrate the possible future capability of AM. From the literature mentioned in the previous sections, it is clear that FGM manufacture by AM is not yet a mature technology. Some of the challenges that are clear to the authors and others (e.g., Muller et al., 2014) are listed as follows:

1. Prominent among the challenges is the determination of the optimal spatial distribution of phases. This will require understanding of how the microstructure, phase composition, and material properties will vary across the build (see challenge 2). Furthermore, models of the FGM structure response to external loading (mechanical, thermal, etc.) will be required to include the effect of any variation in material properties to determine the optimum distribution of the constituent phases. The none discrete nature of the material properties presents new challenges for determining the optimum component geometry as well as composition distribution. For example, deciding the optimum geometry of an I-beam where the material has homogeneous mechanical strength is much easier than when the material strength can be varied smoothly between two different values. Moreover, designing structures to take advantages of FGM requires new CAD methods.
2. Predicting the material properties of the manufactured part, and in particular, how these vary with spatial position, is more challenging than with standard single component or composite structures. The relationship between the material composition and properties must be known in order to determine the best distribution of material. This will require a detailed understanding of not only how changes to the material composition affect the properties but also how the different phases respond to changes in the composition. Although the intention would always be to avoid such situations, weak bonding between different phases may result in a less strong material than predictions.
3. Selection of the materials to be used in the FGM structure is undoubtedly more complex in comparison to manufacturing with a single alloy. In addition to deciding the ideal material for the conditions experienced by the component during service, the two (or more) material must be compatible with each other and lead to a smooth transition from one to the other. Within the transition from one intended phase to another phase, it is possible a tertiary phase, or perhaps even more phases, to form which may be nonoptimum for the material requirements. The brittle phases formed during AM, as mentioned previously, that lead to the failure of the build from cracking is one such example. However, even if the build was to succeed, the extra phases may reduce the material properties to below the required levels. Thus, there must be significant confidence in the material selection procedure and the phases that will be generated within the FGM.
4. The theoretical and actual material composition and gradient have often been found to differ. All the manufacturing methods must be better understood to know why these variations

occur. Possible reasons for the incorrect distributions of the constituent materials include macrosegregation of solutes during solidification or simply poor process control. A good solution (or at least a partial solution on the way to a full solution) to this problem would be the *in situ* monitoring of the phases and chemistry during the build. This would allow both a deeper understanding of the manufacturing method and a method to identify components with potential problems prior to the completion of manufacturing and perhaps allow correction of the issue during the manufacture.

5. It seems likely that, at least in the short to medium term, the material properties will not be as predictable and consistent as those of traditional engineering alloys. If poor control of the material gradient is intrinsic in the manufacture of FGM by AM, it must be decided by how much the material composition can differ before a part is judged to be deficient. It may be necessary to redefine what is meant by engineering material to include a much greater allowable tolerance in the permissible material properties with spatial position. Of course, the reduction in efficiency (e.g., component mass) due to the partially stochastic nature of FGM must be offset by the advantages gained by the use of an FGM, rather than a monolithic material, for there to be any reason to use FGM. Hence, in some situations where the theoretical analysis alone suggests FGM would be beneficial, they may be found to be impractical once the allowances have been made for the possible errors and tolerances during manufacture.

7.6 Summary

FGMs are materials where the phases present are graded with spatial position within the structure. This allows the material properties to be smoothly graded to allow better adaptation of the material to its intended environment. Modern FGMs, which often consist of metal and ceramic phases, are considered to have been first developed in the 1980s in Japan. However, if biological structures are examined, then it can be seen that FGM have existed in nature for millions of years. Furthermore, humans have been taking advantage of subtle changes in composition for hundreds of years prior to the Japanese metal ceramic FGM. Although AM is only one of the possible techniques that can be utilized to manufacture FGM, it offers new ways to manufacture these novel materials. In addition to the direct exploitation of the graded material properties, e.g., for thermal or corrosion protection, or tailored mechanical properties, FGMs can be used to gain more insight into the effect of alloying elements on microstructures by allowing a range of compositions to be quickly manufactured and studied.

Despite the multiple advantages offered by FGM, there are still many challenges to be overcome. Principal among these is the determination of the optimum spatial distribution of the phases. The manufacture of FGM will also require a new generation of CAD software which allows heterogeneous material definitions. Predicting the material properties of the intended material gradient is difficult, but can be achieved with various modeling techniques mentioned in this chapter. However, it must be acknowledged that the theoretical and actual manufactured material distribution and gradient are likely to be quite different, and therefore, the material properties may be less predictable than when using traditional monolithic alloys. Until the material gradient can be controlled more precisely, e.g., with *in situ* monitoring, the tolerances within the material properties must be expanded to allow for the somewhat stochastic nature of FGM.

7.7 Questions

1. What are some of the benefits of the thermomechanical properties of FGM in comparison to single material structures?
2. Are FGMs only found in engineering components?
3. For how long are humans known to have been exploiting the benefits of FGM?
4. What are the major challenges facing FGM produced by AM?
5. List eight ways to predict the material properties of composite materials.
6. Which mathematically idealized models of material properties would engineers commonly use to model the fracture mechanics of FGM?
7. Which mathematically idealized models of material properties would engineers commonly use for stress analysis of FGM?
8. Of blown powder and powder-bed AM systems, which is the easiest to alter the material composition with position?
9. What is the simplest way to model the Young's modulus of a discretely graded lattice structure subject to loading?
10. Why are lattice structures a good use of FGM for medical implants?

References

Ataollahi Oshkour, A. et al., 2014. Comparison of various functionally graded femoral prostheses by finite element analysis. *The Scientific World Journal*, 2014, Article ID 807621, 17 p.

Banerjee, R. et al., 2003. Microstructural evolution in laser deposited compositionally graded α/β titanium-vanadium alloys. *Acta Materialia*, 51(11), pp. 3277–3292.

Banhart, J., 2001. Manufacture, characterisation and application of cellular metals and metal foams. *Progress in Materials Science*, 46(6), pp. 559–632.

Benveniste, Y., 1987. A new approach to the application of Mori-Tanaka's theory in composite materials. *Mechanics of Materials*, 6(2), pp. 147–157.

Bruet, B.J.F. et al., 2008. Materials design principles of ancient fish armour. *Nature Materials*, 7(9), pp. 748–756.

Colin, C. et al., 1993. Processing of functional-gradient WC-Co cermets by powder metallurgy. *International Journal of Refractory Metals and Hard Materials*, 12(3), pp. 145–152.

Dial, L.C. et al., 2015. Dual phase magnetic material component and method of forming, Patent US20150115749.

Drake, J.T. et al., 1993. Finite element analysis of thermal residual stresses at graded ceramic–metal interfaces. Part II. Interface optimization for residual stress reduction. *Journal of Applied Physics*, 74(2), p. 1321.

Drenchev, L. et al., 2003. Numerical simulation of macrostructure formation in centrifugal casting of particle reinforced metal matrix composites. Part 1: Model description. *Modelling and Simulation in Materials Science and Engineering*, 11(4), p. 635.

Erdogan, F., 1995. Fracture mechanics of functionally graded materials. *Composites Engineering*, 5(7), pp. 753–770.

Gao, J.W. and Wang, C.Y., 2000. Modeling the solidification of functionally graded materials by centrifugal casting. *Materials Science and Engineering A*, 292(2), pp. 207–215.

Giannakopoulos, A.E. et al., 1995. Elastoplastic analysis of thermal cycling: Layered materials with compositional gradients. *Acta Metallurgica et Materialia*, 43(4), pp. 1335–1354.

Giannakopoulos, A.E. and Suresh, S., 1997. Indentation of solids with gradients in elastic properties: Part II. Axisymmetric indentors. *International Journal of Solids and Structures*, 34(19), pp. 2393–2428.

Güntner, A. and Sahm, P.R., 1999. Graded metal matrix composites produced by a multi-pouring method with controlled mold filling. *Materials Science Forum*, 308–311, pp. 187–192.

Guo, H. et al., 2003. Laminated and functionally graded hydroxyapatite/yttria stabilized tetragonal zirconia composites fabricated by spark plasma sintering. *Biomaterials*, 24(4), pp. 667–675.

Hashin, Z., 1962. The elastic moduli of heterogeneous materials. *Journal of Applied Mechanics*, 29(1), pp. 143–150.

Hashin, Z., 1968. Assessment of the self consistent scheme approximation: Conductivity of particulate composites. *Journal of Composite Materials*, 2(3), pp. 284–300.

Hashin, Z., 1979. Analysis of properties of fiber composites with anisotropic constituents. *Journal of Applied Mechanics*, 46, pp. 543–550.

Hashin, Z. and Rosen, B.W., 1962. The elastic moduli of heterogeneous materials. *Journal of Applied Mechanics*, 29(63), pp. 143–150.

Hashin, Z. and Shtrikman, S., 1963. A variational approach to the theory of the elastic behaviour of multiphase materials. *Journal of the Mechanics and Physics of Solids*, 11(2), pp. 127–140.

Hill, R., 1965. A self-consistent mechanics of composite materials. *Journal of the Mechanics and Physics of Solids*, 13, pp. 213–222.

Hofmann, D.C. et al., 2008. Designing metallic glass matrix composites with high toughness and tensile ductility. *Nature*, 451(7182), pp. 1085–1089.

Hofmann, D.C. et al., 2014a. Compositionally graded metals: A new frontier of additive manufacturing. *Journal of Materials Research*, 29(17), pp. 1899–1910.

Hofmann, D.C. et al., 2014b. Developing gradient metal alloys through radial deposition additive manufacturing. *Scientific Reports*, 4, p. 5357.

Inoue, T., 2010. Tatara and the Japanese sword: The science and technology. *Acta Mechanica*, 214(1–2), pp. 17–30.

Itoh, Y. et al., 1996. Design of tungsten/copper graded composite for high heat flux components. *Fusion Engineering and Design*, 31(4), pp. 279–289.

Jasim, K.M. et al., 1993. Metal–ceramic functionally gradient material produced by laser processing. *Journal of Materials Science*, 28(10), pp. 2820–2826.

Jedamzik, R. et al., 2000. Functionally graded materials by electrochemical processing and infiltration : Application to tungsten/copper composites. *Journal of Materials Science*, 35(2), pp. 477–486.

Jha, D.K. et al., 2012. Higher order shear and normal deformation theory for natural frequency of functionally graded rectangular plates. *Nuclear Engineering and Design*, 250, pp. 8–13.

Jha, D.K. et al., 2013. A critical review of recent research on functionally graded plates. *Composite Structures*, 96, pp. 833–849.

Khabbaz, R.S. et al., 2009. Nonlinear analysis of FGM plates under pressure loads using the higher-order shear deformation theories. *Composite Structures*, 89(3), pp. 333–344.

Kieback, B. et al., 2003. Processing techniques for functionally graded materials. *Materials Science and Engineering A*, 362(1–2), pp. 81–105.

Lin, X. et al., 2006. Microstructure and phase evolution in laser rapid forming of a functionally graded Ti–Rene88DT alloy. *Acta Materialia*, 54(7), pp. 1901–1915.

Mahamood, R.M. et al., 2012. Functionally graded material : An overview. *World Congress on Engineering*, III, pp. 2–6.

Mori, T. and Tanaka, K., 1973. Average stress in matrix and average elastic energy of materials with misfitting inclusions. *Acta Metallurgica*, 21(5), pp. 571–574.

Mortensen, A. and Suresh, S., 1995. Functionally graded metals and metal–ceramic composites: Part 1. Processing. *International Materials Reviews*, 40(6), pp. 239–265.

Muller, P. et al., 2014. Toolpaths for additive manufacturing of functionally graded materials (FGM) parts. *Rapid Prototyping Journal*, 20(6), pp. 511–522.

Mumtaz, K.A. and Hopkinson, N., 2007. Laser melting functionally graded composition of Waspaloy* and Zirconia powders. *Journal of Materials Science*, 42(18), pp. 7647–7656.

Murr, L.E. et al., 2012. Next generation orthopaedic implants by additive manufacturing using electron beam melting. *International Journal of Biomaterials*, Article ID 245727, 14 p.

Niendorf, T. et al., 2014. Functionally graded alloys obtained by additive manufacturing. *Advanced Engineering Materials*, 16(7), pp. 857–861.

Nogata, F. and Takahashi, H., 1995. Intelligent functionally graded material: Bamboo. *Composites Engineering*, 5(7), pp. 743–751.

Oxman, N. et al., 2011. Functionally graded rapid prototyping. *Innovative Developments in Virtual and Physical Prototyping*, pp. 483–490.

Parthasarathy, J. et al., 2011. A design for the additive manufacture of functionally graded porous structures with tailored mechanical properties for biomedical applications. *Journal of Manufacturing Processes*, 13(2), pp. 160–170.

Rajan, T.P.D. et al., 2010. Characterization of centrifugal cast functionally graded aluminum–silicon carbide metal matrix composites. *Materials Characterization*, 61(10), pp. 923–928.

Reibold, M. et al., 2006. Materials: Carbon nanotubes in an ancient Damascus sabre. *Nature*, 444(7117), p. 286.

Reiter, T. et al, 1997. Micromechanical models for graded composite materials. *Journal of the Mechanics and Physics of Solids*, 45(8), pp. 1281–1302.

Sahasrabudhe, H. et al., 2014. Stainless steel to titanium bimetallic structure using LENS™. *Additive Manufacturing*, 5, pp. 1–8.

Sampath, S. et al., 1995. Thermal spray processing of FGMs. *MRS Bulletin*, 20(1), pp. 27–31.

Savitha, U. et al., 2015. Chemical analysis, structure and mechanical properties of discrete and compositionally graded SS316–IN625 dual materials. *Materials Science and Engineering: A*, 647, pp. 344–352.

Schulz, U. et al., 2003. Graded coatings for thermal, wear and corrosion barriers. *Materials Science and Engineering: A*, 362(1–2), pp. 61–80.

Schwendner, K.I. et al., 2001. Direct laser deposition of alloys from elemental powder blends. *Scripta Materialia*, 45(10), pp. 1123–1129.

Sobczak, J.J. and Drenchev, L., 2013. Metallic functionally graded materials: A specific class of advanced composites. *Journal of Materials Science and Technology*, 29(4), pp. 297–316.

Song, C. et al., 2005. In situ multi-layer functionally graded materials by electromagnetic separation method. *Materials Science and Engineering: A*, 393(1–2), pp. 164–169.

Suresh, S. and Mortensen, A., 1997. Functionally graded metals and metal-ceramic composites: Part 2. Thermomechanical behaviour. *International Materials Reviews*, 42(3), pp. 85–116.

Suresh, S. et al., 1997. Spherical indentation of compositionally graded materials: Theory and experiments. *Acta Materialia*, 45(4), pp. 1307–1321.

Tarlochan, F., 2013. Functionally graded material: A new breed of engineered material. *Journal of Applied Mechanical Engineering*, 02(02), p. e115.

van Grunsven, W. et al., 2014. Fabrication and mechanical characterisation of titanium lattices with graded porosity. *Metals*, 4(3), pp. 401–409.

Wadsworth, J., 2015. Archeometallurgy related to swords. *Materials Characterization*, 99, pp. 1–7.

Watanabe, Y. et al., 2001. Evaluation of three-dimensional orientation of Al3Ti platelet in Al-based functionally graded materials fabricated by a centrifugal casting technique. *Acta Materialia*, 49(5), pp. 775–783.

Williamson, R.L. et al., 1993. Finite element analysis of thermal residual stresses at graded ceramic–metal interfaces. Part I. Model description and geometrical effects. *Journal of Applied Physics*, 74(2), p. 1310.

Wu, C.-P. et al., 2010. Three-dimensional static behavior of functionally graded magneto-electro-elastic plates using the modified Pagano method. *Mechanics Research Communications*, 37(1), pp. 54–60.

Zaragoza, G. and Goodall, R., 2013. Metal foams with graded pore size for heat transfer applications. *Advanced Engineering Materials*, 15(3), pp. 123–128.

Zhang, Y. et al., 2001. Rapid prototyping and combustion synthesis of TiC/Ni functionally gradient materials. *Materials Science and Engineering: A*, 299(1–2), pp. 218–224.

Zlotnikov, I. et al., 2005. Combustion synthesis of dense functionally graded B4C reinforced composites. *Materials Science Forum*, 492–493, pp. 685–692.

Chapter 8

Applications of Laser-Based Additive Manufacturing

Bashir Khoda
North Dakota State University

Taniya Benny
Binghamton University (SUNY)

Prahalad K. Rao and Michael P. Sealy
University of Nebraska–Lincoln

Chi Zhou
University at Buffalo

Contents

8.1 Introduction

This chapter identifies biomedical, aerospace, and automotive applications of laser-based additive manufacturing (LBAM). Examples that express key advantages of LBAM over traditional manufacturing methods are presented. Applications that improve biocompatibility, efficiency, safety, ergonomics, and performance are discussed. As new applications of LBAM are discovered, LBAM will play an ever-increasing role in the production of high-value goods.

8.2 Biomedical Applications

Additive manufacturing (AM), also known as three-dimensional (3D) printing, is one of the preferred fabrication methods for biomedical applications due to its high degree of fidelity and anatomical accuracy. Biomedical applications were categorized into four major areas: (1) tissue scaffolds, (2) cell printing, (3) prosthetics, and (4) implants. Various 3D printing techniques such as extrusion-based systems [1–3], selective laser sintering (SLS) [4], powder-bed printing, vat polymerization, material jetting, and binder jetting technique have been used to manufacture various biomedical products. Depending upon the characteristics of a product and its application, one 3D printing technique will be more advantageous than the other in regard to material use (Figure 8.1) [5] and performance (Table 8.1) [6]. This section highlights biomedical applications using AM. Processes include Stereolithography Apparatus (SLA), SLS, and selective laser melting (SLM).

8.2.1 Tissue Scaffolds

One of the primary tissue engineering strategies to reconstruct or repair damaged tissue is porous structure scaffolds. Scaffolds restore mechanical, biological, and chemical functions by stimulating the mimicry of tissue regeneration processes that will result in *ex vivo* tissues or organ-on-a-chip. The

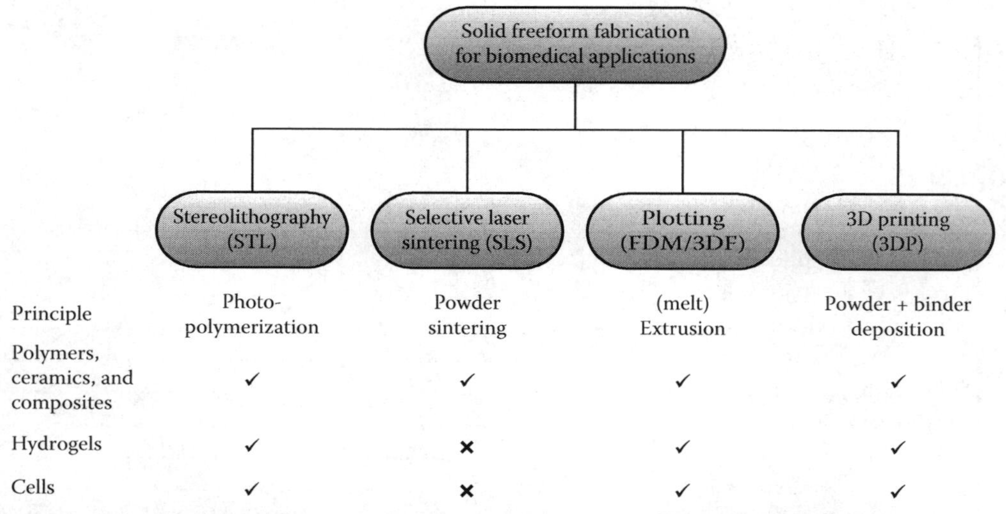

Figure 8.1 Overview of additive processing technologies applied to biomedical applications. (From Melchels, F.P.W. et al., *Biomaterials*, 31, 6121–6130, 2010.)

Table 8.1 Capabilities of Bioprinting Technologies

	Throughput	*Droplet Size*	*Spatial Resolution*	*Single-Cell Control*
Valve-based bioprinting	Medium	100 μm–1 mm	Medium	Medium
Laser-induced bioprinting (LIFT, BioLP, MAPLE)	Medium	>20 μm	Medium	Medium
Laser-guided bioprinting	Low	>10 μm	High	High
Inkjet bioprinting	High	50–300 μm	Medium	Low
Acoustic bioprinting	High	10–500 μm	Medium	High

Source: Tasoglu, S., Demirci, U., *Trends Biotechnol.*, 31, 10–19, 2013.

3D scaffold structures serve as a temporary structural support for isolated cells to grow, provide nutrient to new tissues, facilitate the healing process, restore tissue function, and minimize the wound scar. Far from being a passive component, both the porous architecture and the scaffold material play a significant role in guided tissue regeneration by preserving tissue volume, providing temporary mechanical function, and delivering bioactive agents [7]. The scaffold structure should meet certain criteria to serve specific but often conflicting functions, such as facilitating the migration of cells, providing necessary support (structural integrity) for the cell to proliferate and differentiate, removing the waste, stimulating vascularization, and controlled temporal degradation or integration. The success of tissue regeneration mainly depends on scaffold fabrication technology for generating reliable, fully integrated, complex, three-dimensional, and controlled porous microstructures [8,9].

The fundamental agility of bioadditive processes comes from the layer-based building technique of the digitized model [10]. One of the early systematic approaches for porous scaffold design was proposed and defined as computer-aided tissue engineering (CATE) [11]. The proposed computational algorithm was based upon the unit cell architecture design that combines computer-aided design (CAD) and finite element analysis (FEA), as shown in Figures 8.2 and 8.3. The proposed CATE approach allowed for tailoring and evaluating mechanical properties [12], porosity, permeability, and anatomical shape (biomimetic) of porous tissue scaffolds.

Controllable variation in porosity with pore continuity and connectivity was also proposed for engineered tissue scaffold design [13]. The effectiveness of their design was ensured by the proposed path and process planning of the complex porous internal architecture, as shown in Figure 8.4. Another approach for smooth tissue scaffold structure representation was proposed by fitting the triply periodic minimal surface (TPMS) unit cell within the complex anatomical boundary following the distance field methodology (see Figure 8.5) [14].

A multiscale optimization technique was used to design the scaffold's structure for enhanced tissue regeneration [15]. The elastic properties of the scaffold were computed for different size scale using a homogenization technique. The optimized scaffold structure was fabricated with an EOS P100 Formiga laser sintering system using polycaprolactone (PCL) and 4%–6% hydroxyapatite, as shown in Figure 8.6.

The fabricated structures were analyzed for mechanical properties (Young's modulus) using a universal tensile testing machine. The results demonstrated discrepancies between experimental data and numerical simulations. The authors anticipated the discrepancies could be the result of the highly anisotropic layer-based additive processes, which emphasizes the research gap in design for AM processes.

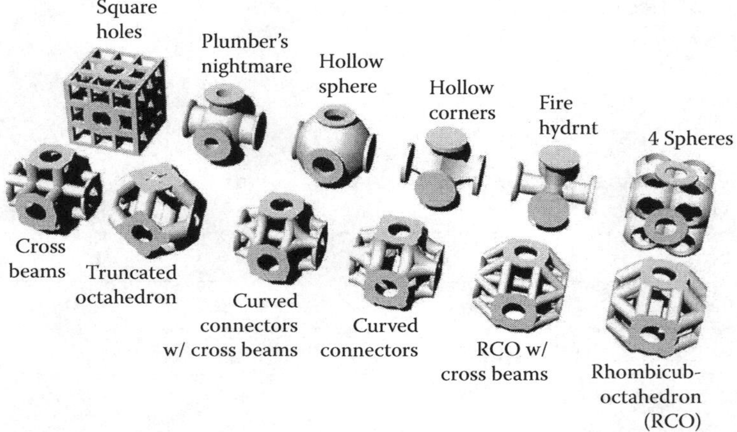

Figure 8.2 Unit block with 20% material volume. (From Wettergreen, M.A. et al., *Comput. Aided Des.*, 37, 1141–1149, 2005.)

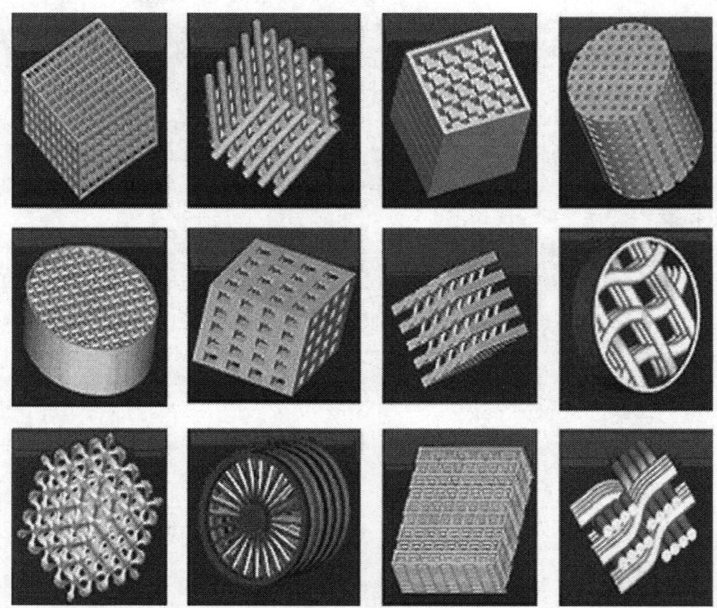

Figure 8.3 Library of designed cellular 3D scaffolds. (From Fang, Z. et al., *Comput. Aided Des.*, 37, 65–72, 2005.)

A parametric design of a scaffold structure was proposed to achieve the desired mechanical properties by considering the strut geometry and the material concentration [16]. The porous microarchitecture of the scaffold was fabricated using a custom-built projection stereolithography (PSL) system. Collagen-based gelatin methacrylate (GelMA) was used as the biomaterial, which was cured under a UV light source from a digital light processing (DLP) chip. Endothelial cells

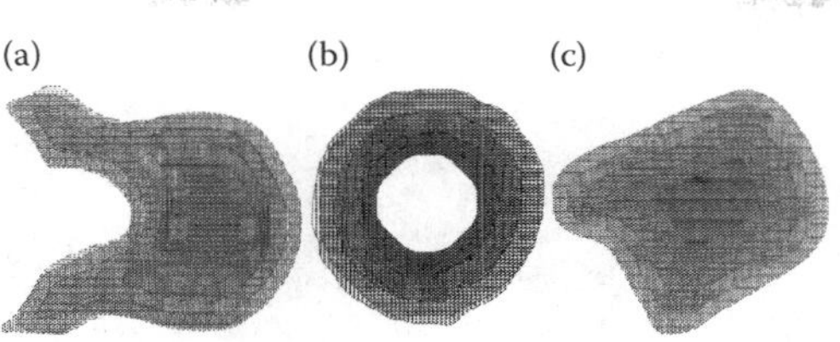

Figure 8.4 Controllable variational porosity in scaffold architecture: (a) vertebra, (b) aorta, and (c) femur. (From Khoda, A. et al., *J. Biomech. Eng.*, 133, 011001, 2011.)

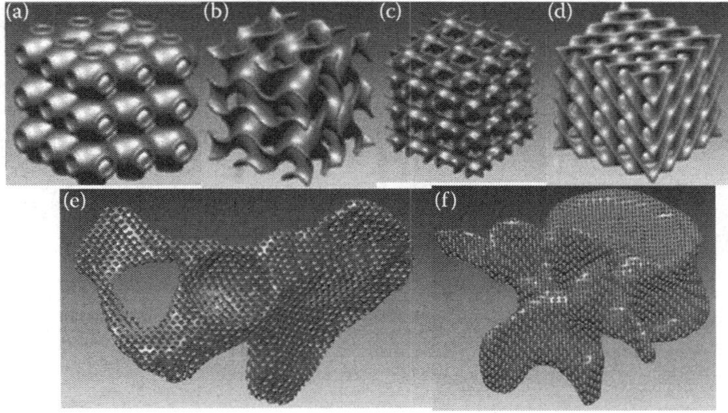

Figure 8.5 TPMS-based unit cell libraries: (a) P surface, (b) D surface, (c) G surface, and (d) I-WP surface; D-surface internal architecture for (e) iliac and (f) spine bone scaffolds. (From Yoo, D.J., *Biomaterials*, 32, 7741–7754, 2011.)

Figure 8.6 Multiscale unit cell topology optimization process fabricated by SLS. (From Coelho, P.G. et al., *Med. Eng. Phys.*, 37, 287–296, 2015.)

were seeded on the fabricated scaffold, which enabled the cells to attach and proliferate on its surface, as shown in Figure 8.7.

A bioinspired biphasic osteochondral scaffold was manufactured with a hybrid ceramic stereolithography (CSL) and gelcasting process (see Figure 8.8) [17]. The engineered scaffolds were composed of beta-tricalcium phosphate (β-TCP) and type-I collagen for osteochondral tissue engineering (TE). The scaffolds fabricated in this study were porous. Pore sizes ranged from 700 to 900 μm and

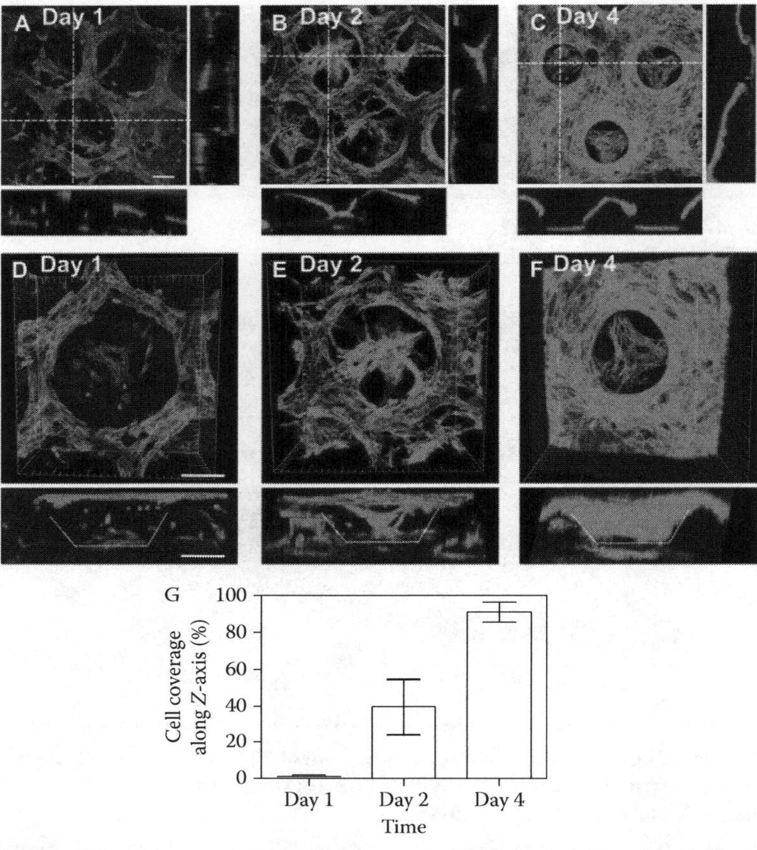

Figure 8.7 3D confocal images showing scaffold coverage by the HUVEC-GFP cells (a through c) and cell penetration into the porous structure (d through f) as a function of time (days) [16]. Cell coverage of 91% ± 5% along *Z*-axis is also reported (g). Scale bar corresponds to 100 μm.

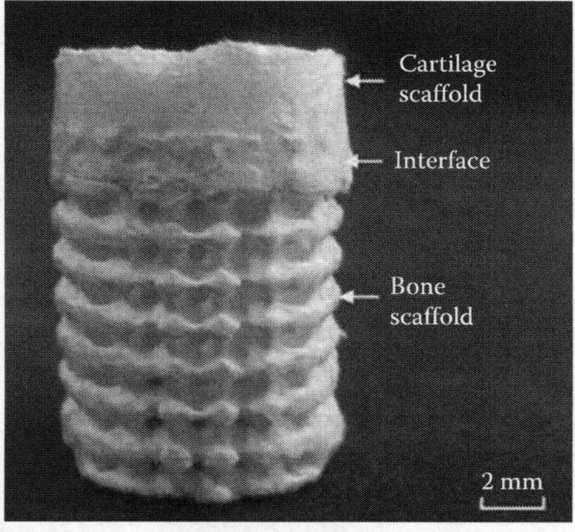

Figure 8.8 A β-TCP-collagen osteochondral scaffold produced by a hybrid ceramic stereolithography and gelcasting process. (From Bian, W. et al., *Rapid Prototyping J.*, 18, 68–80, 2012.)

interconnected pores ranged from 200 to 500 μm. An *in vitro* experiment showed cell attachment with many synapses growing into the micropores and stabilized tightly on the surface. In fact, 60% surface coverage was achieved by Day 7, and a cell colony was formed.

Unlike the previous research, multimaterial scaffolds were designed and fabricated with a He–Cd laser (325 nm wavelength)-based SLA machine [18]. Two biocompatible *polyethylene glycol* (PEG)-based photopolymers were used as the primary materials to prepare the photoreactive solutions in their modified equipment: (1) PEG dimethacrylate PEG-dma MW 1 K and (2) PEG diacrylate PEG-da MW 3.4 K. Cell seeding experiments revealed specific localization of cells in the regions patterned with bioactive PEG containing the cell adhesion ligand RGDS, as shown in Figure 8.9, in both polymers.

Three-dimensional nanocomposite scaffolds based on calcium phosphate (Ca-P)/poly(hydroxybutyrate–co-hydroxyvalerate) (PHBV) and carbonated hydroxyapatite (CHAp)/poly(L-lactic acid) (PLLA) nanocomposite microspheres were successfully fabricated using SLS (see Figure 8.10) [19]. They performed *in vitro* biological evaluation with a human osteoblast-like cell line (SaOS-2), which showed high cell viability and normal morphology and phenotype after 3- and 7-day culture on printed scaffolds.

8.2.2 Cell Printing

Incorporation of living cells during additive processing adds another advantage over other scaffold fabrication approaches. Laser-based bioprinting (LBB) is primarily performed with two different mechanisms: laser-guided direct writing (LGDW) (weakly focused beam) and modified laser-induced forward transfer (LIFT) (weakly focused laser pulses) to deposit with or without cell laden bioink. In LGDW, the laser beam is weakly focused into a liquid suspension of cells, and the photons from a laser beam trap and guide cells by exploiting the differences in refractive indexes of cells and cell media, as shown in Figure 8.11a. The force of the light moves the cells from the suspension onto a translating receiving substrate making the computer-controlled desired pattern.

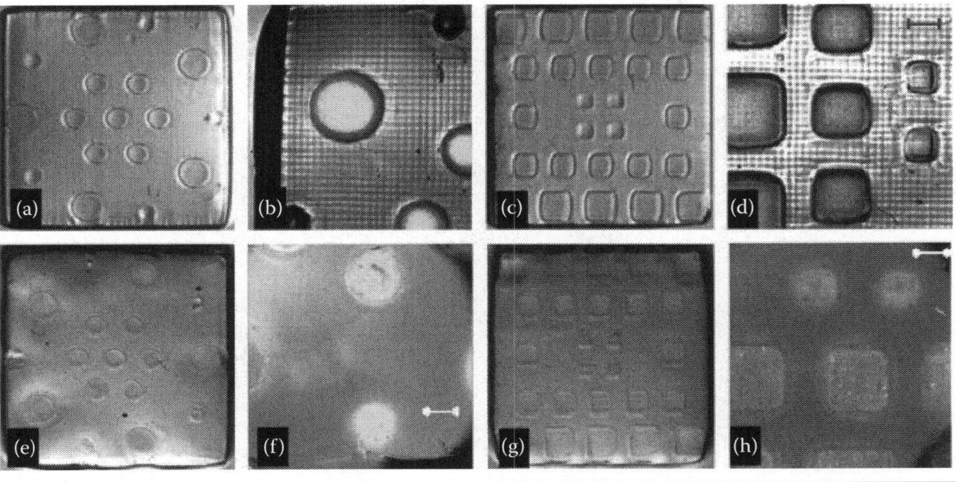

Figure 8.9 (a through d) Scaffolds fabricated with polyethylene glycol dimethacrylate (PEG-dma MW 1 K); (e through h) scaffolds fabricated with polyethylene glycol diacrylate (PEG-da MW 3.4 K) as the main material. (From Arcaute, K. et al., *Acta Biomater.*, 6, 1047–1054, 2010.)

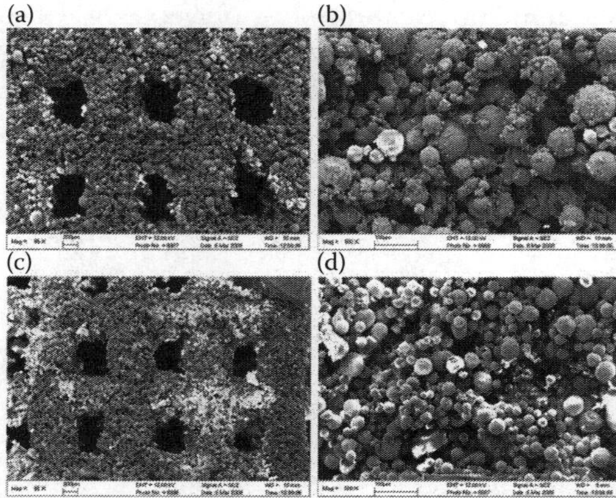

Figure 8.10 SEM images of sintered scaffolds: (a, b) Ca-P/PHBV and (c, d) CHAp/PLLA (magnification 500×) [19]. Scale bar corresponds to 200 μm in (a, c) and 100 μm in (b, d).

In contrast, a donor substrate with a cell-encapsulating film and a receiving substrate are placed in parallel, as shown in Figure 8.11b. A pulsed laser beam is transmitted through the donor substrate and propels cells from the ribbon to the receiving substrate. Absorbing film-assisted laser-induced forward transfer (AFA-LIFT), biological laser processing (BioLP), and matrix-assisted pulsed laser evaporation direct writing (MAPLE DW) are minor variations of the LIFT process.

Nahmais et al. [20] developed a three-stage LGDW system design using IR LD and a simple set of optics, as shown in Figure 8.12a. The system patterned healthy endothelial cells that

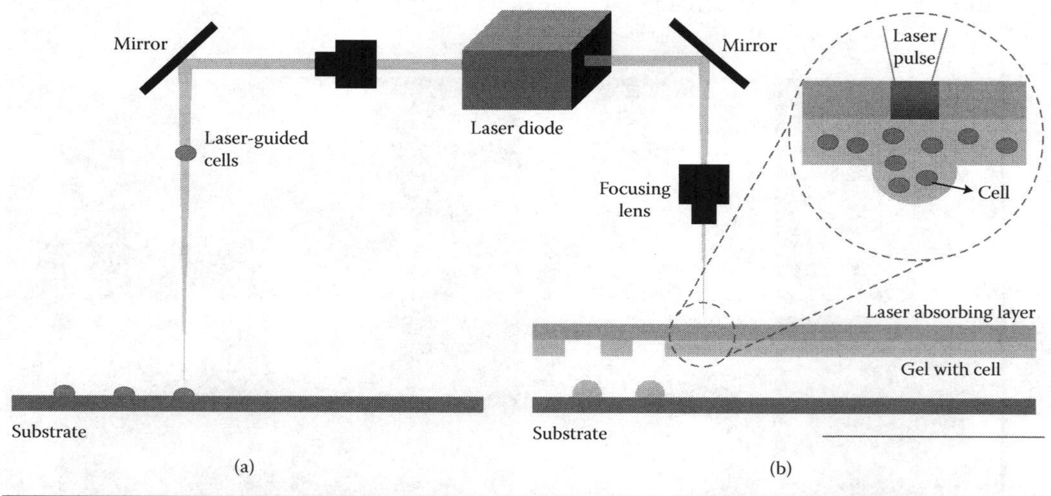

Figure 8.11 Sketch of the laser printing setup. (a) Laser-guided direct cell printing where the laser is focused into a cell suspension and the force due to the difference in refractive indexes moves the cells onto an acceptor substrate. (b) The cell-hydrogel compound is propelled forward as a jet by the pressure of a laser-induced vapor bubble. (From Tasoglu, S., Demirci, U., *Trends Biotechnol.*, 31, 10–19, 2013.)

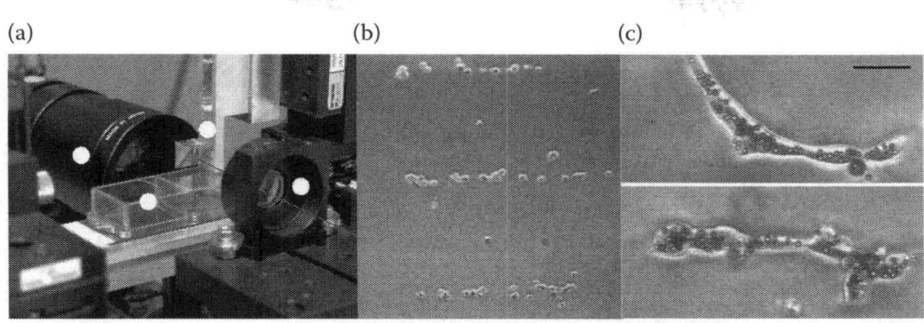

Figure 8.12 (a) An LGDW system; (b) three lines of umbilical vein endothelial cells written on a Matrigel; and (c) a high-resolution image of two vascular structures. (From Nahmias, Y., Odde, D.J., *Nat. Protocols*, 1, 2288–2296, 2006.)

retain their ability to proliferate and their capacity to form vascular structures on Matrigel, as shown in Figure 8.12b and c. One of the key challenges is overheating from excessive exposure to thermal energy from the laser light onto the cell biolayer. To overcome this challenge, near-IR wavelengths (700–1000 nm) have been used, making LGDW unlikely to cause mutations or trigger apoptosis, although wavelength optimization remains an open issue. Other living cells such as embryonic chick neurons, multipotent adult progenitor cells (MAPCs), and bacteria have been patterned with LGDW with no apparent reports of deleterious effects on cellular viability and function.

High cell density and microscale organization were achieved by a laser-assisted cell printer (LIFT) at a laser pulse repetition rate of 5 kHz [21]. A solid Nd:YAG laser source with a 1064 nm wavelength was placed in a five-axis carousel system and focused on concentrated bioink (10^8 cells/mL) to print mammalian cells individually, as shown in Figure 8.13.

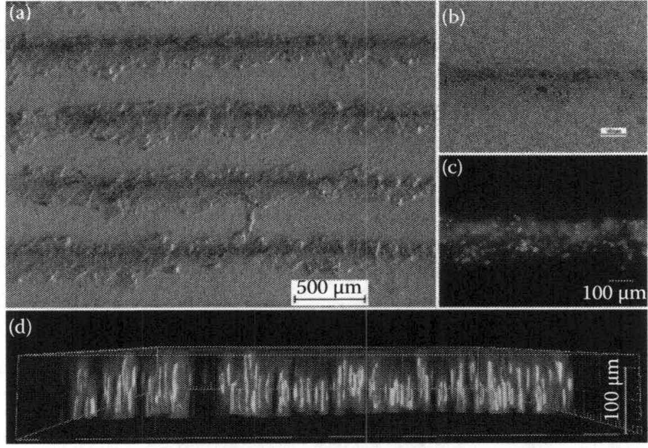

Figure 8.13 A bioink containing endothelial cell line Eahy926 with a concentration of 6×10^7 cells/ml printed on a layer of fibrinogen. (a) Phase contrast optical microscope image, scale bar: 500 µm. (b) Phase contrast optical microscope image, scale bar: 100 µm. (c) Live/dead assay staining of (b), fluorescence microscope image, scale bar: 100 µm. (d) Orthogonal view of the surface representation of the *z*-stack of (c). (From Guillotin, B., *Biomaterials*, 31, 7250–7256, 2010.)

8.2.3 Prosthetics

Prosthetics are artificial devices that modify and/or reinforce less functioning body parts or completely replace missing ones. According to the American Academy of Orthotists and Prosthetists' reports, in 1995, there were ~1.6 million users of prosthesis. By the year 2020, the number of users is expected to increase by 50% to a total of 2.4 million patients [22]. Growing consumers means a larger market; however, it also requires a technology revolution to realize "mass customization." A simple cost–benefit analysis shows that traditional manufacturing technology is not capable of meeting this need. This has led to new opportunities for AM and a new era in manufacturing. In the field of prosthetics, orthodontic braces and hearing aids are two of the most commonly used applications of LBAM technology.

Align Technology is a pioneer in the business of manufacturing customized braces (Figure 8.14). What makes Align Technology unique among its competitors is not only the ability to produce custom braces for each patient but also the ability for mass production of custom braces. The aligner fabrication starts with taking x-ray pictures and impressions of a patient's teeth. Based on these, the orthodontist prescribes a detailed treatment plan with exact movement for each tooth. The impression is then sent to Align Technology, where its geometric information is extracted by a laser scanner, and a tessellation model of the teeth is reconstructed accordingly. The technician from Align Technology can modify the tessellation model by placing each tooth into its desired position based on the specific treatment plan. Then, the aligner mode is built in an SLA-7000 machine from 3D Systems Inc. according to the modified tessellation model. The SLA-7000 is a commercial SLA apparatus, and it adopts laser as the energy source to solidify the liquid resin. In this application, Align Technology uses SLA-7000 to create the aligner mode in a layer-by-layer fashion, and a customized, clear aligner is then fabricated by thermal forming of a sheet of smooth, BPA-free plastic over the aligner mode [23]. It has been reported that each SLA-7000 machine can produce 100 aligner molds per day, with a total daily production of 40,000 unique braces.

Figure 8.15 shows a Siemens LASR hearing aid. In the early 2000s, two major hearing aid competitors, Siemens and Phonak, started to collaborate. They worked together to apply SLS in producing shells for hearing aids. They tried to use AM technologies to replace traditional

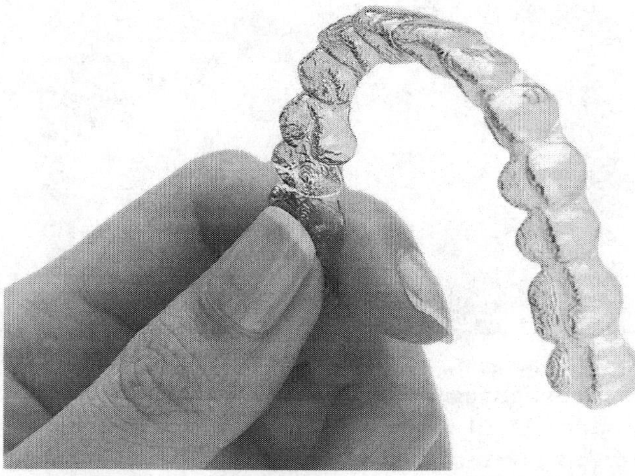

Figure 8.14 Clear tooth aligner from Align Technology.

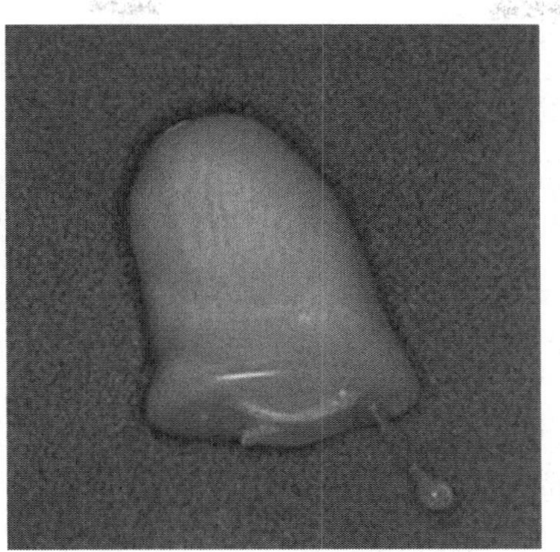

Figure 8.15 Siemens LASR hearing aid and shell. (From Gibson, D.W.R.I., Stucker, B., *Additive Manufacturing Technologies: Rapid Prototyping to Direct Digital Manufacturing,* Springer, New York, 2010.)

injection molding practices. To do so, they first included solid modeling CAD systems to the production process. Impressions taken from customers' ears are scanned by a laser digitizer, and then the resultant point cloud is converted to a tessellation model. This tessellation shell model is exported as an STL file that can be processed by an SLS machine. The impression transformation process is shown in Figure 8.16. Since each tessellation shell mode is patient specific, the hearing aid shell offers more comfort for the user. Siemens never stopped its technology development, and in the mid-2000s, they also investigated SLA technology to complement the SLS capacity. Records show that Siemens' customer satisfaction has been improved dramatically since AM technologies were adopted in the plant [23].

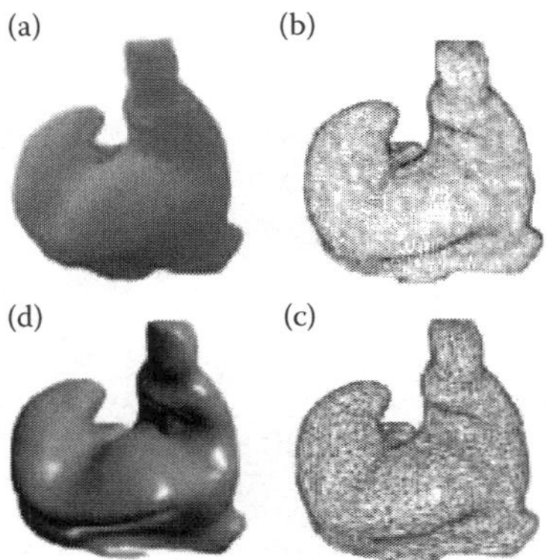

Figure 8.16 Impression transformation: (a) original, talc-coated ethyl methacrylate impression; (b) scanned point cloud image; (c) transformation to contiguous polygon surface representation; and (d) virtual impression (STL file). (From Sullivan, R.F., *Hear. J.*, 60, 21–28, 2007.)

8.2.4 Implants

Metallic implants are widely used to restore the lost structure and functions of human bone. Current metal implants are designed to be bioinert in nature but often demonstrate a significantly higher stiffness than natural bone that leads to an undesirable biological response. The mechanical mismatch between bone and a metal implant often causes stress shielding. An implant shields bone from carrying loads. As a consequence of the reduced load, bone will begin to resorb and be at higher risk for refracture. Porous metallic structures can minimize the discrepancies and achieve stable long-term implant functionality [25]. Patient-specific porous implant fabrication with bioadditive techniques brings novel possibilities, including guided tissue regeneration, load-bearing implants, osseointegration, and on-demand and on-site manufacturing. Both novel design approaches and fabrication technologies are required to achieve balanced mechanical and functional performance in an implant.

Complex anatomically shaped implants were fabricated with laser-engineered net shaping (LENS) to meet the desired mechanical and biological performance (see Figure 8.17) [26]. The LENS process was able to fabricate the designed porosity to improve implant performance. For example, stress shielding was reduced through designed porosities of up to 70 vol%. Similarly, the effective modulus of Ti, NiTi, and other alloys was tailored to suit the modulus of human cortical bone by introducing 12–42 vol% porosity. Furthermore, their fabrication process was able to introduce a graded coating of Co–Cr–Mo to minimize the wear-induced osteolysis.

In addition to the metallic implants, LBAM plastic parts are also widely used for patient-specific implantable devices. For example, laser-based SLA can be used to fabricate medical implants with tailored geometrical and anatomical properties. Such implants include cranio-maxillo-facial surgical components [27–31], bone fracture fixation devices [32], components for artificial hips or knees [33], and nerve guidance channels [34,35]. In a laser-based SLA process, some of the major challenges are the mechanical properties and biocompatibility of the photopolymerizable materials [5]. Research work has been done to develop and study the biodegradability and biocompatibility of the biomaterials for such 3D printing processes. Popov and Evseev [36] studied the laser photopolymerization of a mixture of acrylic resin and hydroxyapatite for custom-designed implant fabrication. Supercritical carbon dioxide was used to enhance the biocompatibility of the materials and increase the area of direct contact of the implant surface with regenerated bone tissue. Schuster et al. [37,38] investigated the reactivity and biocompatibility of different types of commercially available acrylates and photoinitiators for bone replacement applications. Matsuda et al. [39]

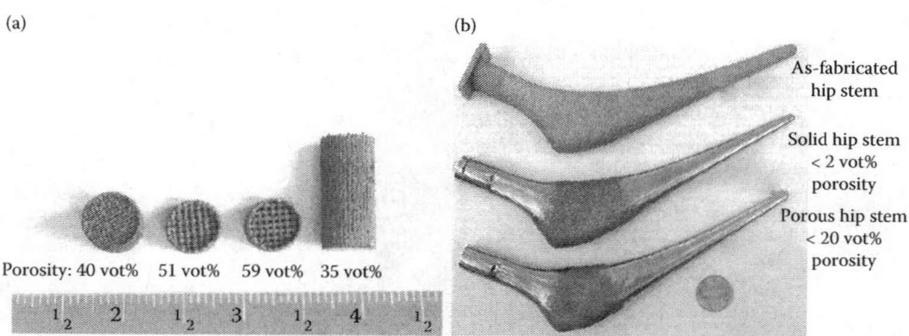

Figure 8.17 LENS processed porous titanium structures: (a) samples with total porosity >50 vol% were fabricated using tool-path–based porosity and (b) net shape functional hip stems with designed porosity fabricated using LENS. (From Bandyopadhyay, A. et al., *J. Mater. Sci.*, 20, 29–34, 2008.)

demonstrated that acrylate-based biodegradable poly(ε-caprolactone-*co*-trimethylene carbonate) processed by SLA showed fast biodegradation and biocompatibility after implantation in rats.

8.3 Aerospace Applications

8.3.1 Introduction

This section focuses on the applications of LBAM in the aerospace industry. Popular applications include repair of turbine blades, fuel injection nozzles, and various spacecraft components [40]. The most widely used AM processes within the aerospace domain are direct metal laser sintering (DMLS), LENS, electron beam additive manufacturing (EBAM), and electron beam freeform fabrication (EBFFF) [41]. Table 8.2, which is reproduced from Ref. [40], shows different categories of AM processes according to American Society for Testing and Materials (ASTM) Standard F2792 [42] contingent on laser-derived energy. Among these, certain AM processes may also employ alternative sources of energy for solidification, for instance, EBAM in the power-bed fusion (PBF) category and EBFFF in direct energy deposition (DED) utilize electron beam as the energy source [40].

AM in general, and LBAM in particular, are attractive to the aerospace industry due to:

1. The ability to produce lightweight, near-net–shaped, hard to machine metal components having complex topologies. AM parts offer the possibility of concomitant reduction in intermediate processing steps, material waste in the form of chips (the so-called buy-to-fly ratio), and energy. The reduction in component weight has a direct influence on the fuel efficiency of the aircraft [43]. For instance,
   - American Airlines Fuel Smart program literature states that, ...*removing just one pound of weight from each aircraft in American's fleet would save more than 11,000 gal of fuel annually* [44]. In an online article by NBC, American Airlines revealed that reduction

Table 8.2 Various Additive Manufacturing Processes

Category	*Processes*	*Application of Laser*
Powder-bed fusion (PBF)	Selective laser melting (SLM), selective laser melting, direct metal laser sintering (DMLS).	Yes
Direct energy deposition (DED)	Direct metal deposition(DMD), direct laser deposition, laser engineered net shaping(LENS)	Yes
Material extrusion	Fused deposition modeling (FDM), fused filament fabrication (FFF), pressure-based extrusion (PBE)	No
Material jetting	Polyjet, Aerosol Jet (AJ)	No
Sheet lamination	Ultrasonic additive manufacturing (UAM) and laminated object manufacturing (LOM)	No
Binder jetting	Three-dimensional printing (3DP)	No
VAT photopolymerization	Stereolithography (SLA)	Yes

Source: Pinkerton, A. J., *Optics Laser Technol.*, 78, 25–32, 2016.

in weight of an onboard food cart by 17 lb saved the company close to $7 million in fuel costs (in 2008 prices) [45].

- EOS of Germany reported that aerospace companies were able to reduce component weight by nearly 50% using LBAM.
- For instance, Dr. Tim Shinbara (Vice President, Association of Manufacturing Technology) presented the follow data (Table 8.3) emphasizing the possibilities of LBAM.

2. The high flexibility on account of low tooling and changeover costs, and relative independence of scale, material, and geometric variability [43]. High-quality LBAM components are capable of withstanding high-temperature and high-pressure conditions of aerospace applications. Integrated with appropriate postprocessing (e.g., hot isostatic pressing), fully dense parts can be produced. Materials that are not amenable to subtractive manufacturing [Inconel, Ti64AlV4, cobalt–chromium (Co–Cr)] can be processed using LBAM processes.
3. The creation of components with complex internal geometries that can substantially magnify the functional boundaries. For instance, Figure 8.18a is an x-ray image of a miniature Inconel turbine blade that contains through channels made using DMLS. The through channels allow for passage of cooling gases, thereby permitting operation at much higher speeds. Likewise, Figure 8.18b shows microcomputed tomography (micro-CT) scans of

Table 8.3 Comparison of Traditional Manufacturing and Laser-Based Additive Manufacturing

Composite Interface Fitting		*Hot Air Mixer*	
Traditional Manufacturing	*Additive Manufacturing*	*Traditional Manufacturing*	*Additive Manufacturing*
~ 500 machining hours	32 build hours	Buy to fly ratio 10–20:1	Buy-to-fly ratio ~ 2:1
16–26-week lead time	4 day lead time	Min. 4 pieces w/2 welds	1 piece with no welding
	60%–70% cost savings		35%–45% cost savings

Source: Adapted from a talk by Tim Shinbara (http://www.aaes.org/sites/default/files/april2013_NAE_Shinbara.pdf).

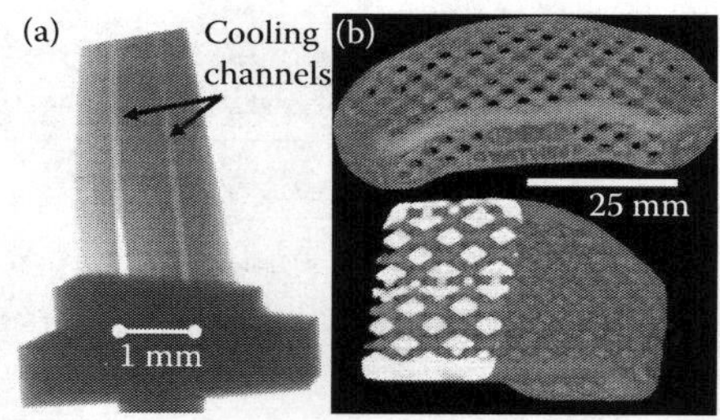

Figure 8.18 (a) X-ray image of a miniature turbine blade made using DMLS and (b) a micro-CT scan courtesy of Binghamton University of a titanium spinal implant crafted using DMLS.

a titanium spinal implant having a complex lattice-like internal structure. Thus, from a broader vista, LBAM reduces product development and design lead times [46]. The ability to make complex geometries is further enhanced by software-based topological optimization—the ability to construct and realize geometries that are difficult, if not impossible to imagine, let alone create, with conventional engineering design principles.

4. Aviation fuel being a leading contributor to greenhouse effect, the adoption of AM for lightweight structural components can mitigate the carbon footprint. In 2014, carbon emissions from aviation fuel were 688 million metric tons. In the coming years, economic growth will inevitably lead to a rapid increase in the number of aircraft, which in turn will further increase the CO_2 emissions. Studies have shown that the use of AM could potentially reduce CO_2 emissions to between 92.1 and 215 million metric tons [47]. Table 8.4, reproduced from Huang et al. [47], shows the reduction in CO_2 emissions in metric tons/aircraft, while taking into consideration the empty mass of the aircraft as 40,622 kg [47].

The widespread adoption of AM technology in both military and commercial aircraft industry was pioneered by Boeing with the implementation of PBF processes (SLS) [48]. Currently, Boeing produces ~20,000 parts using LBAM processes (laser sintering) in both military and commercial aircrafts. The 787 Dreamliner uses 32 different components made by LBAM [49].

General Electric (GE) Aviation has successfully implemented PBF (DMLS) LBAM process for the manufacture of the fuel injector nozzle for their jet engine. The nozzle is 25% lighter and five times more durable than its traditionally manufactured counterparts [50,51]. By the year 2020, GE is planning to produce the majority of its components for its new high-efficiency CFM leading edge aviation propulsion (LEAP) and G90X engines using AM. The utilization of LBAM to produce the nozzles for the LEAP engine, each of which contains 19 nozzles, will further enhance the market for LBAM in the aircraft industry [52–54]. Another leading manufacturer of aircraft components, Lockheed Martin, implements AM techniques to produce brackets. These AM-produced components are one-tenth the weight of their traditionally produced counterparts [50].

In situ fabrication and repair (ISFR) of various parts using LBAM is seen as possibility. This will be imperative to human space explorations. For instance, the ample availability of regolith on

Table 8.4 Carbon Dioxide Emissions in Metric Tons per Aircraft

Component	*Category*	*Al Alloy*	*Ti Alloy*	*Ni Alloy*	*Steel*	*Total*
Body systems	Auxiliary	0–12				0–12
Furnishings	Structural Functional	2–17 12–130				2–17 12–130
Engine	Functional Auxiliary		90–270 7–17	60–270 3–13	1–4 0–5	160–540 10–35
Propulsion systems	Functional		45–160			45–160
Nacelle system	Auxiliary		2–6			2–6
Ref. [47]	Total	14–160	150–450	70–290	1–10	230–900

Source: Huang, R. et al., *Addit. Manuf.*, 135, 1559–1570, 2016.

the moon eliminates the need to carry feedstock such as Ti, Al, and iron (Fe). These elements can be obtained by the molten oxide electrolysis of raw regolith [55].

For effective implementation of LBAM in the aircraft industry, awareness of the various materials–process interactions and thermomechanical phenomena in LBAM is a critical prerequisite. Currently, this knowledge is largely confined to empirical experience. Hence, despite significant progress, repeatability and quality are the main impediments toward adoption of AM parts in the aerospace industry. Therefore, research geared toward in-process quality control of LBAM parts is being conducted on a priority basis by both academia and industry. For instance, Boeing in collaboration with University of Louisville and Integra Services International (Belton, TX) advanced a patented process for the zonal temperature control of build plate for the PBF processes [48]. Steady-state temperature can be established through the incorporation of multizone infrared (IR) heating elements along with online IR imaging [56]. Another research finding by Boeing includes developing a polyamide material that passes flammability tests and can be processed using PBF (SLS) [57].

Recent applications of AM in the aerospace industry include the manufacture of a satellite antenna from titanium parts by Lockheed Martin and a jet-powered UAV by Aurora Flight Sciences Corp. [58]. The UAV weighs 33 lb, has a wingspan of 9 ft, and can fly at speeds of more than 150 mph. The front bearing housing of the PW1500 high-bypass jet engine developed by Pratt & Whitney is made of a titanium alloy using AM techniques. These engines are employed in the C-Series Bombardier Airliners [58].

8.3.2 Powder-Bed Fusion

Common metallic alloys used in aerospace applications of PBF are Al–Si–10Mg, CoCr superalloy, stainless steel, and Ti–6Al–4V. Studies have shown that the aerospace components manufactured by PBF can reduce CO_2 emissions by 40% over the entire life cycle of a part [40]. An emerging application of PBF in aerospace applications is in the manufacture of lightweight cellular lattice structures such as honeycomb and mini truss structures [51,58]. Studies conducted on such structures have shown that they offer excellent thermal and acoustic insulation and good energy absorption characteristics. Figure 8.19 shows an example of a lightweight lattice structure manufactured by DMLS from an AlSiMg alloy at an angle of 45° with respect to build plate [58].

The recent (2014) $50 million investment in an AM facility in Auburn, Alabama by GE Aviation attests to the emergence of AM in this sector. The plant is dedicated to the manufacture of fuel nozzles for the LEAP engine using PBF (DMLS) processes [59]. PBF allowed

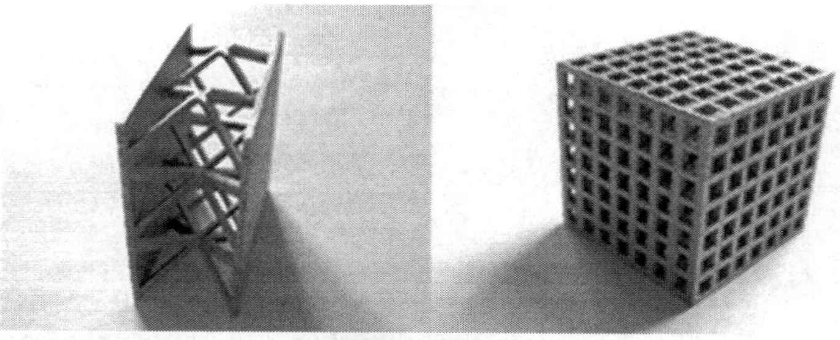

Figure 8.19 Lattice structures manufactured using DMLS. (From Manfredi, D. et al., *Light Metal Alloys Applications*, InTech, 2014.)

Boeing to manufacture the fuel nozzle as a single part that would otherwise be comprised of 20 distinct parts using traditional manufacturing processes. Initially, titanium was used for the fabrication, but the redesigned nozzle was made of a cobalt–chromium alloy [52,60]. The Co–Cr alloy offered better mechanical properties such as resistance to higher temperatures up to 1800 F. Furthermore, a significant reduction in weight (by 25%) was realized [52]. The DMLS processes also allowed incorporation of cooling pathways in the nozzle. The durability of the component was found to be five times more than that of those produced by traditional methods [50,53,61,62].

Another important application of PBF laser-based (DMLS) processing is the manufacture of jet engine combustors using a nickel alloy. Figure 8.20 shows a lightweight jet engine combustor that can withstand high temperatures and provides the same functional characteristics as that of a traditionally manufactured part [63].

Boeing typically uses DMLS followed by hot isostatic pressing to obtain fully dense components for use in mission-critical assemblies [64]. For instance, LBAM processing of Ti–6Al–4V was employed in the manufacture of guidance section housings for missiles; the AIM-9 Sidewinder missile and the guidance section housing are manufactured using PBF LBAM process [48,64].

Airbus Corporation currently uses PBF (DMLS) in the manufacture of nacelle hinge brackets for the A320 aircraft [54,60,65–67]. This component was initially made of cast steel HC101 [65]. Through PBF, the component is now made using a titanium alloy with an accompanying 40% savings in weight [54]. The study was done in order to analyze the energy consumption of the component over a complete life cycle [66].

NASA successfully applied PBF (DMLS) for fabricating fuel injector nozzles at their Marshall Space Flight Center. Figure 8.21 shows the fuel injector nozzle base plate manufactured at NASA [63]. The fuel injector nozzles were built fifteen times faster than those produced during traditional manufacturing practices. The application of DMLS laser additive manufacturing helped in integrating 163 different components into two-piece units of fuel injectors eliminating the need for welding, casting, or bonding. Concurrently, the lead time for production was reduced from 6 to 9 months to 10 days. Switching over to DMLS eliminated the need for further machining processes such as welding, casting, or bonding.

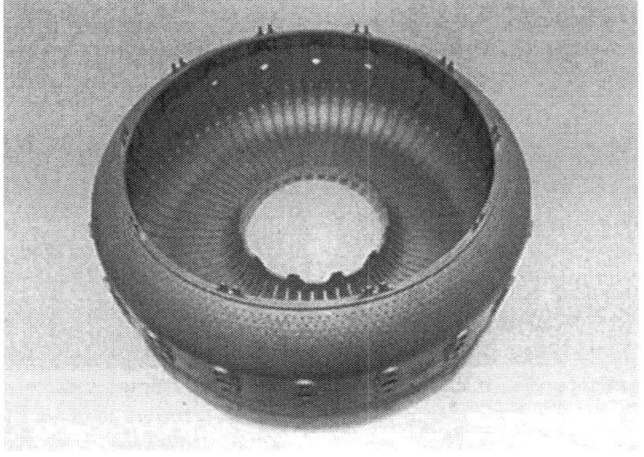

Figure 8.20 Gas turbine engine combustor manufactured by DMLS on an EOS system. (From Weber, A., Additive manufacturing takes off: Aerospace engineers are pushing the boundaries of 3D printing, *Assembly*, 2016.)

Figure 8.21 Fuel injector nozzle manufactured by NASA. (From Weber, A., Additive manufacturing takes off: Aerospace engineers are pushing the boundaries of 3D printing, *Assembly,* 2016.)

One of the many critical parts in the gas turbine is the swirler, which ensures effective recirculation of air in the primary combustion zone, and thus allows the merging of the compressed air with the fuel to maintain the required flame stability [68]. PBF LBAM (DMLS) is used for the manufacture of such swirlers with relative ease compared to traditional manufacturing practices [63]. Figure 8.22 shows a gas turbine swirler made from cobalt–chrome alloy using DMLS [69].

In 2014, SpaceX, a NASA commercial space partner, developed a thrust chamber for its SuperDraco rocket engine using PBF (DMLS). SpaceX SuperDraco engines (also known as

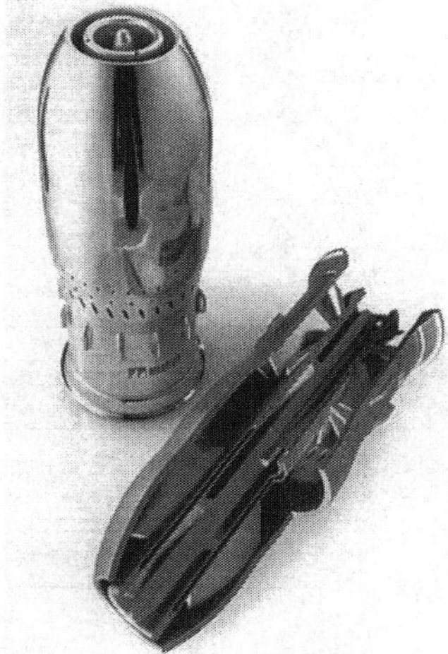

Figure 8.22 Gas turbine swirler made using DMLS. (From A. Weber, Additive manufacturing takes off: Aerospace engineers are pushing the boundaries of 3D printing, *Assembly,* 2016; Udo, B., *Laser Technik J.*, 6, 44–47, 2009.)

Dragon V2 engines) are employed in the SpaceX Rocket Engine that can produce 16,400 lbf thrust, i.e., ~160 times the thrust force produced by the current Draco engine. The added benefits of such components include reduced weight, increased efficiency, and load path optimization [71,72].

PBF LBAM (DMLS) was successfully employed for the manufacture of hot gas discharge ducts for the upper stage of NASA Space Launch System (SLS) [73]. The J-2X gas generator (GG) discharge duct was a result of the collaborative work of NASA Marshall Space Flight Center, Pratt & Whitney Rocketdyne, and Morris technologies. Various tests were conducted on the component to determine operational characteristics in harsh environments. Though the duct was slightly out of tolerance, it was found to be operational. In addition, the mechanical properties and microstructure were found to be consistent. Figure 8.23 shows the J-2X hot gas generator discharge ducts manufactured using DMLS before and after hot-fire tests [73].

Selective laser melting (SLM) is one of the widely adopted PBF processes that utilize a laser to manufacture metallic components in the aircraft industry. Considerable amount of research is being conducted to grow the application of PBF processes that utilize a laser in the aerospace industry. Some of the projects in this domain include the flexible and near-net shape generative manufacturing chains and repair techniques (FANTASIA) project and the TurPro project for engine applications [43].

The aim of the FANTASIA project was to achieve a 40% reduction in repair costs and time from the current values, and it focused on the enhancement of SLM and LMD processes. The main factors responsible for its applications in the aerospace industry are its ability to use nickel- and titanium-based alloys and its capacity for the fabrication of metallic components with unlimited or intricate geometrical features. Applications include manufacturing and repair of various aircraft engine components, such as compressors, casings of combustion chambers, vanes, blades, and Inconel parts such as BLade-Integrated diSKs (BLISKs) [43]. PBF LBAM (SLM) utilizes a laser for processing of multiple layers of metallic powder for the manufacture and repair of distinct aircraft components. For example, the manufacture of a boss for the repair of a plenum case of a turbofan engine is shown in Figure 8.24.

Another application of PBF LBAM (SLM) is the repair of high-pressure turbine (HPT) vanes [43]. An example is the repair of Pratt & Whitney's PW4000 HPT vanes, as shown in Figure 8.25. The repair is done by manufacturing a Mar M247 patch followed by manual TiG welding to replace the crack with a patch.

Figure 8.23 J-2X GG ducts before (left) and after (right) hot-fire tests. (From Betts, E. et al., Using innovative technologies for manufacturing rocket engine hardware, 2011.)

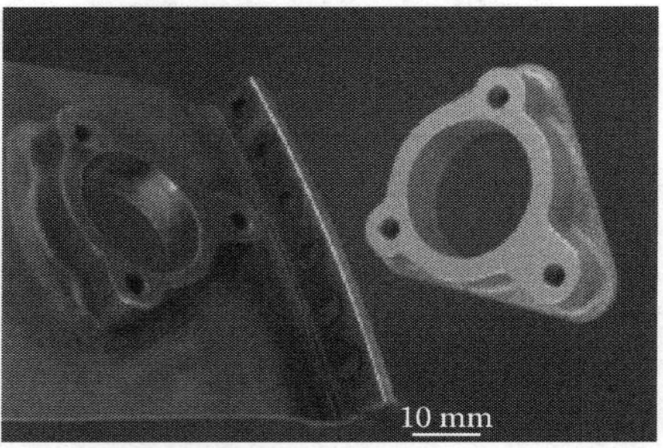

Figure 8.24 Boss manufactured using SLM process. (From Gasser, A. et al., *Laser Technik J.*, 7, 58–63, 2010.)

Figure 8.25 Path produced by SLM on the basis of CAD (left) and repair done on the vane using patch (right) (From Gasser, A. et al., *Laser Technik J.*, 7, 58–63, 2010.)

The vane was found to have several cracks that could not be repaired by traditional techniques. The cracks were remelted with the help of the laser, and then the patch was subjected to hot isostatic pressing at elevated temperatures. The patches for the repair were made by LBAM processing of Inconel 718, and the tensile test conducted on the samples showed an ultimate tensile strength of 1460 MPa and a surface roughness, $R_a < 20\,\mu m$ [43].

The application of PBF LBAM (SLS) also includes the manufacture of prototypes for studying the characteristics of different alloys at different conditions of temperature and pressure in the aerospace industry. LBAM of cermet alloy prototypes to repair the turbine blades reduces the lead time and also eliminates the requirements of pre- and postprocessing and tooling [74]. Pratt and Whitney [75] reported using PBF (EBM) approaches to prototype designs for its PurePower series engines. Engineers can try a wider array of original designs due to the faster prototyping enabled by PBF processes, leading to rapid design innovation [76]. Table 8.5 summarizes specific aerospace applications of PBF processes from the literature.

Table 8.5 Summary of Aerospace Applications That Utilize PBF Processes

Category/LBAM Process	*Title/Authors*	*Material*	*Detail*	*Application(s)*
Powder-bed fusion (DMLS)	Additive manufacturing of aluminum alloys and aluminum matrix composites (AMCs), Manfredi et al. [58].	Aluminum	Additive manufacturing of aluminum alloys AlSi10Mg parts by DMLS	Lattice structure
Powder-bed fusion (SLM)	Additive manufacture of titanium alloy for aircraft components, Uhlmann et al. [77].	TiAl6V4	SLM parameters for the manufacture of Ti alloy components are provided. Microstructure, microhardness, density surface roughness, etc.	Lightweight parts
Powder-bed fusion (SLS)	Processing of titanium net shapes by SLS/HIP, Das et al. [78].	TiAl6V4	Microstructure and mechanical properties of TiAl6V4	AIM-9 missile guidance section housing
Powder-bed fusion (SLS)	Direct laser fabrication of superalloy cement abrasive turbine blade tips, Das et al. [74].	Cermet material, 18% titanium-coated alumina	Process parameters for production, mechanical properties are discussed	Manufacture of prototypes (abrasive blade tips)
Powder-bed fusion (SLS)	Producing metal parts with selective laser sintering/hot isostatic pressing, Das et al. [64].	TiAl64V	Mechanical properties, microstructure, and oxygen levels of TiAl64V component	AIM-9 Sidewinder missile
Powder-bed fusion (EBM)	Fabrication of smart parts using powder-bed fusion additive manufacturing technology, Hossain et al. [79].	TiAl64V	Embedded sensors for *in situ* monitoring of harsh operating conditions	Smart parts

(Continued)

Table 8.5 (Continued) Summary of Aerospace Applications That Utilize PBF Processes

Category/LBAM Process	*Title/Authors*	*Material*	*Detail*	*Application(s)*
Powder-bed fusion	Investigation on additive manufacturing of tungsten carbide-cobalt by selective laser melting, Uhlmann et al. [77].	Tungsten carbide–cobalt (WC–Co)	Density, porosity, microstructure	Manufacture of near-net shape tools
Powder-bed fusion (EBM)	Electron beam melting of Ti–48Al–2Cr–2Nb alloy: Microstructure and mechanical properties investigation, Biamino et al. [80].	Ti–48Al–2Cr–2Nb alloy	Microstructure, porosity, and chemical composition are analyzed after EBM	Aero engine components
Powder-bed fusion (EBM and SLM)	Metal Fabrication by additive manufacturing using laser and electron beam melting technologies, Murr et al. [81].	Cu, alloy 625 (Ni-base alloy), Ti–6Al–4V, 17–4PH stainless steel.	Comparison of SLM and EBM	Cellular lattice structures
Powder-bed fusion (EBM)	Prediction of mechanical properties of EBM Ti–6Al–4V parts using dislocation density-based crystal plasticity framework, Pal et al. [82].	Ti–6Al–4V Gamma TiAl	DDCP-FEM simulations for deformation behavior of TiAl64V parts manufactured using EBM	Turbine blades
Powder-bed fusion (selective EMB)	Selective electron beam melting of Ti–48Al–2Nb–2Cr: Microstructure and aluminum loss, Schwerdtfeger and Korner [83].	Ti–48Al–2Nb–2Cr	Variation of parameters with scan speeds, microstructure	Manufacture of complex shapes (turbine blades, valves for combustion engines)

(Continued)

Table 8.5 (Continued) Summary of Aerospace Applications That Utilize PBF Processes

Category/LBAM Process	*Title/Authors*	*Material*	*Detail*	*Application(s)*
Powder-bed fusion	Powder bed additive manufacturing systems and its applications, Udroiu [84].	Inconel 718-30	Additive manufacturing of metal parts using PBF and DED processes.	Turbine engine case study
Powder-bed fusion (DMLS)	Evaluation of lightweight AlSi10Mg periodic cellular lattice structures fabricated via direct metal laser sintering, Yan et al. [85].	AlSi10Mg	Feasibility to manufacture Al alloy lattice structures at a controlled volume fraction and a particular unit cell size and mechanical properties studied.	Manufacture of lightweight lattice structures
Powder-bed fusion (SLM)	Metal additive manufacturing of a high-pressure micropump, Wits et al. [86].	Ti–6Al–4V	Manufacture of prototype microvalves and leak tests were conducted to check the sealing.	Potential cooling applications for satellites.
Powder-bed fusion (DMLS)	Using innovative technologies for manufacturing rocket engine hardware, Betts et al. [73].	Nickel Alloy 625	Material properties and tolerance conditions	Manufacture of J-2X gas generator (GG) discharge duct
Powder-bed fusion (DMLS)	DMLS—Development history and state of the art, Shellabear and Nyrhilä [87].	Various	Evolution of DMLS process, materials and process developments	

(Continued)

Table 8.5 (Continued) Summary of Aerospace Applications That Utilize PBF Processes

Category/LBAM Process	*Title/Authors*	*Material*	*Detail*	*Application(s)*
LBAM	Laser additive manufacturing: Laser Metal Deposition (LMD) and selective laser melting (SLM) in turbo-engine applications, Gasser et al. [43].	Inconel 718	Turbo engine applications	Repair of various turbo engine component parts.
DMLS	Materials and finishing methods of DMLA manufactured parts, Duleba et al. [88].	CC MP1, In718, SS PH1, Ti6Al4V, 316L SS	Various materials used for DMLS manufacture of metallic parts.	Materials used in aerospace.
Powder-bed fusion (SLM)	On-site additive manufacturing by selective laser melting of composite objects, Fateri et al. [89].	Composites	Reduction of weight for potential space craft by on-site additive manufacturing on Mars/Moon.	Potential space applications of powder-bed fusion AM.
Powder-bed fusion	Laser powder-bed fusion additive manufacturing of metals: Physics, computational, and material challenges, King et al. [90].		Physical-based models for powder-bed fusion (metal) and its application for simulation of the melting and densification of metal powder.	
Powder-bed fusion (laser CUSING)	AM and aerospace: An ideal combination, Nickels [66].	Titanium	Application of powder-bed fusion, advantages (energy benefits), and challenges faced.	Cabin bracket for Airbus A350

8.3.3 Directed Energy Deposition

Direct energy deposition (DED) is primarily employed in the aerospace industry for the repair and remanufacture of aircraft components using metal alloys such as iron-based alloys (e.g., stationary compressor components). The European FANTASIA and TurPro exemplify the repair application focus of DED [41].

Factors that make DED most valuable for the aerospace industry are (1) low heat input, which helps to reduce cracks and (2) rapid cooling of the structures soon after the laser treatment, which facilitates finer microstructure [43]. Recently, researchers have incorporated closed loop feedback control mechanisms to attain high dimensional accuracy/resolution [91]. A common recurring problem in DED, despite precautions, is the presence of residual stresses that can lead to cracking of the components. Residual stresses occur during the solidification of the layers of metal powder.

Components on which DED can be performed involve casings, blades, rotor blades, and BLISKs. Akin to PBF, laser-based DED AM techniques allow the manufacture of complex parts such as turbine blades with internal cooling passages with relative ease [92]. One of the main advantages of DED is that it eliminates the use of a processing gas chamber while carrying out repair [1]. This technology is also utilized in the repair and freeform fabrication of several metallic components in the aerospace industry. The laser beam melting followed by rapid cooling provides fine grain structures, which when combined with excellent mechanical properties is advantageous in aerospace industry. Researchers are actively working on closed feedback control loops that help retain the same quality while reproducing the components [93].

In DED processes, laser power with different spot sizes are utilized depending on the wattage requirements. The laser compatible to blown powder DED processes are Nd:YAG and fiber lasers. The various materials compatible for these processes, especially those within the aerospace industry, are titanium alloys (Ti–6Al–4V, Ti-6-2-4-20), high-grade stainless steel (304, 316, 410, 420, 17–4 PH), nickel alloys (617, 625, 718), and cobalt alloys (#6 Stellite, #21 Stellite) [94,95].

Results show that the arrangement of grains in the Ti–6AL–4V is long and columnar parallel to the deposition direction, developed as a result of heat extraction from the substrate [96]. From the figure, the presence of macro- and micro-heat-affected zones (HAZ) is evident. Micro-HAZ were observed in between the interfaces of LENS deposition and substrates. Macro-HAZ were evident as a result of reheating of the succeeding layers on subsequent deposition. The properties of the DED components also varied depending on the different build orientations. This was due to the microstructure as a result of the presence of micro- and macro-HAZ [96,97]. The surface finish of components varied according to the laser power used. Low-power laser yielded surfaces with a smoother finish than high-power laser produced surfaces.

It is observed that for a given material, the mechanical properties of a printed part vary based on the AM process. Table 8.6, reproduced from Uriondo et al. [41], shows the difference in mechanical properties of Ti–6AL–4V when manufactured by PBF and DED processes.

Figure 8.26 shows the microstructure of Ti–6Al–4V manufactured by DED and PBF [41]. An exceptional case is that of Al alloys. Aluminum alloys are not compatible with DED due to their reflective properties. Further research is being conducted on the use of Tungsten, Tantalum, Rhenium, and Molybdenum [94].

DED processes (particularly, Optomec's LENS system) are employed in the repair of various components of gas turbine engines. One application is the repair of titanium bearing housings. Published results are available for repair done on a low wattage Ti–6Al–4V bearing housing of a gas turbine engine [94]. The bearing housing was found to be out of tolerance, which made it unserviceable. The implementation of DED (LENS) helped return it to operational conditions at

Table 8.6 Mechanical Properties of AM Ti–6Al–4V Parts

Process	*Yield Strength (GPa)*	*Ultimate Tensile Strength (GPa)*	*% Elongation*	*Reference(s)*
EBM	1.1–1.15 0.83 0.735	1.15–1.2 0.915 0.775	16–25 — 2.3	Murr et al. [98] Facchini et al. [99] Koike et al. [100]
DMLS	0.865 1.07 0.835 0.99	0.972 1.2 0.915 1.095	10 11 10.6 8	SLM solutions EOS Facchini et al. [101] Facchini et al. [101]
Wire and Arc AM (WAAM, similar to DED)	803	918	—	Ding and William [102]

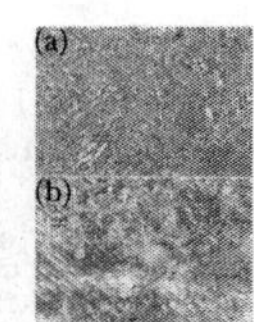

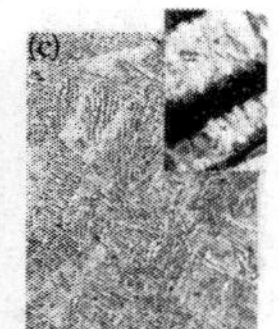

Figure 8.26 Microstructure of Ti–6Al–4V manufactured by (a) EBM, (b) SLM, and (c) dislocation density in α-phase (From Uriondo, A. et al., *Proc. Inst. Mech. Eng.,* 229, 2132–2147, 2015.)

50% of what it would have cost to replace it entirely. The repair was done by melting off multiple layers of Ti–6Al–4V material using a laser beam [94].

Near-net shape, thin-walled structures, and freeform builds can be built using LENS by varying the laser characteristics; e.g., from low wattage/low spot sizes to high wattage/high spot sizes of the laser. Results of freeforms made of stainless steel at different spot sizes (1, 2, 3, and 4 mm) and (right) are available in Ref. [94].

One of the DED applications includes the repair of a compressor seal. This seal is crucial for preventing gas leakages in gas turbine engines [103]. In this application, the compressor seal is made of Inconel 718. When the diameter of labyrinth seal wear exceeds 0.08 in., the efficiency of the engine is reduced and considered not airworthy. Results from repair done on a gas turbine engine using Inconel 718 powder are available in Ref. [94]. The repair not only saved time but also reduced the costs to 45% of a new one [94].

DED is also reported in the repair of HPT shrouds (see Figure 8.27). The function of HPT shrouds is to prevent leakage of air around the tips of turbine blades and thus increase the

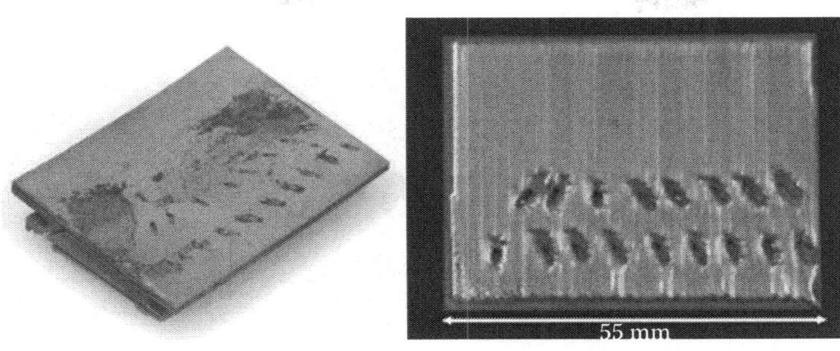

Figure 8.27 HPT shroud before and after LBAM process. (From Gasser, A. et al., *Laser Technik J.*, 7, 58–63, 2010.)

efficiency of gas turbine engines. The shrouds are arranged side by side in the circumferential direction to form a seal. The hot air that passes over its surface in the axial direction is cooled by the presence of cooling holes. These cooling holes reduce overheating of the engine and thus reduce the stress loads on the turbine blades [104]. The first step involved in repairing is scanning of the part using the laser beam. This is done so that the areas that require repair are differentiated from the other regions, and also because it helps to create a defined path for the tool. The main challenge faced is performing the repair without affecting the cooling holes on the surface [43].

Another repair-based application of DED involves the repair of gas turbine engine cases; e.g., Rolls-Royce used DED in the repair of the BR715 HPT case [43]. Special DED nozzles used in metal deposition make the process more flexible with minimum contamination in the components. The material of the turbine case was Nimonic PE16 (nickel-based alloy), and the material used for repair was Inconel 625 [43].

The applications of DED extend to space explorations. One such application includes the fabrication of parts using raw lunar regolith. This demonstrates the potential use of DED for the manufacture of spare parts on the Moon and eliminating the need to carry spare parts. The material used for blown powder DED fabrication of parts is JSC-1AC (regolith) of particle size ≤5 mm. The particle size was reduced to the range of 50–150 μm to ensure its compatibility with blown powder DED applications. Optomec LENS-750 was used for the fabrication of parts using an aluminum alloy as a substrate and Nd-YAG laser with a spot size of 1.65 mm. Vickers hardness test and FE-SEM microscopy conducted on these structures with a 200 g load revealed a hardness of 500 ± 18 Hv structure with fine microstructural characteristics [55].

Cylindrical structures were fabricated from raw regolith (JSC-1AC) using direct energy deposition (LENS). An Nd:YAG laser using 50 W power, 20 mm/s scan speed, and 12.36 g/min of feed rate was used for the direct melting of the raw material [55].

Laser cladding that works with the same principle as that of DED utilizes CO_2 lasers for the manufacture of different components in the aeronautical industry at Aeromet [105,106]. One such application includes the manufacture of a burn-resistant Ti alloy for turbine blades at Rolls Royce. The components produced by laser cladding exhibit higher tensile strength and other mechanical properties compared to wrought alloys. Commonly used materials in this process are cobalt, nickel, aluminum, and titanium alloys [105].

Table 8.7 further summarizes the aerospace-related applications of DED in the literature.

Table 8.7 Summary of Aerospace Applications That Utilize DED Processes

Category/LBAM Process	*Title/authors*	*Material*	*Detail*	*Application(s)*
Direct energy deposition (LENS)	Laser engineered net shaping (LENS): A tool for direct fabrication of metal parts, Atwood et al. [107].		Fabrication of injection mold tools with internal cooling passages	
Direct energy deposition (LENS)	Laser engineered net shaping advances additive manufacturing and repair, Mudge and Wald [94].		LENS (blown powder DED) process and its applications	Repair of bearing hosing, compressor seal, shaft, and manufacture of freeform structures
EBM and LENS	Microstructure, static properties, and fatigue crack growth mechanisms in Ti–6Al–4V fabricated by additive manufacturing: LENS and EBM, Zhai et al. [96].	Ti–6Al–4V	Powder-bed fusion and directed energy deposition and its applications	
LENS	First demonstration on direct laser fabrication of lunar regolith parts, Balla et al. [55].	Lunar regolith	Experimentation of the manufacture of regolith structures using	Manufacture of regolith structures.
EBM	Case Study: Additive manufacturing of aerospace brackets, Dehoff et al. [108].	Ti–6Al–4V	Additive manufacturing of BALD brackets using Electron Beam Melting	BALD bracket
AM	Additive manufacturing technology: Potential implications for US manufacturing competitiveness, Ford [51].		Additive manufacturing applications in various industrial sectors in the United States.	Aerospace applications.

(Continued)

Table 8.7 (Continued) Summary of Aerospace Applications That Utilize DED Processes

Category/LBAM Process	*Title/authors*	*Material*	*Detail*	*Application(s)*
AM	A critical analysis of additive manufacturing technologies for aerospace applications, Angrish [109].	Metals and nonmetals	AM applications in aerospace and its advantages and disadvantages.	
AM	Additive manufacturing of metals: A review, Herderick [110].	Metal alloys	Additive manufacture of metallic components.	
LBAM	Sensing, modeling and control for laser-based additive manufacturing, Hu et al. [111].		Improving the performance of LBAM process. Aero apps not mentioned.	
AM	Additive manufacturing of aerospace alloys for aircraft structures, Kobryn et al. [112].	Ti6Al4V	Manufacture of metallic parts, applications and challenges.	Potential aerospace applications: Rib-web structures, turbine engine cases, engine blades and vanes.
LBAM	Laser-based manufacturing of metals, Kumar et al. [93].		LBAM techniques, comparisons, applications, and challenges.	
LBAM	Laser additive manufacturing (AM): Classification, processing philosophy and Metallurgical Mechanisms, Gu [113].			
LBAM	Improving solid freeform fabrication by laser-based additive manufacturing, Hu et al. [114].		Closed loop control on LBAM processes.	

(Continued)

Table 8.7 (Continued) Summary of Aerospace Applications That Utilize DED Processes

Category/LBAM Process	*Title/authors*	*Material*	*Detail*	*Application(s)*
EBM	Characterization of titanium aluminide alloy components fabricated by additive manufacturing using EBM, Murr et al. [115].	Ti aluminide alloys	Fabrication of Ti aluminide products using EBM.	
LBAM (DMLS)	Rapid Manufacturing of metal components by laser forming, Santos et al. [105].	Titanium, nickel, and aluminum alloys	Aero apps mentioned. Laser-based additive manufacturing of metallic components.	Aerospace components by Aeromet.
AM	Additive Manufacturing for aerospace. Schiller [50].		Additive manufacturing applications in aerospace, cost evaluation and challenges.	Fuel nozzle by GE aviation, liquid engine propulsion injector plate (NASA).
LBAM	Laser additive manufacturing: LMD & SLM in turbo engine applications, Gasser et al. [43].	Inconel 718(SLM) Ti–6Al–4V (LMD)	Laser-based additive manufacturing applications in the repair of aero engine components	Repair of aero engine components.
AM	Analyzing product lifecycle costs for a better understanding of cost derivers in additive manufacturing, Lindemann et al. [116].	Sample aerospace part	Cost analysis for parts during its entire lifecycle (life cycle–based approach)	Weight reduction in aircraft and cost savings.
AM	Additive manufacturing and its societal impact: A literature review, Huang et al. [117].		The societal impact of additive manufacturing is discussed	

(Continued)

Table 8.7 (Continued) Summary of Aerospace Applications That Utilize DED Processes

Category/LBAM Process	*Title/authors*	*Material*	*Detail*	*Application(s)*
AM	Additive manufacturing takes off, Weber [63].		Discusses the recent applications of AM.	Recent applications in aerospace industry.
AM	Aircraft engine maker opens additive manufacturing lab, Ann [75].	Nickel and Ti alloys	Discusses the recent Am applications by Pratt & Whitney	Prototypes for PurePower engines.
AM	Economics of additive manufacturing for end usable metal parts, Atzeni et al. [118].		Comparison of the costs for manufacturing using high pressure die casting and PBF (SLS).	Case study of aircraft landing gear for DFM.
AM	Aviation finds that extra dimension, Brookes [119].		Various additively manufactured components exhibited at the Paris Air Show 2015.	Recent applications of AM
SFF	Solid freeform fabrication: An enabling technology for future space missions, Taminger et al. [120].		Application of SFF techniques for space missions	Potential aerospace such as on-orbit construction of space structures, manufacture of spare parts, online monitoring, and repair of structures

(Continued)

Table 8.7 (Continued) Summary of Aerospace Applications That Utilize DED Processes

Category/LBAM Process	*Title/authors*	*Material*	*Detail*	*Application(s)*
AM	Energy and emissions saving potential of additive manufacturing: The case of lightweight aircraft components, Huang et al. [47].		Estimation of the energy saving and carbon dioxide emission reduction using additive manufactured components.	Reduction in emission of carbon dioxide and fuel savings.
LBAM	Laser additive manufacturing and Bionics: Redefining lightweight design, Emmelmann et al. [121].		Optimizing the design for manufacture of lightweight structures	Lightweight aircraft structures
AM	Could 3D printing change the world? Campbell et al. [122].		Discusses current as well as futuristic applications of additive manufacturing	
AM	Making sense of 3D printing: Creating a map of additive manufacturing products and services, Conner et al. [52].		Evaluation of products that can be manufactured using AM: Customization, complexity, and production quantity	Decision making for manufacture of components between traditional and AM methods
AM	3D opportunity: Additive manufacturing, parts to performance, innovation and growth, Cotteleer [61].		Additive manufacturing: Materials used, challenges, solutions, and its market	Titanium propellant tank and Bleed air leak detect bracket (BALD) by Lockheed Martin

(Continued)

Table 8.7 (Contiuned) Summary of Aerospace Applications That Utilize DED Processes

Category/LBAM Process	*Title/authors*	*Material*	*Detail*	*Application(s)*
AM	Thinking ahead the future of additive manufacturing, Gausemeier et al. [69].		Applications of additive manufacturing in various industries	Swirler made of Cobalt Chrome (Co–Cr) MP1 for gas turbine
AM	Metal additive manufacturing: A review, Frazier [123].	Ti–6Al–4V IN-625, 718	Discusses various additive manufacturing systems, challenges faced economics, and environmental impact	
AM	Additive layer manufacturing: State of the art in industrial applications through case studies, Petrovic et al. [124].		Various industrial applications and case studies	Lightweight lattice structures
AM	Additive manufacturing in aerospace: Examples and research outlook, Lyons [48].	PAEK materials, polyamides	Various aerospace applications, challenges, and current research	Air ducts
AM	Layer by layer, Freedman [49].		Current applications and challenges to overcome	Aerospace applications: Boeing and GE aviation
AM	Additive manufacturing, Babu et al. [67].		Design of parts for additive manufacturing; Topology optimization	

(Continued)

Table 8.7 (Continued) Summary of Aerospace Applications That Utilize DED Processes

Category/LBAM Process	*Title/authors*	*Material*	*Detail*	*Application(s)*
AM	Is additive manufacturing truly the future? Winkless [125].	Titanium, polymer composites	Applications in aerospace industry	Additive manufacturing for repair of components manufactured by Rolls Royce
AM	Metallic additive manufacturing: State of the art review and prospects, Vayre et al. [126].		Various LBAM techniques, cost evaluation, and their environmental impact.	
AM	Environmental aspects of laser-based and conventional tool and die manufacturing, Morrow et al. [91].		DMD-based manufacturing of tools, molds, and dies and their environmental impact.	Manufacture of tools for various industries.
AM	Additive manufacturing technologies used for superalloys processing, Udroiu [84].	Inconel 718 CoCr	Additive manufacturing of metal parts.	Aerospace applications: Stator of gas turbine, turbine blade.
AM	The present and future of additive manufacturing in the aerospace sector: A review of important aspects, Uriondo et al. [41].		Manufacture of metal parts using EMB and SLM. Various aerospace alloys used and the mechanical properties.	
AM	Nondestructive evaluation of additive manufacturing, Waller et al. [71].	Inconel 625 Ti–6Al–4V	AM applications in aerospace industry (NASA)	Manufacture of thrust chamber

8.4 Automotive Applications

The automotive industry was early to adopt commercialized AM machines and accounts for ~16.1% of revenue in the AM industry [127]. The primary uses were prototyping parts to test form and fit as well as rapid product development. Large-scale producers in the automotive industry have not been able to incorporate AM into mass production parts. Unlike the aerospace industry, AM technology is cost prohibitive for most production-level automotive applications because the production volume is significantly higher. Instead, AM has found a niche in the automotive sector for low production, high-quality parts that demand performance, and weight reduction. Several AM processes have been incorporated into producing automotive parts, such as SLS, SLM, SLA, binder jetting, fused deposition modeling (FDM), and electron beam melting for a variety of applications [128,129]. LBAM of automotive parts was limited until metal AM became more mainstream. Prior to commercial metal systems, LBAM was primarily restricted to SLS and SLA. Printing metal opened up new possibilities for functional parts. LBAM of automotive parts remains a small niche market mainly found on high-end custom-built vehicles, such as the Bloodhound SSC, student formula cars, or luxury sports vehicles. The advantage of LBAM is that production and performance problems can be solved that were impossible with traditional manufacturing technologies. This section identifies several LBAM applications around the automotive industry.

Nearly every major AM system manufacturer is involved in LBAM for the automotive industry. Concept Laser uses their LaserCUSING technology to produce engine components, body work, structural parts, and interior components [130]. EOS has been extensively involved in using SLS on parts for the student Formula race cars [131–133]. Stratasys has used SLA and SLS to fabricate prototypes, investment casting patterns, and final automotive parts. One of the latest projects from Stratasys is teaming up with Equus Automotive to fabricate interior and exterior parts of Equss' model 770 muscle car [134]. Renishaw's SLM system has been used on the Bloodhound Supersonic Car [135]. As more 3D printing companies develop counterparts in the automotive industry, manufacturing competitiveness and the products available to consumers will continue to advance and push the technological limits [51].

One of the most impressive feats in the automotive world is to break the land speed record. Renishaw, based in the United Kingdom, contributed to development of the Bloodhound Supersonic Car (Figure 8.28), which will attempt to break the world's land speed record and

Figure 8.28 Bloodhound supersonic car with a nose cone printed on Renishaw's AM250. (From Green, S., Bloodhound diary: Riding four 'gyroscopes', BBC News, http://www.bbc.com/news/science-environment-36302946, 2016.)

exceed 1000 mph [135,136]. Renishaw's AM250 printed the nose cone out of Ti–6Al–4V. Printing the nose cone allows for a more complex surface curvature and a hollow taper that reduces weight and improves aerodynamic performance. Furthermore, a honeycomb structure can be printed on internal surfaces for strength.

AM allows for custom solutions when complexity poses a challenge for traditional manufacturing methods. Koenigsegg, a Swedish car company, ran into difficulties when they were unable to cast a turbocharger housing for one of its vehicles due to the complexity of the part (Figure 8.29) [137]. To fix this problem, Koenigsegg turned to AM and used SLM to craft the part out of stainless steel for its One:1 supercar. Due to the low volume of production required, this turned out to be a cost-effective solution for the company.

The first 3D-printed electric motorcycle, known as Light Rider, debuted in May 2016 and was developed by APWorks, a subsidiary of Airbus Group (see Figure 8.30) [138]. The motorcycle frame was printed with and EOS M290 and EOS M400 and capitalizes on freedom of design with its bionic frame structure. This example shows how AM can allow the frame's design to match the stress/load capacity needed for real-world driving scenarios. The frame is 30% lighter than conventionally

Figure 8.29 Koenigsegg's patented twisted chamber turbocharger printed by SLM. (From Virgil, R., Koenigsegg 3D printed variable turbocharger, CarVerse, http://www.carverse.com/news/3d-printed-turbocharger/, 2014.)

Figure 8.30 The first 3D-printed electric motorcycle, known as Light Rider, by APWorks. (From APWorks, Light Rider, Taufkirchen, Germany, http://www.apworks.de, 2016.)

manufactured electric motorcycles and made of a proprietary high-strength, corrosion-resistant aluminum alloy known as Scalmalloy. With a 6 kW (8 hp) motor, the 35 kg (77.2 lb) motorcycle can reach speeds of 45 km/h in 3 s. Apart from the frame, other applications involving motorbikes include a custom-built dashboard to house electronics in a 125GP Honda racing motorbike [139]. The dashboard was printed from Windform XT material using SLS.

LBAM has also found a need in the student Formula race car competitions held all over the world. The University of Warwick's Formula race car team extensively used AM to reduce their vehicle's weight by 10% [140]. The team laser sintered titanium splines on a drive shaft, an Inconel exhaust header, and alumide powder for the plenum chamber and inlet runners of the induction system (Figure 8.31). Switching from a hardened steel drive shaft to a carbon fiber and Ti–6Al–4V shaft reduced the weight by 70% and led to better use of the power generated by the engine. The freedom of design afforded by AM allowed the Inconel header to vary in wall thickness from 0.5 to 5.0 mm. The advantage of this capability was a 40% reduction in weight without sacrificing performance in terms of handling high temperatures and exhaustive vibration. Using the EOSINT P 380 to manufacture a lighter weight induction system made of alumide powder provided design flexibility while maintaining a rigid structure. The incorporation of AM to a Formula car helped propel the University of Warwick's team to a top ten finish in the United Kingdom in 2010.

AM has not only had an impact on combustion vehicles but has also led to progress for electric vehicles. In competitions featuring electrically powered vehicles, many important factors affect the performance of the car. The most important consideration is safety, with others including the total weight and size. The use of AM opens up new design possibilities for the construction of complete energy storage system. The Global Formula Team consisting of students from Oregon State University in the United States and Baden-Württemberg Cooperative State University in Germany partnered to build a battery housing for electric formula race cars (see Figure 8.32a) [132]. The volume of the battery housing was reduced by half and the weight by 40%. The efficient design also improved heat dissipation.

As with nearly all electric vehicles, the ability to maintain proper cooling of the engine is of the utmost importance when it comes to performance. The GreenTeam from the University of Stuttgart sought out an innovative way to provide additional cooling to the electric motors of their vehicle. The GreenTeam used an EOS M 290 to develop a cooling jacket made from AlSi10Mg

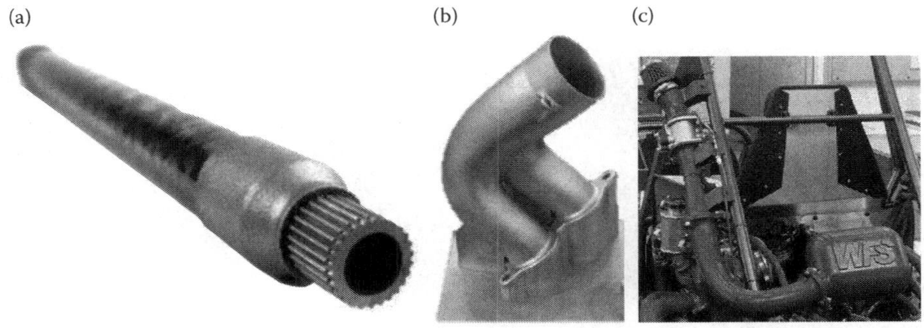

Figure 8.31 Laser sintered parts using various EOS laser sintering systems for a Formula race car: (a) titanium splines on a drive shaft, (b) Inconel exhaust header, and (c) plenum chamber and inlet runners of an induction system made from alumide powder. (From I. I. C. Ltd, Race car's weight cut by 10% using laser sintered titanium, Inconel and Alumide, *Powder Metallurgy Review*, http://www.ipmd.net/articles/001239.html, 2011.)

Figure 8.32 Additive manufacturing in electric formula vehicles produced by EOS laser sintering systems: (a) Global Formula Racing battery housing [132] and (b) University of Stuttgart cooling jacket. (From GmbH, E., Automotive: Formula student—Improved cooling thanks to aluminum jacket with internal helix-structure produced with DMLS, http://www.eos.info/press/customer_case_studies/additive-manufacturing-of-water-cooled-electric-motor-component, 2014.)

(see Figure 8.32b) [131]. The cooling jacket helped improve the performance and efficiency of the engine. The innovative design incorporated built-in cooling channels that improved performance by 37% in the 2013/2014 season. AM helped propel the Stuttgart team to first place at the Formula Student China competition.

The world's first 3D-printed race car, known as Areion, was a joint effort by Materialise based out of Belgium and KU Leauven's Group T Formula team [141]. The main shell of the car was created using one of Materialise's Mammoth Stereolithography machines, which has a build volume of 2100 mm × 680 mm × 800 mm (Figure 8.33). From start to finish, the car body was completed in 3 weeks. By printing the car, students were able to incorporate innovative aspects into the

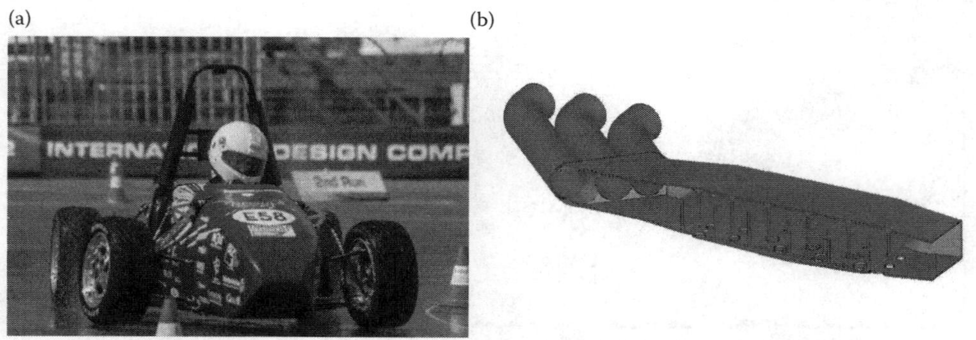

Figure 8.33 Areion formula car printed using Materialise's Mammoth Stereolithography machines: (a) external body and (b) side pods. (From Materialise, The Areion by Formula Group T: The world's first 3D printed race car, http://www.materialise.com/cases/the-areion-by-formula-group-t-the-world-s-first-3d-printed-race-car, 2016.)

Figure 8.34 Ergonomic and lightweight car seat produced by an EOS sintering system. (From GmbH, E., Hanover Fair 2016: EOS is part of the innovative shared stand emotional engineering in lightweight design, http://www.eos.info/press/eos-at-hanover-fair-2016-lightweight-design, 2016.)

design. For example, the nose of the car was printed with a shark skin texture. Although the effect on drag may have been negligible in this case, the idea of printing skin textures onto vehicles may become a mainstream approach to improve fuel efficiency in the future. In addition, complex cooling channels were printed into the side pods for ideal air flow into the engine.

AM is also well suited for internal and ergonomic components of vehicles. Figure 8.34 shows a polymer-based conceptual car seat printed by an EOS system that epitomizes lightweight design with ergonomic form [133]. The ability to produce complex geometries using AM opens up the possibility of performance enhancement by integrating functions like ventilation and posture support.

The previously mentioned examples demonstrate how LBAM has found a need in the automotive industry. Applications that improve efficiency, safety, ergonomics, and performance were discussed. These examples show how AM has an ever-encroaching presence in how goods are produced. As process efficiency improves, LBAM will play a vital role in high-volume production of custom goods.

References

1. A. B. Khoda and B. Koc, Designing controllable porosity for multifunctional deformable tissue scaffolds, *Journal of Medical Devices*, vol. 6, p. 031003, 2012. doi:10.1115/1.4007009.
2. A. K. M. Khoda, I. T. Ozbolat, and B. Koc, Designing heterogeneous porous tissue scaffolds for additive manufacturing processes, *Computer-Aided Design*, vol. 45, pp. 1507–1523, 2013. doi:10.1016/j.cad.2013.07.003.
3. W. Sun, B. Starly, A. Darling, and C. Gomez, Computer-aided tissue engineering: Application to biomimetic modelling and design of tissue scaffolds, *Biotechnology and Applied Biochemistry*, vol. 39, pp. 49–58, 2004. doi:10.1042/BA20030109
4. H. Sahasrabudhe, R. Harrison, C. Carpenter, and A. Bandyopadhyay, Stainless steel to titanium bimetallic structure using LENS™, *Additive Manufacturing*, vol. 5, pp. 1–8, 2015. doi:10.1016/j.addma.2014.10.002.
5. F. P. W. Melchels, J. Feijen, and D. W. Grijpma, A review on stereolithography and its applications in biomedical engineering, *Biomaterials*, vol. 31, pp. 6121–6130, 2010. doi:10.1016/j.biomaterials.2010.04.050.

6. S. Tasoglu and U. Demirci, Bioprinting for stem cell research, *Trends in Biotechnology*, vol. 31, pp. 10–19, 2013. doi:10.1016/j.tibtech.2012.10.005.
7. S. Hollister, Porous scaffold design for tissue engineering, *Nature Material*, vol. 4, pp. 518–526, 2005.
8. D. W. Hutmacher, J. T. Schantz, C. X. F. Lam, K. C. Tan, and T. C. Lim, State of the art and future directions of scaffold-based bone engineering from a biomaterials perspective, *Journal of Tissue Engineering and Regenerative Medicine*, vol. 1, pp. 245–260, 2007.
9. A. Khoda, I. T. Ozbolat, and B. Koc, A functionally gradient variational porosity architecture for hollowed scaffolds fabrication, *Biofabrication*, vol. 3, pp. 1–15, 2011. doi:10.1088/1758-5082/3/3/034106.
10. A. Bandyopadhyay, S. Bose, and S. Das, 3D printing of biomaterials, *MRS Bulletin*, vol. 40, pp. 108–115, 2015. doi:10.1557/mrs.2015.3.
11. M. A. Wettergreen, B. S. Bucklen, B. Starly, E. Yuksel, W. Sun, and M. A. K. Liebschner, Creation of a unit block library of architectures for use in assembled scaffold engineering, *Computer-Aided Design*, vol. 37, pp. 1141–1149, 2005. doi:10.1016/j.cad.2005.02.005.
12. Z. Fang, B. Starly, and W. Sun, Computer-aided characterization for effective mechanical properties of porous tissue scaffolds, *Computer-Aided Design*, vol. 37, pp. 65–72, 2005. doi:10.1016/j.cad.2004.04.002.
13. A. Khoda, I. T. Ozbolat, and B. Koc, Engineered tissue scaffolds with variational porous architecture, *Journal of Biomechanical Engineering*, vol. 133, p. 011001, 2011.
14. D. J. Yoo, Porous scaffold design using the distance field and triply periodic minimal surface models, *Biomaterials*, vol. 32, pp. 7741–7754, 2011. doi:10.1016/j.biomaterials.2011.07.019.
15. P. G. Coelho, S. J. Hollister, C. L. Flanagan, and P. R. Fernandes, Bioresorbable scaffolds for bone tissue engineering: Optimal design, fabrication, mechanical testing and scale-size effects analysis, *Medical Engineering & Physics*, vol. 37, pp. 287–296, 2015. doi:10.1016/j.medengphy.2015.01.004.
16. R. Gauvin, Y.-C. Chen, J. W. Lee, P. Soman, P. Zorlutuna, J. W. Nichol, H. Bae, S. Chen, and A. Khademhosseini, Microfabrication of complex porous tissue engineering scaffolds using 3D projection stereolithography, *Biomaterials*, vol. 33, pp. 3824–3834, 2012. doi:10.1016/j.biomaterials.2012.01.048.
17. W. Bian, D. Li, Q. Lian, X. Li, W. Zhang, K. Wang, and Z. Jin, Fabrication of a bio-inspired β-tricalcium phosphate/collagen scaffold based on ceramic stereolithography and gel casting for osteochondral tissue engineering, *Rapid Prototyping Journal*, vol. 18, pp. 68–80, 2012. doi:10.1108/13552541211193511.
18. K. Arcaute, B. Mann, and R. Wicker, Stereolithography of spatially controlled multi-material bioactive poly(ethylene glycol) scaffolds, *Acta Biomaterialia*, vol. 6, pp. 1047–1054, 2010. doi:10.1016/j.actbio.2009.08.017.
19. B. Duan, M. Wang, W. Y. Zhou, W. L. Cheung, Z. Y. Li, and W. W. Lu, Three-dimensional nanocomposite scaffolds fabricated via selective laser sintering for bone tissue engineering, *Acta Biomaterialia*, vol. 6, pp. 4495–4505, 2010. doi:10.1016/j.actbio.2010.06.024.
20. Y. Nahmias and D. J. Odde, Micropatterning of living cells by laser-guided direct writing: Application to fabrication of hepatic-endothelial sinusoid-like structures, *Nature Protocols*, vol. 1, pp. 2288–2296, 2006.
21. B. Guillotin, A. Souquet, S. Catros, M. Duocastella, B. Pippenger, S. Bellance, R. Bareille, M. Rémy, L. Bordenave, J. Amédée, and F. Guillemot, Laser assisted bioprinting of engineered tissue with high cell density and microscale organization, *Biomaterials*, vol. 31, pp. 7250–7256, 2010. doi:10.1016/j.biomaterials.2010.05.055.
22. American Academy of Orthotists and Prosthetists, O&P Trends & Statistics. Weblink of American Academy: https://www.oandp.org/
23. D. W. R. I. Gibson, B. Stucker, *Additive Manufacturing Technologies Rapid Prototyping to Direct Digital Manufacturing*. New York: Springer, 2010.
24. R. F. Sullivan, Scan/print vs. invest/pour shell-making technologies for CIC hearing aid fittings, *The Hearing Journal*, vol. 60, pp. 21–28, 2007.
25. A. Bandyopadhyay, F. Espana, V. K. Balla, S. Bose, Y. Ohgami, and N. M. Davies, Influence of porosity on mechanical properties and in vivo response of Ti6Al4V implants, *Acta Biomaterialia*, vol. 6, pp. 1640–1648, 2010. doi:10.1016/j.actbio.2009.11.011.
26. A. Bandyopadhyay, B. V. Krishna, W. Xue, and S. Bose, Application of laser engineered net shaping (LENS) to manufacture porous and functionally graded structures for load bearing implants, *Journal of Materials Science: Materials in Medicine*, vol. 20, pp. 29–34, 2008. doi:10.1007/s10856-008-3478-2.

27. H. Anderl, D. Zur Nedden, W. Mu, K. Twerdy, E. Zanon, K. Wicke, and R. Knapp, CT-guided stereolithography as a new tool in craniofacial surgery, *British Journal of Plastic Surgery*, vol. 47, pp. 60–64, 1994.
28. J. Cui, L. Chen, X. Guan, L. Ye, H. Wang, and L. Liu, Surgical planning, three-dimensional model surgery and preshaped implants in treatment of bilateral craniomaxillofacial post-traumatic deformities, *Journal of Oral and Maxillofacial Surgery*, vol. 72, pp. 1138.e1–11.8.e14, 2014.
29. P. S. D'Urso, D. J. Effeney, W. J. Earwaker, T. M. Barker, M. J. Redmond, R. G. Thompson, and F. H. Tomlinson, Custom cranioplasty using stereolithography and acrylic, *British Journal of Plastic Surgery*, vol. 53, pp. 200–204, 2000.
30. H. Klein, W. Schneider, G. Alzen, E. Voy, and R. Günther, Pediatric craniofacial surgery: Comparison of milling and stereolithography for 3D model manufacturing, *Pediatric Radiology*, vol. 22, pp. 458–460, 1992.
31. M. Robiony, I. Salvo, F. Costa, N. Zerman, M. Bazzocchi, F. Toso, C. Bandera, S. Filippi, M. Felice, and M. Politi, Virtual reality surgical planning for maxillofacial distraction osteogenesis: The role of reverse engineering rapid prototyping and cooperative work, *Journal of Oral and Maxillofacial Surgery*, vol. 65, pp. 1198–1208, 2007.
32. R. Petzold, H.-F. Zeilhofer, and W. Kalender, Rapid prototyping technology in medicine—Basics and applications, *Computerized Medical Imaging and Graphics*, vol. 23, pp. 277–284, 1999.
33. E. De Momi, E. Pavan, B. Motyl, C. Bandera, and C. Frigo, Hip joint anatomy virtual and stereolithographic reconstruction for preoperative planning of total hip replacement, in *International Congress Series*, vol. 1281, pp. 708–712, 2005.
34. K. Arcaute, B. K. Mann, and R. B. Wicker, Fabrication of off-the-shelf multilumen poly(ethylene glycol) nerve guidance conduits using stereolithography, *Tissue Engineering Part C: Methods*, vol. 17, pp. 27–38, 2010.
35. K. Arcaute, L. Ochoa, B. Mann, and R. Wicker, Stereolithography of PEG hydrogel multi-lumen nerve regeneration conduits, in *ASME 2005 International Mechanical Engineering Congress and Exposition Manufacturing Engineering and Materials Handling, Parts A and B* Orlando, FL, USA, November 5 – 11, 2005, pp. 161–167. http://proceedings.asmedigitalcollection.asme.org/proceeding.aspx?articleid=1582732
36. V. Popov, A. Evseev, A. Ivanov, V. Roginski, A. Volozhin, and S. Howdle, Laser stereolithography and supercritical fluid processing for custom-designed implant fabrication, *Journal of Materials Science: Materials in Medicine*, vol. 15, pp. 123–128, 2004.
37. M. Schuster, C. Turecek, B. Kaiser, J. Stampfl, R. Liska, and F. Varga, Evaluation of biocompatible photopolymers I: Photoreactivity and mechanical properties of reactive diluents, *Journal of Macromolecular Science, Part A*, vol. 44, pp. 547–557, 2007.
38. M. Schuster, C. Turecek, A. Mateos, J. Stampfl, R. Liska, and F. Varga, Evaluation of biocompatible photopolymers II: Further reactive diluents, *Monatshefte für Chemie-Chemical Monthly*, vol. 138, pp. 261–268, 2007.
39. T. Matsuda and M. Mizutani, Liquid acrylate-endcapped biodegradable poly (ε-caprolactone-co-trimethylene carbonate). II. Computer-aided stereolithographic microarchitectural surface photoconstructs, *Journal of Biomedical Materials Research*, vol. 62, pp. 395–403, 2002.
40. A. J. Pinkerton, [INVITED] Lasers in additive manufacturing, *Optics & Laser Technology*, vol. 78, Part A, pp. 25–32, 2016. doi:10.1016/j.optlastec.2015.09.025.
41. A. Uriondo, M. Esperon-Miguez, and S. Perinpanayagam, The present and future of additive manufacturing in the aerospace sector: A review of important aspects, *Proceedings of the Institution of Mechanical Engineers, Part G: Journal of Aerospace Engineering*, vol. 229, pp. 2132–2147, 2015.
42. Standard, A.S.T.M. *F2792. 2012. Standard Terminology for Additive Manufacturing Technologies.* West Conshohocken, PA: ASTM International. http://web.mit.edu/2.810/www/files/readings/AdditiveManufacturingTerminology.pdf, 2012.
43. A. Gasser, G. Backes, I. Kelbassa, A. Weisheit, and K. Wissenbach, Laser additive manufacturing, *Laser Technik Journal*, vol. 7, pp. 58–63, 2010.
44. S. Fletcher, P. Norman, S. Galloway, and G. Burt, Fault detection and location in DC systems from initial di/dt measurement, in *Euro Tech Con Conference*, Manchester, UK, 2012.

45. S. James. And you think you're trying to save gas...: Meet the man in charge of cutting American Airlines' fuel bill. http://www.nbcnews.com/, 2012.
46. I. Gibson, D. W. Rosen, and B. Stucker, *Additive Manufacturing Technologies: Rapid Prototyping to Direct Digital Manufacturing*. New York: Springer, 2010.
47. R. Huang, M. Riddle, D. Graziano, J. Warren, S. Das, S. Nimbalkar, J. Cresko, and E. Masanet, Energy and emissions saving potential of additive manufacturing: The case of lightweight aircraft components, *Journal of Cleaner Production*, vol. 135, pp. 1559–1570, 2016.
48. B. Lyons, Additive manufacturing in aerospace: Examples and research outlook, *The Bridge*, vol. 44, pp. 13–19, 2014.
49. D. H. Freedman, Layer by layer, *Technology Review*, vol. 115, pp. 50–53, 2012.
50. G. Schiller, Additive manufacturing for Aerospace, in *2015 IEEE Aerospace Conference*, Big Sky, MT, USA, pp. 1–8, March 7–14, 2015.
51. S. L. Ford, Additive manufacturing technology: Potential implications for US manufacturing competitiveness, *Journal of International Commerce and Economics*, pp. 2–35, 2014. https://ssrn.com/abstract=2501065.
52. B. P. Conner, G. P. Manogharan, A. N. Martof, L. M. Rodomsky, C. M. Rodomsky, D. C. Jordan, and J. W. Limperos, Making sense of 3-D printing: Creating a map of additive manufacturing products and services, *Additive Manufacturing*, vol. 1, pp. 64–76, 2014.
53. J. Waller, R. Saulsberry, B. Parker, K. Hodges, E. Burke, K. Taminger, D. Chimenti, and L. Bond, Summary of NDE of additive manufacturing efforts in NASA, in *AIP Conference Proceedings*, vol. 1650, pp. 51-62, 2015. http://aip.scitation.org/doi/abs/10.1063/1.4914594
54. I. Gibson, D. Rosen, and B. Stucker, *Additive Manufacturing Technologies: 3D Printing, Rapid Prototyping, and Direct Digital Manufacturing*. New York: Springer, 2014.
55. V. Krishna Balla, L. B. Roberson, G. W. O'Connor, S. Trigwell, S. Bose, and A. Bandyopadhyay, First demonstration on direct laser fabrication of lunar regolith parts, *Rapid Prototyping Journal*, vol. 18, pp. 451–457, 2012.
56. C. S. Huskamp, Methods and systems for controlling and adjusting heat distribution over a part bed, Google Patents, US 7,515,986 B2, 2009.
57. R. B. Booth, B. C. Thornton, D. L. Vanelli, and M. L. Gardiner, Methods and systems for fabricating fire retardant materials, Google Patents, US 20080153947 A1, 2012.
58. D. Manfredi, F. Calignano, M. Krishnan, R. Canali, E. P. Ambrosio, S. Biamino, D. Ugues, M. Pavese, and P. Fino, Additive manufacturing of Al alloys and aluminium matrix composites (AMCs), W. A. Monteiro (Ed.), *Light Metal Alloys Applications*. Croatia: InTech, 250, 2014.
59. S. Bland, New report: 3D printing opportunities in the metal powder industry, *Metal Powder Report*, vol. 70, p. 40, 2015.
60. S. Bland and B. Conner, Mapping out the additive manufacturing landscape, *Metal Powder Report*, vol. 70, pp. 115–119, 2015.
61. M. Cotteleer and J. Joyce, 3D opportunity: Additive manufacturing paths to performance, innovation, and growth, *Deloitte Review*, vol. 14, pp. 5–19, 2014.
62. P. Timothy, *3D Printing: Opportunities, Challenges and Policy Implications of Additive Manufacturing*. GAO, United States Government Accountability Office, 2015.
63. A. Weber, Additive manufacturing takes off in Aerospace Industry, Assembly, January 2014. http://www.assemblymag.com/articles/93176-additive-manufacturing-takes-off-in-aerospace-industry
64. S. Das, M. Wohlert, J. J. Beaman, and D. L. Bourell, Producing metal parts with selective laser sintering/hot isostatic pressing, *JoM*, vol. 50, pp. 17–20, 1998.
65. M. Tomlin and J. Meyer, Topology optimization of an additive layer manufactured (ALM) aerospace part, in *Proceeding of the 7th Altair CAE Technology Conference*, UK, 2011, pp. 1–9. http://www.pfonline.com/cdn/cms/uploadedFiles/Topology-Optimization-of-an-Additive-Layer-Manufactured-Aerospace-Part.pdf
66. L. Nickels, AM and aerospace: An ideal combination, *Metal Powder Report*, vol. 70, pp. 300–303, 2015.
67. S. Babu and R. Goodridge, Additive manufacturing, *Materials Science and Technology*, 2015.
68. A. H. Lefebvre, *Gas Turbine Combustion*. Boca Raton, FL: CRC Press, 1998.

69. I. J. Gausemeier, Thinking ahead the future of additive manufacturing—Analysis of promising industries, Heinz Nixdorf Institute, University of Paderborn Product Engineering, Paderborn, Germany, p. 14, 2011.
70. B. Udo, Laser-sintering in the aerospace industry is on the rise, *Laser Technik Journal*, vol. 6, pp. 44–47, 2009.
71. J. M. Waller, B. H. Parker, K. L. Hodges, E. R. Burke, J. L. Walker, and E. R. Generazio, Nondestructive evaluation of additive manufacturing, state-of-the discipline report, NASA, vol. 17, 2014.
72. R. W. Appleton, Additive manufacturing overview for the United States marine corps, R W Appleton & Company, Inc, 2014.
73. E. Betts, D. Eddleman, D. Reynolds, and N. Hardin, *Using Innovative Technologies for Manufacturing Rocket Engine Hardware*. Huntsville: Engineering Research and Consulting, Inc, 2011.
74. S. Das, T. P. Fuesting, G. Danyo, L. E. Brown, J. J. Beaman, and D. L. Bourell, Direct laser fabrication of superalloy cermet abrasive turbine blade tips, *Materials & Design*, vol. 21, pp. 63–73, 2000.
75. T. R. Ann, Aircarft engine maker opens additive manufacturing lab, *Design News*, vol. 68, 2013.
76. D. Jeff, Materials move aerospace additive manufacturing from prototypes to finished products, *Design News*, vol. 68, 2013.
77. E. Uhlmann, R. Kersting, T. B. Klein, M. F. Cruz, and A. V. Borille, Additive manufacturing of titanium alloy for aircraft components, *Procedia CIRP*, vol. 35, pp. 55–60, 2015.
78. S. Das, M. Wohlert, J. J. Beaman, and D. L. Bourell, Processing of titanium net shapes by SLS/HIP, *Materials & Design*, vol. 20, pp. 115–121, 1999.
79. M. S. Hossain, J. A. Gonzalez, R. M. Hernandez, M. A. I. Shuvo, J. Mireles, A. Choudhuri, Y. Lin, and R. B. Wicker, Fabrication of smart parts using powder bed fusion additive manufacturing technology, *Additive Manufacturing*, vol. 10, pp. 58–66, 2016.
80. S. Biamino, A. Penna, U. Ackelid, S. Sabbadini, O. Tassa, P. Fino, M. Pavese, P. Gennaro, and C. Badini, Electron beam melting of Ti–48Al–2Cr–2Nb alloy: Microstructure and mechanical properties investigation, *Intermetallics*, vol. 19, pp. 776–781, 2011.
81. L. E. Murr, S. M. Gaytan, D. A. Ramirez, E. Martinez, J. Hernandez, K. N. Amato, P. W. Shindo, F. R. Medina, and R. B. Wicker, Metal fabrication by additive manufacturing using laser and electron beam melting technologies, *Journal of Materials Science & Technology*, vol. 28, pp. 1–14, 2012.
82. D. Pal, N. Patil, and B. E. Stucker, Prediction of mechanical properties of electron beam melted Ti6Al4V parts using dislocation density based crystal plasticity framework, in *Proceedings of the Solid Freeform Fabrication Symposium*, Austin, TX, pp. 6–8, 2012.
83. J. Schwerdtfeger and C. Körner, Selective electron beam melting of Ti–48Al–2Nb–2Cr: Microstructure and aluminium loss, *Intermetallics*, vol. 49, pp. 29–35, 2014.
84. R. Udroiu, Tehnologia Inovativa, Vol 3-4, 2014. https://www.researchgate.net/profile/
85. C. Yan, L. Hao, A. Hussein, S. L. Bubb, P. Young, and D. Raymont, Evaluation of light-weight AlSi10Mg periodic cellular lattice structures fabricated via direct metal laser sintering, *Journal of Materials Processing Technology*, vol. 214, pp. 856–864, 2014.
86. W. W. Wits, S. J. Weitkamp, and J. van Es, Metal additive manufacturing of a high-pressure micro-pump, *Procedia CIRP*, vol. 7, pp. 252–257, 2013.
87. M. Shellabear and O. Nyrhilä, DMLS-Development history and state of the art, in *Proceedings of the 4th Laser Assisted Netshape Engineering 4 LANE*, pp. 21–24, 2004.
88. B. Duleba, F. Greškovič, and J. Sikora, Materials and finishing methods of DMLS manufactured parts, *Transfer Inovácií*, vol. 21, pp. 143–148, 2011.
89. M. Fateri and M. Khosravi, On-site additive manufacturing by selective laser melting of composite objects, in *Concepts and Approaches for Mars Exploration*, Houston TX, pp. 12–14, 2012.
90. W. King, A. Anderson, R. Ferencz, N. Hodge, C. Kamath, S. A. Khairallah, and A. M. Rubenchik, Laser powder bed fusion additive manufacturing of metals; physics, computational, and materials challenges, *Applied Physics Reviews*, vol. 2, p. 041304, 2015.
91. W. Morrow, H. Qi, I. Kim, J. Mazumder, and S. Skerlos, Environmental aspects of laser-based and conventional tool and die manufacturing, *Journal of Cleaner Production*, vol. 15, pp. 932–943, 2007.
92. H. Lipson and M. Kurman, *Fabricated: The New World of 3D Printing*. Indianapolis: John Wiley & Sons, 2013.

93. S. Kumar and S. Pityana, Laser-based additive manufacturing of metals, *Advanced Materials Research*, vol. 227, pp. 92–95, 2011.
94. R. P. Mudge and N. R. Wald, Laser engineered net shaping advances additive manufacturing and repair, *Welding Journal-New York*, vol. 86, p. 44, 2007.
95. M. Peters, J. Kumpfert, C. H. Ward, and C. Leyens, Titanium alloys for aerospace applications, *Advanced Engineering Materials*, vol. 5, pp. 419–427, 2003.
96. Y. Zhai, D. A. Lados, and J. L. LaGoy, Additive manufacturing: Making imagination the major limitation, *JOM*, vol. 66, pp. 808–816, 2014.
97. B. Baufeld, E. Brandl, and O. Van der Biest, Wire based additive layer manufacturing: Comparison of microstructure and mechanical properties of Ti–6Al–4V components fabricated by laser-beam deposition and shaped metal deposition, *Journal of Materials Processing Technology*, vol. 211, pp. 1146–1158, 2011.
98. L. E. Murr, E. V. Esquivel, S. A. Quinones, S. M. Gaytan, M. I. Lopez, E. Martinez et al., Microstructures and mechanical properties of electron beam-rapid manufactured Ti–6Al–4V biomedical prototypes compared to wrought Ti–6Al–4V, *Materials Characterization*, vol. 60, pp. 96–105, 2009.
99. L. Facchini, E. Magalini, P. Robotti, and A. Molinari, Microstructure and mechanical properties of Ti–6Al–4V produced by electron beam melting of pre-alloyed powders, *Rapid Prototyping Journal*, vol. 15, pp. 171–178, 2009.
100. M. Koike, K. Martinez, L. Guo, G. Chahine, R. Kovacevic, and T. Okabe, Evaluation of titanium alloy fabricated using electron beam melting system for dental applications, *Journal of Materials Processing Technology*, vol. 211, pp. 1400–1408, 2011.
101. L. Facchini, E. Magalini, P. Robotti, A. Molinari, S. Höges, and K. Wissenbach, Ductility of a Ti–6Al–4V alloy produced by selective laser melting of prealloyed powders, *Rapid Prototyping Journal*, vol. 16, pp. 450–459, 2010.
102. J. Ding, Thermo-mechanical analysis of wire and arc additive manufacturing process, 2012.
103. M. P. Boyce, *Gas Turbine Engineering Handbook*. Oxford: Elsevier, 2011.
104. T. Nijdam and R. V. Gestel, Service experience with single crystal superalloys for high pressure turbine shrouds, 2011.
105. E. C. Santos, M. Shiomi, K. Osakada, and T. Laoui, Rapid manufacturing of metal components by laser forming, *International Journal of Machine Tools and Manufacture*, vol. 46, pp. 1459–1468, 2006.
106. T. Wohlers and T. Gornet, History of additive manufacturing, Wohlers Report: Additive manufacturing and 3D printing state of the industry annual worldwide progress report, 2011.
107. C. Atwood, M. Ensz, D. Greene, M. Griffith, L. Harwell, D. Reckaway et al., *Laser Engineered Net Shaping (Lens): A Tool for Direct Fabrication of Metal Parts*. 1998, 7th International Congress on Applications of Lasers and Elector-Optics, Orlando, FL; 11/16-19/1998
108. R. R. Dehoff, C. Tallman, C. E. Duty, W. H. Peter, Y. Yamamoto, W. Chen et al., Case study: Additive manufacturing of aerospace brackets, *Advanced Materials and Processes*, vol. 171, 2013.
109. A. Angrish, A critical analysis of additive manufacturing technologies for aerospace applications, in *2014 IEEE Aerospace Conference*, 2014, Orlando, FL; 11/16-19/1998 pp. 1–6.
110. E. Herderick, Additive manufacturing of metals: A review, *Materials Science & Technology*, pp. 1413–1425, 2011.
111. D. Hu and R. Kovacevic, Sensing, modeling and control for laser-based additive manufacturing, *International Journal of Machine Tools and Manufacture*, vol. 43, pp. 51–60, 2003. doi:10.1016/S0890-6955(02)00163-3
112. P. Kobryn, N. Ontko, L. Perkins, and J. Tiley, Additive manufacturing of aerospace alloys for aircraft structures, DTIC Document2006.
113. D. Gu, Laser additive manufacturing (AM): Classification, processing philosophy, and metallurgical mechanisms, in *Laser Additive Manufacturing of High-Performance Materials*. New York: Springer, 2015, pp. 15–71.
114. D. Hu, H. Mei, and R. Kovacevic, Improving solid freeform fabrication by laser-based additive manufacturing, *Proceedings of the Institution of Mechanical Engineers, Part B: Journal of Engineering Manufacture*, vol. 216, pp. 1253–1264, 2002.

115. L. E. Murr, S. Gaytan, A. Ceylan, E. Martinez, J. Martinez, D. Hernandez et al., Characterization of titanium aluminide alloy components fabricated by additive manufacturing using electron beam melting, *Acta Materialia*, vol. 58, pp. 1887–1894, 2010.
116. C. Lindemann, U. Jahnke, M. Moi, and R. Koch, Analyzing product lifecycle costs for a better understanding of cost drivers in additive manufacturing, in *23th Annual International Solid Freeform Fabrication Symposium–An Additive Manufacturing Conference*, Austin, TX, August 6–8, 2012.
117. S. H. Huang, P. Liu, A. Mokasdar, and L. Hou, Additive manufacturing and its societal impact: A literature review, *The International Journal of Advanced Manufacturing Technology*, vol. 67, pp. 1191–1203, 2013.
118. E. Atzeni and A. Salmi, Economics of additive manufacturing for end-usable metal parts, *The International Journal of Advanced Manufacturing Technology*, vol. 62, pp. 1147–1155, 2012.
119. K. J. Brookes, Aviation finds that extra dimension, *Metal Powder Report*, vol. 5, pp. 239–244, 2015.
120. K. Taminger, R. A. Hafley, and D. L. Dicus, Solid freeform fabrication: An enabling technology for future space missions, 2002.
121. C. Emmelmann, P. Sander, J. Kranz, and E. Wycisk, Laser additive manufacturing and bionics: Redefining lightweight design, *Physics Procedia*, vol. 12, pp. 364–368, 2011.
122. T. Campbell, C. Williams, O. Ivanova, and B. Garrett, Could 3D printing change the world? Technologies, potential, and implications of additive manufacturing, Atlantic Council, Washington, DC, 2011.
123. W. E. Frazier, Metal additive manufacturing: A review, *Journal of Materials Engineering and Performance*, vol. 23, pp. 1917–1928, 2014.
124. V. Petrovic, J. Vicente Haro Gonzalez, O. Jorda Ferrando, J. Delgado Gordillo, J. Ramon Blasco Puchades, and L. Portoles Grinan, Additive layered manufacturing: Sectors of industrial application shown through case studies, *International Journal of Production Research*, vol. 49, pp. 1061–1079, 2011.
125. L. Winkless, Is additive manufacturing truly the future? *Metal Powder Report*, vol. 70, pp. 229–232, 2015.
126. B. Vayre, F. Vignat, and F. Villeneuve, Metallic additive manufacturing: State-of-the-art review and prospects, *Mechanics & Industry*, vol. 13, pp. 89–96, 2012.
127. T. T. Wohlers and T. Caffrey, Wohlers report 2015: 3D printing and additive manufacturing state of the industry annual worldwide progress report, Wohlers Associates, 2015.
128. J. Gausemeier, N. Echterhoff, M. Kokoschka, and M. Wall, Thinking ahead the future of additive manufacturing—Analysis of promising industries, Study for the Direct Manufacturing Research Center, Paderborn, Germany, 2011.
129. I. Gauscmeier, Thinking ahead the future of additive manufacturing-innovation roadmapping of required of required advancements, DMRC study Part-3, University of Paderborn, pp. 48–51, 2013.
130. C. L. A. F. D. c. i. v. construction. http://www.concept-laser.de.
131. E. GmbH, Automotive: Formula student—Improved cooling thanks to aluminum jacket with internal helix-structure produced with DMLS. http://www.eos.info/press/customer_case_studies/additive-manufacturing-of-water-cooled-electric-motor-component, 2014.
132. E. GmbH, Automotive: Formula student racing-additive manufactured lightweight battery housing for an electric-powered race car. http://www.eos.info/press/customer_case_studies/formula_student, 2013.
133. E. GmbH, Hanover Fair 2016: EOS is part of the innovative shared stand emotional engineering in lightweight design. http://www.eos.info/press/eos-at-hanover-fair-2016-lightweight-design, 2016.
134. Stratasys, The new American muscle car. https://www.stratasysdirect.com/case-studies/new-american-muscle-car/, 2016.
135. Renishaw, Bloodhound SSC turns to Renishaw for 3D printing expertise. http://www.renishaw.com/en/bloodhound-ssc-turns-to-renishaw-for-3d-printing-expertise--21224, 2013.
136. A. Green, Bloodhound Diary: Riding four 'gyroscopes', BBC News. http://www.bbc.com/news/science-environment-36302946, 2016.
137. R. Virgil, KOENIGSEGG 3D printed variable turbocharger, CarVerse. http://www.carverse.com/news/3d-printed-turbocharger/, 2014.

138. APWorks, Light Rider, Taufkirchen, Germany. https://www.lightrider.apworks.de/en, 2016.
139. L. Cevolini, G. Testoni, and S. Davis, Rapid manufacturing and continued development of highly stressed fibre-reinforced plastic parts: Motorbike dash assembly made by Windform XT and SLS Technology, *CRP Technology*, pp. 1–11, 2008.
140. I. I. C. Ltd, Race car's weight cut by 10% using laser sintered titanium, Inconel and Alumide, *Powder Metallurgy Review*. http://www.ipmd.net/articles/001239.html, 2011.
141. Materialise, The Areion by Formula Group T: The world's first 3D printed race car. http://www.materialise.com/cases/the-areion-by-formula-group-t-the-world-s-first-3d-printed-race-car, 2016.

Chapter 9

Economics of Additive Manufacturing

Douglas S. Thomas
National Institute of Standards and Technology

Contents

CHAPTER OUTLINE

In this chapter, we discuss:

- Additive manufacturing's contribution to the US and global economy
- Costs and benefits associated with additive manufacturing
- Cost models associated with additive manufacturing
- Adoption and diffusion of additive manufacturing

9.1 Introduction

Learning Objectives

- Review the topics of this chapter
- Discuss the different aspects of additive manufacturing economics

There are three aspects to the economics of additive manufacturing. The first involves measuring the size of additive manufacturing. This includes measuring the value of the goods produced using this technology in the context of the total economy. The second aspect involves measuring the costs and benefits of using this technology. It includes understanding when and why additive manufacturing is more cost effective than traditional manufacturing methods. It also involves understanding other advantages such as manufacturing new products that might not be possible with traditional methods. The last aspect of additive manufacturing economics is the adoption and diffusion of this technology. Additive manufacturing is significantly different from traditional methods; thus, determining when and how to take advantage of the benefits of additive manufacturing is a challenge. Manufacturers have to invest resources into understanding whether additive manufacturing makes sense for their business. If they adopt this technology, they will have to train or hire employees who are able to utilize it. This chapter is not intended to be a comprehensive review of these issues, but rather it provides an introduction into each of the topics.

Section 9.2 discusses measuring the size of the additive manufacturing industry in the context of the total economy. It discusses output, value-added, and standard industry classification systems. Section 9.3 discusses categories of costs and benefits for additive manufacturing. It includes a discussion of ill-structured costs such as inventory and transportation. It also discusses well-structured costs such as materials and energy use. Section 9.4 presents the use of these costs in the context of the two primary cost models for additive manufacturing as well. This section discusses the cost of traditional manufacturing compared to additive manufacturing. Section 9.5 discusses the adoption and diffusion of additive manufacturing as well. It includes a discussion about the circumstances where this technology tends to be cost effective. It also discusses five methods for measuring economic performance: present value of net benefits, present value of net savings, benefit-to-cost ratio, savings-to-investment ratio, and the adjusted internal rate of return. This section concludes with an overview of the trends in the adoption of additive manufacturing.

When possible, this chapter will utilize studies and research on metal parts; however, due to the limited amount of research on additive manufacturing economics, there are some instances of other types of materials that are discussed and used as illustrations. It is also important to note that some of the economic issues discussed in this chapter are about current capabilities of additive manufacturing, while others are about potential future capabilities. For example, there is some discussion regarding this technology's ability to produce assembled products in one build; however, the current state of technology has some limitation on this ability. Since this is a new technology that is rapidly changing, the current abilities may advance significantly in a short period of time. For this reason, it is important to consider the potential changes the technology may facilitate rather than limiting the discussion to only the current state of technology.

9.2 Additive Manufacturing and the Global Economy

Learning Objectives

- Understand metrics for measuring economic activity
- Understand standard industry classification systems
- Understand additive manufacturing's current contribution to the global economy

Occasionally, there are exaggerated claims and criticisms made about the economics of new technologies and their current status or impact on the economy. Claims might even utilize economic data, but may not fairly represent the economic status. For this reason, it is important to understand standard measures of economic activity and how additive manufacturing fits into the global economy using these measures.

9.2.1 Measuring Economic Activity

In 2013, there was \$9.5 trillion in global manufacturing value added of the total \$53.7 trillion value added (also referred to as gross domestic product or GDP) for all industries, according to the United Nations Statistics Division. The top five manufacturing countries accounted for \$5.5 trillion or 57.6% of the manufacturing value added: the United States (18.8%), China (18.5%), Japan (10.5%), Germany (7.0%), and France (2.8%) (UNSD 2015). Value added is the increase in the value of output at a given stage of production; that is, the value of output minus the cost of inputs from other firms (Dornbusch et al. 2000). The primary elements that remain after subtracting inputs are taxes, compensation to employees, and gross operating surplus; thus, the sum of these also equals value added. Gross operating surplus is used to calculate profit, which is gross operating surplus less the depreciation of capital such as buildings and machinery.

$$\begin{aligned}\text{Value added} &= \text{Output} - \text{Inputs from other firms} \\ &= \text{Taxes} + \text{Compensation} + \text{Gross operating surplus.}\end{aligned} \tag{9.1}$$

A nation's economy is typically measured as GDP, which is the sum of the value added for all industries. The distinction between output and value added is important, as output (also called shipments) double counts the value of production. This double counting occurs because the data is collected from each establishment or factory location, including locations producing intermediate parts. An intermediate product might leave a factory and the revenue that is generated for that location is counted as part of the nation's output. It is delivered to another factory, but when it leaves this factory as part of a finished product, the revenue is counted as output again. For example, a car radio might be produced at an electronics factory and shipped to an automotive factory. The revenue to the electronics factory is counted as part of the nation's output when it leaves the electronics factory. The revenue from the second factory is also counted as output when the assembled car leaves that factory, and this value of output includes all of the components in the automobile, including the radio. So, when the output from the assembly factory is added to the output from the electronics factory to calculate national or global output, there is a double counting of the radio. Value added takes the output from the assembly factory and subtracts off the cost of all of the intermediate components made at other locations such as the radio. Thus, when the value for the radio is added to the value for the assembly, it is not double counting as it has been subtracted from the output counted at the assembly factory.

9.2.2 Additive Manufacturing's Share of GDP

Calculating additive manufacturing's contribution to value added has some challenges. The value added and output data that is collected and distributed is typically organized by standard industry classifications such as the International Standard Industrial Classification system (ISIC) or the North American Industry Classification System (NAICS). NAICS is the standard system used by federal statistical agencies classifying business establishments in the United States. NAICS was jointly developed by the US Economic Classification Policy Committee, Statistics Canada, and Mexico's Instituto Nacional de Estadistica y Geografia and was adopted in 1997. NAICS codes are categorized at varying levels of detail, as seen in Table 9.1. This table presents the lowest level of detail, which is the two-digit NAICS, where there are 20 categories. Additional detail is attained by adding more digits; thus, three digits provide more detail than the two digits, and the four digits provide more detail than the three digits. The maximum is six digits, as illustrated for automobile manufacturing (NAICS 336111) and light truck and utility manufacturing (NAICS 336112). Sometimes a two-, three-, four-, or five-digit code is followed by zeros, which do not represent categories. They are null or placeholders. For example, the code 336000 represents NAICS 336. Other industry classifications, such as the ISIC, have similar designs where more digits represent more details and zeros represent placeholders.

When most individuals ask how much additive manufacturing contributes to the economy, they either want to know the value added of the goods being produced using this technology, or the output of the goods using this technology. Output would, essentially, be the revenue received for the parts produced using additive manufacturing technology, while value added would include

Table 9.1 North American Industry Classification System, Two-Digit Codes

Sector	*Description*
11	Agriculture, forestry, fishing, and hunting
21	Mining, quarrying, and oil and gas extraction
22	Utilities
23	Construction
31–33	Manufacturing
336	Transportation equipment manufacturing
3361	Motor vehicle manufacturing
33611	Automobile and light duty motor vehicle manufacturing
336111	Automobile manufacturing
336112	Light truck and utility manufacturing
42	Wholesale trade
44–45	Retail trade
48–49	Transportation and warehousing
51	Information
52	Finance and insurance
53	Real estate and rental and leasing
54	Professional, scientific, and technical services
55	Management of companies and enterprises
56	Administrative and support, and waste management and remediation services
61	Educational services
62	Health care and social assistance
71	Arts, entertainment, and recreation
72	Accommodation and food services
81	Other services (except public administration)
92	Public administration

the output less the cost of inputs such as the materials and machinery. Unfortunately, the standard classification systems tend to be oriented around the good or service being produced at a particular location. For example, there are ISIC codes for computers and peripheral equipment, furniture, motor vehicles, and motor vehicle parts. Additive manufacturing is a process that occurs within each of these categories; therefore, it is not separated into its own category. Fortunately, Wohler Associates, Inc. collects and publishes data on additive manufacturing (Wohlers 2015).

Wohlers estimates the 2014 revenue from additive manufacturing worldwide to be $4.103 billion; however, the estimate that is most consistent with the definitions previously discussed is Wohlers' estimate for service providers. It states that there was "an estimated $1.307 billion from the sale of parts produced by additive manufacturing systems in 2014" (Wohlers 2015). In order to estimate the value added, we would need to subtract off the materials, machinery, and other intermediate goods that were purchased for production. Wohlers estimates that material sales amounted to $640 million in 2014; thus, an estimate of global value added for additive manufacturing can be estimated by taking the $1.307 billion less the $640 million for materials, totaling $667 million. This equates to 0.01% of total global manufacturing value added. An estimate for each category of production is provided in Table 9.2. These figures, however, are for all material types and not just metal parts. Unfortunately, there is not a breakout for metal parts; however, Wohlers (2015) provides an estimate of $48.7 million spent on metal materials for additive manufacturing, which amounts to approximately 7.6% of the $640 million spent on all materials shown in Table 9.2. Assuming that the proportion is the same for shipments and value added, there were $99.5 million in metal part shipments (1307 × 7.6%) and $50.8 million in value added (667 × 7.6%).

Example 1

Table 9.3 shows economic data on the US economy for 2013. The data is in billions of dollars and includes intermediate purchases (i.e., cost of inputs), compensation, taxes, gross operating surplus, and total output. The value added for each industry is the sum of compensation, taxes, and gross operating surplus. It can also be calculated as total output less total intermediate purchases. The total value added for the United States in 2013 is the sum of the row of value added or $16768 billion.

Table 9.2 Global Additive Manufacturing Shipments and Value Added

Category	Percent of Total AM Made Products[a] (%)	Shipments of AM Products ($millions, 2014)[b]	AM Materials ($millions 2014)	AM Value Added ($millions, 2014)[b]
Motor vehicles	16.1	210.4	103.0	107.4
Aerospace	14.8	193.4	94.7	98.7
Industrial/business machines	17.5	228.7	112.0	116.7
Medical/dental	13.1	171.2	83.8	87.4
Government/military	6.6	86.3	42.2	44.0
Architectural	3.2	41.8	20.5	21.3
Consumer products/electronics, academic institutions, and other	28.7	375.1	183.7	191.4
Total	100.0	1307.0	640.0	667.0

Note: Numbers may not add up to total due to rounding.

[a] Value is from Wohlers (2015).

[b] It is assumed that the share of revenue for each category is same for the materials.

Table 9.3 US Economic Data, 2013 ($Billions)

NAICS	*11–22*	*23*	*31–33*	*42–45*	*48–49*	*51–53*	*54–81, Other*
NAICS Description	*Agriculture, Forestry, Mining, and Utilities*	*Construction*	*Manufacturing*	*Wholesale and Retail Trade*	*Transportation and Warehousing*	*Information, Finance, Insurance, Real Estate, and Rentals*	*Professional and Business Services*
Total intermediate	553	515	3,912	1,044	515	2,345	4,069
Employee compensation	216	390	932	997	283	1,048	4,987
Taxes	100	8	85	402	28	324	157
Gross operating surplus	627	222	1,011	571	171	2,789	1,421
Total industry output	1,496	1,135	5,940	3,014	996	6,506	10,634
Value added	943	620	2,028	1,970	481	4,161	6,565

9.3 Categories of Costs and Benefits for Additive Manufacturing

Learning Objectives

- Understand ill-structured costs
- Understand well-structured costs
- Learn the seven categories of waste from lean manufacturing

The manufacturing supply chain is a system that moves products between the suppliers and the consumers. Many costs are hidden in the supply chain. Additive manufacturing may, potentially, have significant impacts on the design and size of this system, reducing its associated costs (Reeves 2008). The costs of production can be categorized in two ways: (1) "well-structured" costs such as labor, material, and machine costs and (2) "ill-structured" costs such as those associated with build failure, machine setup, and inventory (Young 1991). Well-structured costs are easier to track and measure than ill-structured costs. As a result, there tends to be more focus on well-structured costs of additive manufacturing than ill-structured costs; however, some of the more significant benefits and cost savings in additive manufacturing may be hidden in the ill-structured costs.

Concepts from lean manufacturing, a method for eliminating waste within a production system, might prove useful when examining the costs and benefits of additive manufacturing. Additive manufacturing may impact a significant number of lean categories for various types of products, reducing some of lean manufacturing categories of waste. An important concept of lean manufacturing is the identification of waste, which is classified into seven categories:

1. *Overproduction:* occurs when more is produced than is currently required by customers
2. *Transportation:* does not make any change to the product and is a source of risk to the product
3. *Rework/Defects:* result in wasted resources or extra costs correcting the defect
4. *Overprocessing:* occurs when more work is done than is necessary
5. *Motion:* unnecessary motion results in unnecessary expenditure of time and resources
6. *Inventory:* similar to that of overproduction and results in the need for additional handling, space, people, and paperwork to manage extra product
7. *Waiting:* when workers and equipment are waiting for material and parts, these resources are being wasted

An examination of automobile production in the United State provides an example of the significance of some of these costs. Wholesale and retail trade, transportation, and warehousing along with transportation and material moving occupations, which are activities that do not alter the final product, contribute 18.1% of the total value added for new automobile production (Thomas and Kandaswamy 2015).

9.3.1 Ill-Structured Costs

9.3.1.1 Inventory

Suppliers often suffer from high inventory and distribution costs and the resources spent storing products could be used elsewhere if the need for inventory were reduced. For example, in the spare parts industry, a specific type of part is infrequently ordered; however, when a spare part is ordered, it is needed quite rapidly, as delays in delivery leave machinery and equipment idle. Traditional production technologies make it too costly and require too much time to produce parts on demand. The result is a significant amount of inventory built up for infrequently ordered parts (Walter et al. 2004). This inventory is tied up capital for unused products. They occupy physical space, buildings, and land, while requiring rent, utility costs, insurance, and taxes. Meanwhile, the products are deteriorating and becoming obsolete. Additive manufacturing provides the ability to manufacture parts on demand, which reduces the need for maintaining large inventory and eliminates the associated costs.

9.3.1.2 Transportation

Traditional manufacturing often includes production of parts at multiple locations, where an inventory of each part might be stored. The parts are shipped to a facility where they are assembled into a product, as illustrated in Figure 9.1. The manufacturing of each component requires material inventory to be used in the production of the component, and each component is then stored in finished goods inventory until it is needed. Additive manufacturing has the potential for producing multiple parts simultaneously in the same build, making it possible to produce an entire product. This would reduce the need to maintain large inventories for each part of one product. It also reduces the transportation of parts produced at varying locations and reduces the need for just-in-time delivery. It is important to note that the ability to produce an assembled product using additive manufacturing technology may require precision accuracy, which may or may not be available or cost effective at this time.

9.3.1.3 Consumer's Proximity to Production

It has been proposed that there are three alternatives for the diffusion of additive manufacturing technologies. The first involves consumers purchasing additive manufacturing systems or 3D printers and producing products themselves (Neef et al. 2005). The second is a copy shop scenario,

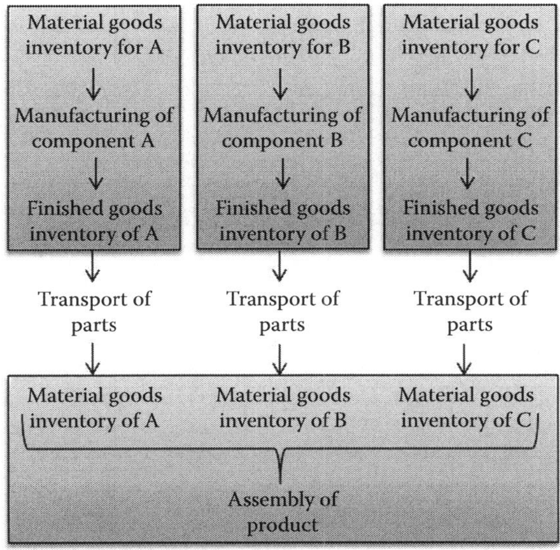

Figure 9.1 Example of traditional manufacturing flow.

where individuals submit their designs to a service provider that produces goods (Neef et al. 2005). The third scenario involves additive manufacturing being adopted by the commercial manufacturing industry, changing the technology of design and production (Neef et al. 2005). A fourth scenario, however, is possible. Additive manufacturing has the potential to bring production closer to the consumer for some products, often called distributed manufacture. Unlike traditional manufacturing methods, additive manufacturing has the potential to produce a final product in one build. It results in limited exposure to hazardous conditions and has little hazardous waste (Huang et al. 2013), making it possible for producing in nonindustrial settings. For example, currently, a more remote geographic area may order automotive parts on demand, which may take multiple days to be delivered. Additive manufacturing might allow some of these parts or products to be produced near the point of use or even onsite (Holmström et al. 2010). Further, localized production combined with simplified processes may begin to blur the line between manufacturers, wholesalers, and retailers, as each could potentially produce products in their facilities. The current quality and accuracy of additive manufacturing systems might make this infeasible in the short run; however, technological advancements may change this situation.

Khajavi et al. (2014) compared the operating cost of centralized additive manufacturing production and distributed production, where production is in close proximity to the consumer. This analysis examined the production of spare parts for the air-cooling ducts of the environmental control system for the F-18 Super Hornet fighter jet, which is a well-documented instance where additive manufacturing has already been implemented. The expected total cost per year for centralized production was \$1.0 million and \$1.8 million for distributed production. Inventory obsolescence cost, initial inventory production costs, inventory carrying costs, and spare parts transportation costs are all reduced for distributed production; however, significant increases in personnel costs and the initial investment in additive manufacturing machines make it more expensive than centralized production. Increased automation and reduced machine costs are needed for this scenario to be cost effective. It is also important to note that this analysis examined the manufacture of a relatively simple component with little assembly. One potential benefit of additive manufacturing might be to produce an assembled product rather than individual components. Research by Holmström et al. (2010), which also examines spare parts in the aircraft industry, concurs that currently, on-demand centralized production of spare parts is the most likely approach to succeed; however, if additive manufacturing develops into a widely adopted process, the distributed approach becomes more feasible (Holmström et al. 2010).

9.3.1.4 Supply Chain Management

Supply chain management includes purchasing, operations, and distribution. Purchasing involves sourcing product suppliers. Operations involve demand planning, forecasting, and inventory, while distribution involves the movement of products. Reducing the need for these activities can result in a reduction in costs. Some large businesses and retailers largely owe their success to the effective management of their supply chain. They use technology to innovate the way they track inventory and restock shelves resulting in reduced costs. Walmart, for example, cut links in the supply chain, making the link between their stores and the manufacturers more direct. It also began vender managed inventory (VMI), where manufacturers were responsible for managing their products in Walmart's warehouses. It also advanced its communication and collaboration network. Management of the supply chain can be a factor that drives a company to market leadership. Additive manufacturing may have significant impacts on the manufacturing supply chain, reducing the need for supply chain management. This technology has the potential to bring manufacturers closer to consumers, reducing the links in the supply chain.

9.3.1.5 Vulnerability to Supply Disruption

If additive manufacturing reduces the number of links in the supply chain and brings production closer to consumers, it will result in a reduction in the vulnerability to disasters and disruptions. Every factory and warehouse in the supply chain for a product is a potential point where a disaster or disruption can stop or hinder the production and delivery of a product. A smaller supply chain with fewer links means there are fewer points for potential disruption. Additionally, if production is brought closer to consumers, it will result in a more decentralized production with many facilities producing a few products rather than a few facilities producing many products. A single facility can be disrupted by one localized disaster, while multiple facilities would require multiple disasters or one mega disaster. Moreover, disruptions in decentralized production might result in only localized impacts rather than regional or national impacts. Figure 9.2 provides an illustration of the difference between traditional manufacturing and distributed manufacturing. In traditional manufacturing, material resource providers deliver materials to the manufacturers of parts and components, who might deliver parts and components to each other and then to an assembly plant. From there, the assembled product is delivered to a retailer or a distributor. A disruption at any of the points in manufacturing or assembly may result in a disruption of deliveries to all the retailers or distributors if there is not redundancy in the system. Additive manufacturing with localized production does not have the same vulnerability. First, there may not be any assembly of parts or components. Second, a disruption to manufacturing does not impact all of the retailers and distributors.

9.3.2 Well-Structured Costs

9.3.2.1 Material Cost

Metal and plastic are the primary materials used in additive manufacturing. Currently, the cost of material for additive manufacturing can be quite high when compared to traditional manufacturing. Atzeni and Salmi (2012) showed that the material costs for a selected metal part made from aluminum alloys was €2.59 per part for traditional manufacturing and €25.81 per part for additive manufacturing using selective laser sintering. Thus, the additive manufacturing material was nearly ten times more expensive (Atzeni and Salmi 2012).

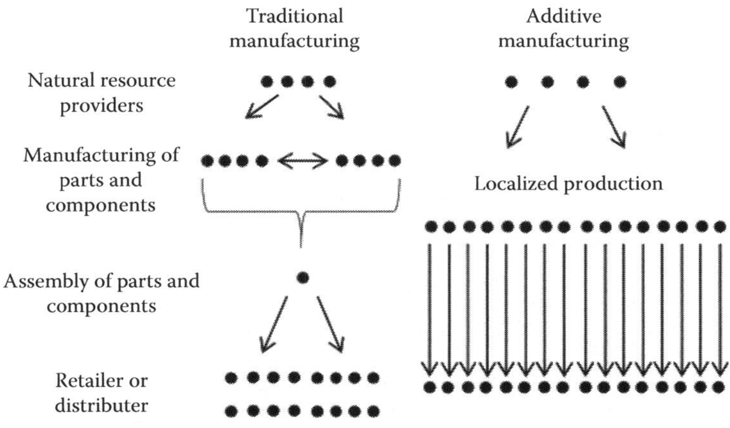

Figure 9.2 Example of traditional supply chain compared to the supply chain for additive manufacturing with localized production.

Other research on metal parts confirms that material costs are a major cost driver for this technology as shown in Figure 9.3, which presents data for a sample part made of stainless steel. For this example, four cost factors are varied, and the production quantity is a little less than 200 for the base case. This analysis provides insight into identifying the largest costs of additive manufacturing. The first cost factor that is varied is the building rate, which is the speed at which the additive manufacturing system operates. In this example, it is measured in cubic centimeters per hour. The second factor that is varied is the machine utilization measured as the number of hours per year that the machine is operated. The third factor is the material cost, and the last factor is the machine investment costs, which include items related to housing, using, and maintaining the additive manufacturing system. Among other things, this includes energy costs, machine purchase, and associated labor costs to operate the system. The base model has a build rate of 6.3 cm³/h, a utilization of 4500 h/year, a material cost of 89 €, and a machine investment cost of 500,000 €. For comparison, the base case is shown at the far left of the figure. On average, the machine costs are accounted for 62.9% of the cost estimates, as shown in Figure 9.3. This cost was the highest even when building rate was more than tripled and other factors were held constant. This cost was the highest in all but one case, where material costs were increased to 600 €/kg. The second highest cost is of the materials, which, on average, accounted for 18.0% of the costs; however, it is important to note that this cost is likely

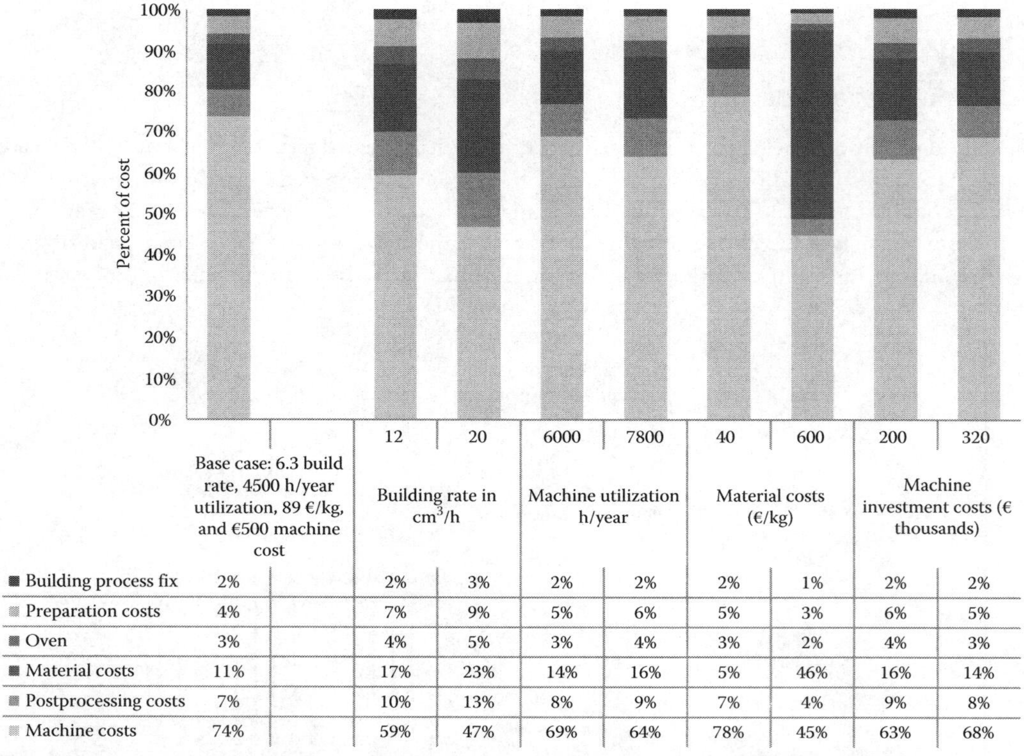

	Base case: 6.3 build rate, 4500 h/year utilization, 89 €/kg, and €500 machine cost	Building rate in cm³/h: 12	20	Machine utilization h/year: 6000	7800	Material costs (€/kg): 40	600	Machine investment costs (€ thousands): 200	320
■ Building process fix	2%	2%	3%	2%	2%	2%	1%	2%	2%
■ Preparation costs	4%	7%	9%	5%	6%	5%	3%	6%	5%
■ Oven	3%	4%	5%	3%	4%	3%	2%	4%	3%
■ Material costs	11%	17%	23%	14%	16%	5%	46%	16%	14%
■ Postprocessing costs	7%	10%	13%	8%	9%	7%	4%	9%	8%
■ Machine costs	74%	59%	47%	69%	64%	78%	45%	63%	68%

Figure 9.3 Cost distribution of additive manufacturing of metal parts by varying factors. (Modified from Lindemann C., et al., *Proceedings of the 2012 Solid Freeform Fabrication Symposium—An Additive Manufacturing Conference,* Austin, TX, 2012.)

to decrease as more suppliers enter the field (Lindemann et al. 2012). Additionally, improved utilization of materials may improve costs, as many additive manufacturing systems deposit more material than is needed in the final product. Systems also require a very narrow specification in powder properties, which drives up the cost of material. Postprocessing, preparation, oven heating, and building process fix were approximately 8.4%, 5.4%, 3.3%, and 1.9% of the costs, respectively.

The material costs for additive manufacturing are significant; however, technologies can often be complementary, where two technologies are adopted alongside each other and the benefits are greater than if they were adopted individually. One example is computer-aided design (CAD) and computer-aided manufacturing (CAM), as both are needed to be utilized for the other to be valuable. Additive manufacturing and the raw materials that are used may be a condition where they are complementary (Baumers 2012). All additive manufacturing requires raw materials, and according to Stoneman (2002) this may create a feedback loop (Stoneman 2002). Increasing adoption of additive manufacturing may lead to a reduction in raw material cost through economies of scale. The reduced cost in raw material might then propagate further adoption of additive manufacturing. There may also be economies of scale in raw material costs if particular materials become more common rather than a plethora of different materials.

9.3.2.2 Machine Cost

In addition to material costs, machine cost is also one of the most significant costs involved in additive manufacturing. The average selling price of an industrial additive manufacturing system was $87,140 in 2014 as against $90,370 in 2013 (Wohlers 2012). Between 2001 and 2010, the price decreased by 57% after adjusting for inflation (Wohlers 2012). While the trends in machine costs are generally downward, large differences remain between the costs for polymer-based systems and metal-based systems, and the tremendous growth in sales of low-cost, polymer-based systems during this time has strongly influenced the average selling price of additive manufacturing systems. The machine cost estimates for Lindemann et al. (2012) ranged from 45% to 78% of the cost of a metal part, as shown in Figure 9.3.

9.3.2.3 Build Envelope and Envelope Utilization

The size of the build envelope* and the utilization of the envelope both have an impact on the cost of an additive manufactured product. The size of the build envelope has two impacts. First, products can only be built to the size of the build envelope, meaning that it might not be possible to build some products using additive manufacturing technologies without enlarging the build envelope. The second impact of the build envelope is related to utilizing the total amount of build capacity. A significant efficiency factor lies in the ability to exhaust the available build space. For example, Baumers et al. (2011) examined the impact of capacity utilization on energy using six different machines (Arcam—A1, MTT Group—SLM 250, EOS GmbH—EOSINT M 270, Concept Laser GmbH—M3 Linear, StratasysInc—FDM 400 mc, and EOS GmbH—EOSINT P390) and four different materials (titanium, stainless steel, and two kinds of polymers). As shown in Figure 9.4, the full build case, where the build envelope is fully utilized, uses less energy per kilogram deposited than one single part being produced for all machines shown.

* The build envelope is the maximum area for part production in an additive manufacturing system.

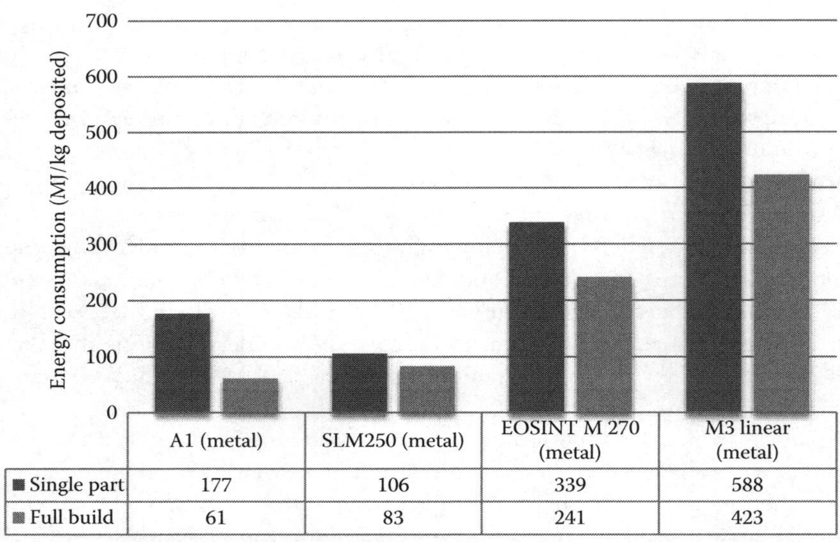

	A1 (metal)	SLM250 (metal)	EOSINT M 270 (metal)	M3 linear (metal)
■ Single part	177	106	339	588
■ Full build	61	83	241	423

Figure 9.4 Energy consumption per kilogram deposited. (Modified from Baumers, M., et al., *22nd Annual International Solid Freeform Fabrication Symposium—An Additive Manufacturing Conference*. Austin, TX, 2011.)

9.3.2.4 Build Time

Build time is a significant component in regard to estimating the cost of additive manufacturing and a number of software packages are available for estimating build time (Ruffo 2006b; Campbell et al. 2008). There tends to be two approaches to estimating build time: (1) detailed analysis and (2) parametric analysis (Di Angelo and Di Stefano2011). Detailed analysis utilizes knowledge about the inner workings of a system, while parametric analysis utilizes information on process time and characteristics such as layer thickness. Build time estimations tend to be specific to the system and material being used. Although this is an important factor in the cost of additive manufacturing, the details of build time are beyond the scope of this chapter.

9.3.2.5 Energy

Some cost studies for additive manufacturing, such as Hopkinson and Dickens (2003), included an examination of energy consumption, but they did not include energy in their reporting, as it contributed less than one percent to the final cost (Hopkinson and Dickens 2003). Energy consumption, however, is an important factor in considering the cost of additive manufacturing compared to other methods of manufacturing, especially in terms of examining the costs from cradle to grave. Energy studies on additive manufacturing, however, tend to focus only on the energy used in material refining and by the additive manufacturing system itself.

As discussed previously, Baumers (2012) examined energy consumption among a number of machines. The results shown in Figure 9.4 provide the results for energy consumption among machines producing metal parts. Morrow et al. (2007) compare direct metal deposition to conventional tool and die manufacturing. This work identifies that energy consumption is driven by the solid-to-cavity volume ratio. At low ratios, the additive manufacturing process of direct metal deposition minimizes energy, while at high ratios computer numeric controlled milling minimizes energy consumption. Other studies have shown that the orientation of the part being made can also impact the energy being consumed.

9.3.2.6 Labor

Labor associated with the additive manufacturing process tends to be a small portion of the cost. Labor might include removing the finished product or refilling the raw material. For a polymer-based product, Hopkinson and Dickens estimate labor at 2% of the cost (Hopkinson and Dickens 2003), while Ruffo et al. estimate it at 2% and 3% (Ruffo 2006a). However, it is important to note that the additional labor is built into the material cost and machine cost, as these items also require labor to produce.

9.3.2.7 Postprocessing

In most cases, there is a need for postprocessing of a part that is produced using additive manufacturing. This might include hot isostatic pressing where heat and pressure are applied to metals or other traditional manufacturing processes to produce parts with proper dimensions and finishes. Thus, in many instances, these parts are a highbred of additive manufacturing and traditional manufacturing processes.

9.3.2.8 Product Enhancements and Quality

Although the focus of this section is the costs of additive manufacturing, it is important to note that there are product enhancements and quality differences due to using this technology. There is more geometric freedom with additive manufacturing and it creates more flexibility; however, there are limitations, as some designs require support structures and means for dissipating heat in production (Baumers 2012). Complexity does not increase the cost of production as it does with traditional methods. With the exception of the design cost, each product produced can be customized at little or no expense, although these types of parts often need support structure and finishing processes that can be difficult and add costs. There is a significant need for custom products in the medical sector for replacement joint implants, dental work, and hearing aids among other things (Doubrovski et al. 2011). There is also the possibility of customers designing their own products or customizing them. One concern with additive manufacturing, however, is quality assurance. Currently, there is a need for standard methods to evaluate and ensure accuracy, surface finish, and feature detail to achieve desired part quality.

Example 2

Build orientation can have a significant impact on energy consumption. Mognol et al. (2006) examined the impact of part orientation for three systems: Stratasys FDM 3000, 3D Systems Thermojet, and EOS EOSINT M250 Xtended. They examined 18 positions for a single part. Due to the change in the position of the part, the energy consumed could increase between 75% and 160% depending on the system, as illustrated in Figure 9.5. This figure also illustrates that the position for one system may have low-energy consumption, but for another system it might not have a low consumption (Mognol et al. 2006).

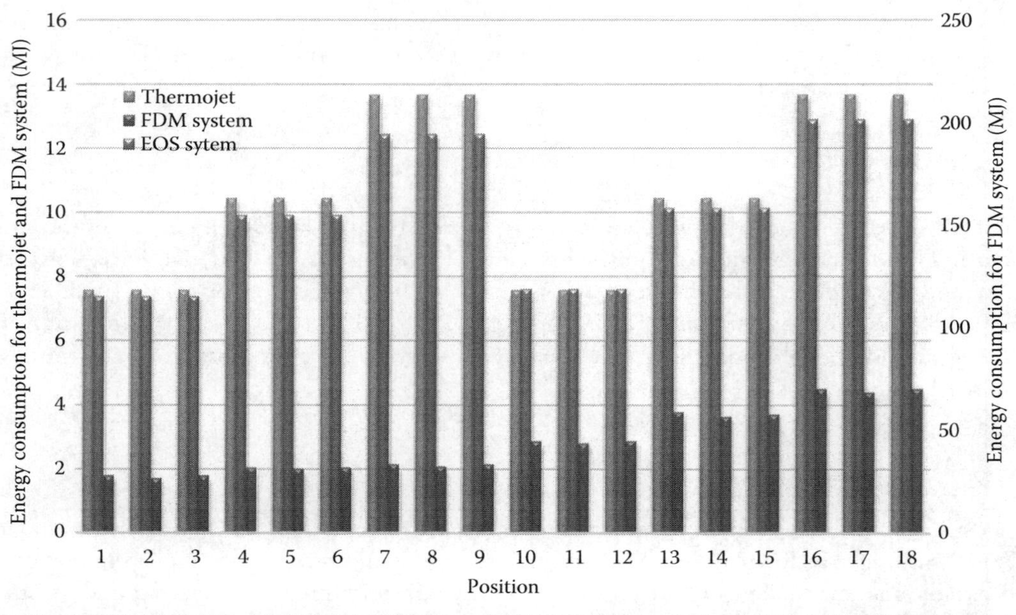

Figure 9.5 Energy consumption by position. (From Mognol, P., et al., *Rapid Prototyping J.*, 12(1), 26–34.)

9.4 Cost Models and Economic Performance of Additive Manufacturing

Learning Objectives

- Understand the cost models associated with additive manufacturing
- Understand the cost advantage and total advantage of additive manufacturing

The following sections discuss cost models and economic performance of additive manufacturing. Section 9.4.1 discusses estimating the cost of producing a product using additive manufacturing. Section 9.4.2 discusses more details into the source of these costs, such as those built into the supply chain.

9.4.1 Models for Estimating the Cost of Additive Manufacturing Production

There are two cost models that receive significant attention in additive manufacturing: (1) Hopkinson and Dickens (2003) and (2) Ruffo et al. (2006a). Although they are not focused on metal parts, they provide approaches for examining costs (Hopkinson and Dickens 2003; Ruffo 2006a; Baumers 2012). The cost of additive manufactured parts is calculated by Hopkinson and Dickens based on calculating the average cost per part and three additional assumptions: (1) the system produces a single type of part for 1 year, (2) it utilizes maximum volumes, and (3) the machine operates for 90% of the time. The analysis includes labor, material, and machine costs.

Other factors such as power consumption and space rental were considered, but they contributed less than one percent of the costs; therefore, they were not included in the results. The average part cost is calculated by dividing the total cost by the total number of parts manufactured in a year. Costs can be broken into machine costs, labor costs, and material costs. The costs are calculated for two parts, a lever and a cover, using stereolithography, fused deposition modeling, and laser sintering. A cost breakout for the lever is provided in Table 9.4, which shows that in this analysis, laser sintering was the cheapest additive manufacturing process for this product. Machine cost was the major contributing cost factor for stereolithography and fused deposition modeling, while the material cost was the major contributor for laser sintering.

Hopkinson and Dickens estimate an annual machine cost per part where the machine cost completely depreciates after 8 years; that is, it is the sum of depreciation cost per year (calculated as machine and ancillary equipment cost divided by 8) and machine maintenance cost per year divided by production volume. The result is a machine cost per part that is constant over time, as shown in Figure 9.6.

The cost of additive manufactured parts is calculated by Ruffo et al. using an activity-based cost model, where each cost is associated with a particular activity. They produce the same lever that Hopkinson and Dickens produced using selective laser sintering. In their model, the total cost of a build (C), is the sum of raw material costs and indirect costs. The raw material costs are the product of price ($P_{material}$), measured in euros per kilogram, and the mass in kg (M). The indirect costs are calculated as the product of total build time (T) and a cost rate ($P_{indirect}$). The total cost of a build is then represented as:

$$C = P_{material} \times M + P_{indirect} \times T. \tag{9.2}$$

The cost per part is calculated as the total cost of a build (C) divided by the number of parts in the build. In contrast, Ruffo et al. indicate that the time and material used are the main variables in the costing model. It was assumed that the machine worked 100 h/week for 50 weeks/year (57% utilization).

There are three different times that are calculated in Ruffo et al.'s model: (1) "time to laser scan the section and its border in order to sinter"; (2) "time to add layers of powder"; and (3) "time to heat the bed before scanning and to cool down slowly after scanning, adding layers of powder or just waiting time to reach the correct temperature." The sum of these times is the build time (T), and the resulting cost model along with the Hopkinson and Dickens model is shown along with Hopkinson and Dickens model in Figure 9.6. Ruffo et al.'s model has a jagged saw tooth shape to it, which occurs when an additional unit of production requires a new build or new line or layer within a build. Each time one of these is added (i.e., a build, line, or layer), average costs increase irregularly from raw material consumption and process time. At 1600 parts, the cost of the lever is estimated at €2.76 per part when compared to Hopkinson and Dickens' €2.20 for laser sintering. Ruffo et al. also conducted an examination where unused material was recycled. In this examination, the per-unit cost was €1.86.

Many of the cost studies assume a scenario where one part is produced repeatedly; however, one of the benefits of additive manufacturing is the ability to produce different components simultaneously. Therefore, a "smart mix" of components in the same build might achieve reduced costs. In a single part production, the cost per part for a build is the total cost divided by the number of parts; however, the cost for different parts being built simultaneously is more complicated. Ruffo and Hague (2007) compared three costing methodologies for assessing this cost. The results

Table 9.4 Cost Breakout

	Stereolithography	*Fused Deposition Modeling*	*Laser Sintering*
Other			
Number per platform	190	75	1,056
Platform build time	26.8	67.27	59.78
Production rate per hour	7.09	1.11	17.66
Hours per year in operation	7,884	7,884	7,884
Production volume total per year	55,894	8,790	139,269
Machine costs			
Machine and ancillary equipment (€)	1,040,000	101,280	340,000
Equipment depreciation cost per year (€)	130,000	12,660	42,500
Machine maintenance cost per year (€)	89,000	10,560	30,450
Total machine cost per year (€)	219,000	23,220	72,950
Machine cost per part (€)	3.92	2.64	0.52
Labor costs			
Machine operator cost per hour (€)	5.30	5.30	5.30
Setup time to control machine (min)	33	10	120
Postprocessing time per build (min)	49	60	360
Labor cost per build (€)	7.24	6.18	42.37
Labor cost per part (€)	0.04	0.08	0.04
Material costs			
Material per part (kg)	0.0047	0.0035	
Support material per part (kg)		0.0016	
Build material cost per kg (€)	275.20	400.00	54.00
Support material cost per kg (€)		216.00	
Cost of material used in one build (€)			1,725.72
Material cost per part (€)	1.29	1.75	1.63
Total cost per part (€)	5.25	4.47	2.20

Source: Hopkinson, N., Dickens, P.M., *Proc. Inst. Mech. Eng.*, 217(1), 31–39, 2003.

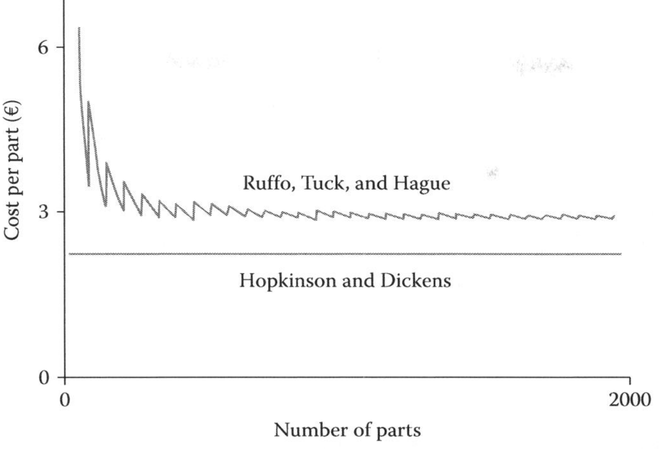

Figure 9.6 Cost model comparison (Ruffo, Tuck, and Hague vs. Hopkinson and Dickens). (Modified from Hopkinson, N., Dickens, P.M., *Proc. Inst. Mech. Eng.*, 217(1), 31–39, 2003; Ruffo, M., et al., *Proc. Inst. Mech. Eng.*, 220(9), 1417–1427, 2006a.)

suggested that one of the models provided a "fair assignment method." The other two were identified as being inappropriate due to the result drastically reducing the estimated cost of larger components at the expense of smaller parts. The model is based on the cost of a part built in high volume. It is similar to the second method in that only the cost variables in γ_i are calculated using a high number of parts rather than a single part. It is represented as

$$\text{Cost}_{P_i} = \frac{\gamma_i^{\infty} \times \text{Cost}_{\text{B}}}{n_i}, \tag{9.3}$$

where

$$\gamma_i^{\infty} = \frac{\text{Cost}_{P_i}^{\infty} \times n_i}{\sum_j \left(\text{Cost}_{P_j}^{\infty} \times n_j\right)} \tag{9.4}$$

$$\text{Cost}_{P_i}^{\infty} = \text{Hypothetical cost of a number, which approaches infinity, of manufactured parts } i. \tag{9.5}$$

$$\text{Cost}_{\text{B}} = \sum \frac{\text{indirect\_costs}}{\text{working\_time}}\left(t_{xy} + t_z + t_{\text{HC}}\right) + \frac{\text{direct\_cost}}{\text{mass\_unit}} m_{\text{B}} \tag{9.6}$$

where

m_{B} = mass of the planned production proportional to the object volumes, and the time to manufacturing the entire build

t_{xy} = time to laser-scan the section and its border to sinter powder

t_z = time to add layers of powder

t_{HC} = time to heat the bed before scanning and to cool down after scanning and adding layers of powder

n_i = the number of parts identified with i

i = an index going from one to the number of parts in the build

j = an index for all parts manufactured in the same bed

This method incorporates time and part volume into the cost. Using a high number of units being produced (i.e., infinite) stabilizes the cost. For example, the cost per part in Figure 9.6 stabilizes at large numbers of units. Given that a large production volume flattens the curve, it is reasonable to use the method outlined in Equations 9.3 and 9.4. The variable Υ_i^{∞} is a weight for the full build cost, while $\text{Cost}_{P_i}^{\infty}$ is the cost for a hypothetical infinite number of parts (i). "Infinite parts" is not applicable in engineering problems; therefore, in practice, "high number of parts" is used instead of "infinite."

9.4.2 Supply Chain and Product Utilization

Many of the cost studies examine costs such as material and machine costs; however, many of the benefits may be hidden in inventory and supply chain costs. For instance, a dollar invested in automotive assembly takes 10.9 days to return in revenue. It spends 7.9 days in material inventory, waiting to be utilized. It spends 19.8 h in production time and another 20.6 h in down time when the factory is closed. Another 1.3 days is spent in finished goods inventory.† Moreover, of the total time used, only 8% is spent in actual production. According to concepts from lean manufacturing, inventory and waiting, which constitute 92% of the automotive assembly time, are two of seven categories of waste. This is just the assembly of an automobile. The production of the engine parts, steering, suspension, power train, body, and others often occurs separately and each has inventories of its own. Additionally, all of these parts are transported between locations. The average shipment of manufactured transportation equipment travels 801 miles (1289 km). For the United States, this amounts to 45.3 billion ton-miles (66.1 ton-km) of transportation equipment being moved annually. Because additive manufacturing can, in some instances now and possibly more in the future, build an entire assembly in one build, it has the potential to reduce the need for some of the transportation and inventory costs, resulting in impacts throughout the supply chain. It is important to note that the ability to produce more complex assemblies, such as those in an automobile, is still developing and involves some speculation about future capabilities. In addition to building complete or partial assemblies, there is also the potential of reducing the size of the supply chain through distributed manufacturing. Therefore, in order to understand the cost difference between additive manufacturing and other processes, it is necessary to examine the costs from raw material extraction to production and through the sale of the final product. This might be represented as:

$$C_{\text{AM}} = \left(\text{MI}_{R,\text{AM}} + \text{MI}_{M,\text{AM}}\right) + \left(P_{E,\text{AM}} + P_{R,\text{AM}} + P_{M,\text{AM}}\right) + \left(\text{FGI}_{E,\text{AM}} + \text{FGI}_{R,\text{AM}} + \text{FGI}_{M,\text{AM}}\right) + \text{WT}_{\text{AM}} + \text{RT}_{\text{AM}} + T_{\text{AM}}, \tag{9.7}$$

where

C_{AM} = Cost of producing an additive manufactured product

† Calculated from data in the Annual Survey of Manufactures and the quarterly survey of plant capacity utilization.

MI = Cost of material inventory for refining raw materials (R) and for manufacturing (M) for additive manufacturing (AM)

P = Cost of the process of material extraction (E), refining raw materials (R), and manufacturing (M), including administrative costs, machine costs, and other relevant costs for additive manufacturing (AM)

FGI = Cost of finished goods inventory for material extraction (E), refining raw materials (R), and manufacturing (M) for additive manufacturing (AM)

$\mathrm{WT}_{\mathrm{AM}}$ = Cost of wholesale trade for additive manufacturing (AM)

$\mathrm{RT}_{\mathrm{AM}}$ = Cost of retail trade for additive manufacturing (AM)

T_{AM} = Transportation cost throughout the supply chain for an additive manufactured product (AM)

This could be compared to the cost of traditional manufacturing, which could be represented as

$$C_{Trad} = \left(\mathrm{MI}_{R,\,Trad} + \mathrm{MI}_{I,\,Trad} + \mathrm{MI}_{A,\,Trad}\right) + \left(P_{E,\,Trad} + P_{R,\,Trad} + P_{I,\,Trad} + P_{A,\,Trad}\right)$$
$$+ (\mathrm{FGI}_{E,Trad} + \mathrm{FGI}_{R,Trad} + \mathrm{FGI}_{I,Trad} + \mathrm{FGI}_{A,Trad}) + \mathrm{WT}_{Trad} + \mathrm{RT}_{Trad} + T_{Trad}, \qquad (9.8)$$

where

C_{Trad} = Cost of producing a product using traditional processes (*Trad*)

MI = Cost of material inventory for refining raw materials (R), producing intermediate goods (I), and assembly (A) for traditional manufacturing (*Trad*)

P = Cost of the process of material extraction (E), refining raw materials (R), producing intermediate goods (I), and assembly (A), including administrative costs, machine costs, and other relevant costs for traditional manufacturing (*Trad*)

FGI = Cost of finished goods inventory for material extraction (E), refining rawmaterials (R), producing intermediate goods (I), and assembly (A) for traditional manufacturing (*Trad*)

WT_{Trad} = Cost of wholesale trade for traditional manufacturing (*Trad*)

RT_{Trad} = Cost of retail trade for traditional manufacturing (*Trad*)

T_{Trad} = Transportation costs throughout the supply chain for a product made using traditional manufacturing (*Trad*)

Currently, there is a better understanding about the cost of the additive manufacturing process cost (P_{AM}) than there is for the other costs of additive manufacturing. Additionally, most cost studies examine a single part or component; however, it is in the final product where additive manufacturing might have significant cost savings. Traditional manufacturing requires numerous intermediate products that are transported and assembled, whereas additive manufacturing can achieve the same final product with fewer component parts and multiple components built either simultaneously or at the same location. For example, consider the future possibility of an entire jet engine housing being made in one build using additive manufacturing compared to an engine housing that has parts made and shipped for assembly from different locations with each location having its own factory, material inventory, finished goods inventory, administrative staff, and transportation infrastructure among other things. Additionally, the jet engine housing might be made using less material, perform more efficiently, and last longer because the design is not limited to the methods used in traditional manufacturing; however, many of these benefits would not be captured in the previously mentioned cost model, as it focuses on production-oriented costs. To capture these benefits, one would need to include a cradle to grave analysis.

Typically, for a company, the goal is to maximize profit; however, at the societal level, there are multiple stakeholders to consider and different costs and benefits. At this level, one might consider the goal to be to minimize resource use and maximize utility, where utility is the usefulness obtained by consumers. Dollar values are affected by numerous factors such as scarcity, regulations, and education costs, among other things, that impact how resources are efficiently allocated. The allocation of resources is an important issue; however, understanding the societal impact of additive manufacturing requires separating resource allocation issues from resource utilization issues.

The factors of production are, typically, considered to be land (i.e., natural resources), labor, capital, and entrepreneurship; however, capital includes machinery and tools, which themselves are made of land and labor. Additionally, a major element in the production of all goods and services is time, as illustrated in many operations management discussions. Therefore, one might consider the most basic elements of production to be land, labor, human capital, entrepreneurship, and time. The human capital and entrepreneurship utilized in producing additive manufactured goods are important, but these are complex issues that are set aside for this discussion. The remaining items, land, labor, and time, constitute the primary cost elements for the production of a product.

It is important to note that there is a trade-off between time and labor (measured in labor hours per hour). For example, one hundred people can build a house quicker than one person. It is also important to note that there is also a trade-off between time/labor and land (i.e., natural resources). For example, a machine can reduce both the time and the number of people needed for production, but utilizes more energy. The triangular plane in Figure 9.7 represents, at a given point in time, possible combinations of land (i.e., natural resources), labor, and time, required for producing a manufactured good. It is important to note that this figure only illustrates that a trade-off exists between time, labor, and natural resources, and the relationship is not actually linear as shown in the figure. For some products, it may be a set of alternatives represented by points; while others may have a sliding scale such as the building of a house. Since there are many possible scenarios, a simple plane is used for this discussion. Moving anywhere along this plane is simply an alteration of resource use, trading off one item for another. This combination is, typically, determined by the least cost combination of inputs.

A company can maximize profit by altering resources and/or by reducing the resources needed for production. Moving along the plane in Figure 9.7 may result in a more efficient allocation of resources for a firm and for society; however, it does not reduce the combination of resources needed for production. For this to occur there needs to be a shift in the plane. In the past, efficiency

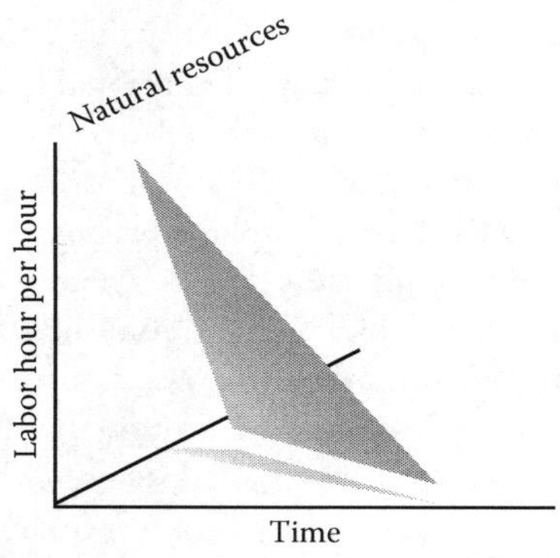

Figure 9.7 Primary elements needed to produce a manufactured product.

improvements have been driven by utilizing natural resources to produce more goods per person, but fewer goods per unit of natural resources. For example, one could fasten two pieces of wood together using a nail and a hammer or a nail gun driven by compressed air. The hammer and nail method requires more labor but less natural resources than the nail gun, as the gun requires additional materials (e.g., metal, plastic, and rubber) and electricity. In recent years, there has been increasing concern about the sustainability of the consumption of resources and the impact on the environment. Therefore, when examining the cost and benefits of a product or process from a societal perspective, it becomes apparent that one needs to measure the natural resources, labor, and time needed for production. If additive manufacturing results in a reduction in the resources needed for production, then the plane in Figure 9.7 will shift, ideally toward the origin.

In addition to production, manufactured goods are also produced to serve a designated purpose. For example, automobiles transport objects and people; cell phones facilitate communication; and monitors display information. Each item produced is designed for some purpose. In the process of fulfilling this purpose, more resources are expended in the form of land, labor, and time. Additionally, a product with a short life span results in more resources being expended to reproduce the product. The disposal of the old product may also result in expending further resources. Additive manufactured products may provide product enhancements, new abilities, or an extended useful life. For example, auto parts might be produced lighter than is possible using traditional methods, which would increase gas mileage. The total advantage of an additive manufactured good is the difference in the use of land, labor, and time expended on production, utilization, and disposal combined with the utility gained from the product compared to that of traditional manufacturing methods. This can be represented as the following:

$$\mathrm{TA}_L = \left(L_{\mathrm{AM},P} + L_{\mathrm{AM},U} + L_{\mathrm{AM},D}\right) - \left(L_{T,P} + L_{T,U} + L_{T,D}\right) \tag{9.9}$$

$$\mathrm{TA}_{LB} = \left(\mathrm{LB}_{\mathrm{AM},P} + \mathrm{LB}_{\mathrm{AM},U} + \mathrm{LB}_{\mathrm{AM},D}\right) - \left(\mathrm{LB}_{T,P} + \mathrm{LB}_{T,U} + \mathrm{LB}_{T,D}\right) \tag{9.10}$$

$$\mathrm{TA}_T = \left(T_{\mathrm{AM},P} + T_{\mathrm{AM},U} + T_{\mathrm{AM},D}\right) - \left(T_{T,P} + T_{T,U} + T_{T,D}\right) \tag{9.11}$$

$$\mathrm{TA}_U = U\left(P_{\mathrm{AM}}\right) - U\left(P_T\right), \tag{9.12}$$

where

TA = the total advantage of additive manufacturing compared to traditional methods for land (L), labor (LB), time (T), and utility of the product (U).

L = the land or natural resources needed using additive manufacturing processes (AM) or traditional methods (T) for production (P), utilization (U), and disposal (D) of the product

LB = the labor hours per hour needed using additive manufacturing processes (AM) or traditional methods (T) for production (P), utilization (U), and disposal (D) of the product

T = the time needed using additive manufacturing processes (AM) or traditional methods (T) for production (P), utilization (U), and disposal (D) of the product

$U(P_{\mathrm{AM}})$ = the utility of a product manufactured using additive manufacturing processes, including the utility gained from increased abilities, enhancements, and useful life.

$U(P_{\mathrm{T}})$ = the utility of a product manufactured using traditional processes, including the utility gained from increased abilities, enhancements, and useful life.

In this case, production includes material extraction, material refining, manufacturing, and transportation among other things. Unfortunately, our current abilities fall short of being able to

Table 9.5 Model Landing Gear Production Costs Compared

	High-Pressure Die-Cast Part	*Selective Laser Sintering Part*
Material cost per part	2.59	25.81
Preprocessing cost per part	—	8.00
Processing cost per part[a]	0.26 + 21,000/*N*	472.50
Postprocessing cost per part	17.90	20.00
Assembly	0.54	—
Total	21.29 + 21,000/*N*	526.31

Source: Atzeni, E., Salmi, A., *Int. J. Adv. Manuf. Technol.*, 62, 1147–1155, 2012.

[a] Includes the mold for high-pressure die casting.
N, the number of parts produced.

measure all of these items for all products; however, it is important to remember that these items must be considered when measuring the total advantage of additive manufacturing. An additional challenge is that land, labor, time, and utility are measured in different units, making them difficult to compare. An additive manufactured product might require more labor, but reduce the natural resources needed. In this instance, there is a trade-off.

Example 3

Atzeni and Salmi (2011) showed that the per assembly cost for a landing gear assembly for a 1:5 scale model of the P180 Avant II by Piaggio Aero Industries S.p.A. (i.e., the machine cost per assembly), with an estimated 5 years of useful life, was €526.31 for the additive manufacturing process of selective laser sintering (see Table 9.5). Compared to high-pressure die-casting, the mold cost and processing cost per part was €21.29 + €21,000/*N*, where *N* is the number of parts produced. For production runs of <42, selective laser sintering was more cost effective than the traditional process of high-pressure die-casting.

9.5 Adoption and Diffusion of Additive Manufacturing

Learning Objectives

- Understand trade-off of flexibility and controllability
- Understand the barriers to the adoption of additive manufacturing
- Understand the Rogers model of technology diffusion

9.5.1 Implementation and Adoption of Additive Manufacturing

The manufacturing industry is oriented toward optimizing operations using traditional methods of production. Additive manufacturing is significantly different from these methods; thus, determining when and how to take advantage of the benefits of additive manufacturing is a challenge. Identifying products that benefit from increased complexity or distributed production along with understanding how additive manufacturing impacts inventory costs is complex and difficult to measure.

In order to create products and services, a firm needs resources, established processes, and capabilities (Kim 2015). Resources include natural resources, labor, and other items needed for production. A firm must have access to resources in order to produce goods and services. The firm must also have processes in place that transform resources into products and services. Two firms may have the same resources and processes in place; however, their products may not be equivalent due to quality, performance, or cost of the product or service. This difference is due to the capabilities of the firm; that is, capabilities are the firm's ability to produce a good or service effectively. Kim and Park (2013) present three entities of capabilities (see Figure 9.8): controllability, flexibility, and integration (Kim 2013).

Controllability is the firm's ability to control its processes. Its primary objective is to achieve efficiency that minimizes cost and maximizes accuracy and productivity. Flexibility is the firm's ability to deal with internal and external uncertainties. It includes reacting to changing circumstances while sustaining few impacts in time, cost, or performance. According to Kim and Park, there is a trade-off between controllability and flexibility; that is, in the short term, a firm chooses combinations of flexibility and controllability, sacrificing one for the other, as illustrated at the bottom of Figure 9.8. Over time, a firm can integrate and increase both flexibility and controllability through technology or knowledge advancement among other things. In addition to the entities of capabilities, there are also categories of capabilities or a chain of capabilities that include basic capabilities, process-level capabilities, system-level capabilities, and performance.

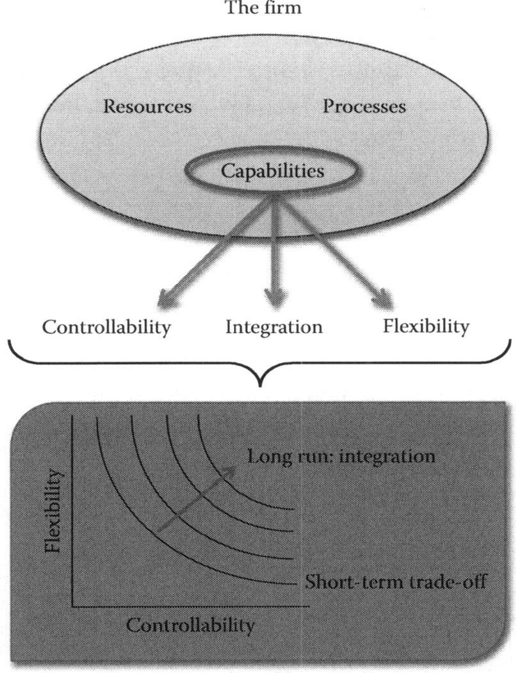

Figure 9.8 Necessities of a firm. (Adapted from Kim, B., Park, C., *Int. J. Prod. Res.*, 51(4), 1258–1278, 2013.)

As seen in Figure 9.9, basic capabilities include overall knowledge and experience of a firm and its employees, including their engineering skills, safety skills, and work ethics among other things. Process-level capabilities include individual functions such as assembly, welding, and other individual activities. System-level capabilities include bringing capabilities together to transform resources into goods and services. The final item in the chain is performance, which is often measured in terms of profit, revenue, or customer satisfaction among other things.

Adopting a new technology, such as additive manufacturing, can have significant impacts on a firm's capabilities. As discussed in the previous sections, in some instances, the cost per unit can be higher for additive manufacturing than for traditional methods. The result is that a firm sacrifices controllability for flexibility; thus, it makes sense for those firms that seek a high flexibility position to adopt additive manufacturing. In some instances, however, additive manufacturing can positively affect controllability. Additive manufacturing can reduce costs for products that have complex designs that are costly to manufacture using traditional methods. As the price of material and systems comes down for additive manufacturing, the controllability associated with this technology will increase, making it attractive to more firms.

In addition to the trade-off between flexibility and controllability, additive manufacturing can also directly impact a firm's chain of capabilities, including the basic, process-level, and system-level capabilities. At the basic level, additive manufacturing requires new knowledge, approaches, and designs. These new knowledge areas can be costly and difficult to acquire. At the process level, a firm that adopts additive manufacturing is abandoning many of its current individual functions to adopt a radically new production method. Former functions might have required significant investment in order to fully develop. Many firms may be apprehensive in abandoning these capabilities for a new process, which itself may require significant investment to fully develop. Finally, additive manufacturing can impact the system-level capability, as it is not only a process that affects the production of individual parts but also the assembly of the parts. All of these changes can make it costly and risky for a business to adopt additive manufacturing technologies and can result in reducing the rate at which this technology is adopted.

9.5.2 Methods for Measuring Economic Performance

Adopting additive manufacturing is an investment where there are costs and benefits. A number of methods of economic evaluation are available to measure the economic performance of an investment, such as investing in the adoption of additive manufacturing. These methods include, but are not limited to, present value of net benefits, present value of net savings, benefit-to-cost ratio, savings-to-investment ratio, and the adjusted internal rate of return. Applicability of each method varies depending on the type of investment decision. The methods described below are based on

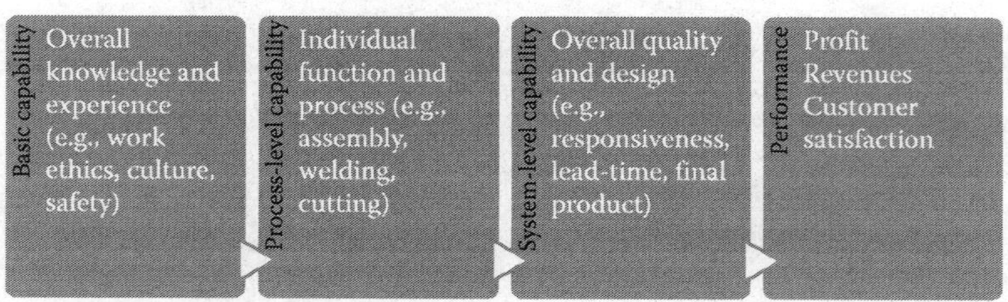

Figure 9.9 Chain of capability. (Adapted from Kim, B., Park, C., *Int. J. Prod. Res.*, 51(4), 1258–1278, 2013.)

ASTM International standard practices(ASTM 2012d), and the National Institute of Standards and Technology (NIST) provides a free software tool for making calculations using the five methods (Rushing and Chapman 2008). The tool and the standards are directed at building economics; however, they are applicable to any investment. Much of the methods, description, and equations below are taken from Chapman et al. (2009). Detailed descriptions of each of the standardized methods are also given in Chapman and Fuller (1996). An in-depth survey, covering these as well as other methods, is discussed in Ruegg and Marshall (1990).

In the following discussion, a number of terms are used in reference to the standardized methods:

a^* = the alternative under analysis;
t = a unit of time, where $-t^a$ is the earliest point for alternative a (i.e., beginning of the study period) before the base year (i.e., $t = 0$) and T is the last point after the base year (i.e., end of the study period);
L = the length of the study period (e.g., $t^a + T$);
$B_t^{a^*}$ = the benefits for alternative a^* in year t;
$I_t^{a^*}$ = the investment costs for alternative a^* in year t;
$C_t^{a^*}$ = the noninvestment costs for alternative a^* in year t;
$\underline{C}_t^{a^*}$ = the combined cost for alternative a^* in year t (i.e., $\underline{C}_t^{a^*} = I_t^{a^*} + C_t^{a^*}$);
$S_t^{a^*}$ = the savings for alternative a^* in year t;
d = the discount rate expressed as a decimal.

Throughout this section, the prefix PV is used to designate dollar denominated quantities in present value terms. The present value is derived by discounting (i.e., using the discount rate) to adjust all benefits, costs, and savings—past, present, and future—to the base year (i.e., $t = 0$). Discounting accounts for the time value of money. Typically, one would prefer income today, rather than getting the same income in the future. The discount rate, which is determined or selected by a firm or an entity, is used to account for this value difference. If one selects 3% as their annual discount rate, for example, this means that $100 today is valued at $103, 1 year from now. The dollar denominated quantities defined above and their associated present value terms are the present value of benefits (PVB), the present value of investment costs (PVI), the present value of noninvestment costs (PVC), the present value of combined costs (PV$\underline{\text{C}}$), and the present value of savings (PVS).

9.5.2.1 Present Value of Net Benefits and Present Value Net Savings

The present value of net benefits (PVNB) method is reliable, straightforward, and widely applicable for finding the economically efficient choice among alternatives (e.g., building systems). It measures the amount of net benefits from investing in a given alternative instead of investing in the foregone opportunity (e.g., some other alternative or maintenance of the *status quo*).

PVNB is computed by subtracting the time-adjusted costs of an investment from its time-adjusted benefits. If PVNB is positive, the investment is economic; if it is zero, the investment is as good as the next best investment opportunity; if it is negative, the investment is uneconomical. Emphasis is on economic efficiency because the method is appropriate for evaluating alternatives that compete on benefits, such as revenue or other advantages that are measured in dollars, in addition to costs.

The present value of net savings (PVNS) method is the PVNB method recast to fit the situation where there are no significant benefits in terms of revenue or the like, but there are reductions in future costs (e.g., reductions in the cost of ownership to consumers).[‡] By treating savings like revenue benefits, the PVNB method may be reformulated as the PVNS method.

The PVNB for a given alternative, $a*$, may be expressed as:

$$\begin{aligned} \mathrm{PVNB}^{a^*} &= \mathrm{PVB}^{a^*} - \mathrm{PV\underline{C}}^{a^*} \\ &= \sum_{t=-t^a}^{T} \frac{(B_t^{a^*} - \underline{C}_t^{a^*})}{(1+d)^t}. \end{aligned} \tag{9.13}$$

If there are no important benefits in terms of revenue, but there are reductions in future costs, then the PVNS for a given alternative, $a*$, may be expressed as:

$$\begin{aligned} \mathrm{PVNS}^{a^*} &= \left(\mathrm{PVS}^{a^*}\right) - \mathrm{PVI}^{a^*} \\ &= \sum_{t=-t^a}^{T} \frac{\left(S_t^{a^*} - I_t^{a^*}\right)}{(1+d)^t}. \end{aligned} \tag{9.14}$$

If the decision maker anticipates revenues from the investment, then use the PVNB measure. If the decision maker expects costs to be reduced, then use the PVNS measure. The PVNS measure is documented in ASTM Standard E1074-09(ASTM 2012c).

9.5.2.2 Benefit-to-Cost Ratio and Savings-to-Investment Ratio

The benefit-to-cost ratio (BCR) and the savings-to-investment ratio (SIR) are numerical ratios whose sizes indicate the economic performance of an investment. The BCR is computed as benefits, net of future noninvestment costs, divided by investment costs. The SIR is savings divided by investment costs. The SIR is the BCR method recast to fit the situation where the investment's primary advantage is lower costs. SIR is to BCR as PVNS is to PVNB.

A ratio less than 1.0 indicates an uneconomic investment; a ratio of 1.0 indicates an investment whose benefits or savings just equal its costs; and a ratio greater than 1.0 indicates an economic project. A ratio of, say, 4.75 means that the investor (e.g., the general public for a public-sector research program) can expect to receive \$4.75 for every \$1.00 invested (e.g., public funds expended), over and above the required rate of return imposed by the discount rate.

The BCR for a given alternative, $a*$, may be expressed as:

$$\begin{aligned} \mathrm{BCR}^{a^*} &= \frac{\left(\mathrm{PVB}^{a^*} - \mathrm{PVC}^{a^*}\right)}{\mathrm{PVI}^{a^*}} \\ &= \frac{\sum_{t=-t^a}^{T} \left(B_t^{a^*} - C_t^{a^*}\right) / (1+d)^t}{\sum_{t=-t^a}^{T} I_t^{a^*} / (1+d)^t} \end{aligned} \tag{9.15}$$

‡ If there are any benefits, say in the form of revenues or other positive cash flows; add them to the cost savings associated with the alternative under analysis.

The SIR for alternative *a** may be expressed as:

$$\text{SIR}^{a^*} = \frac{\text{PVS}^{a^*}}{\text{PVI}^{a^*}} = \frac{\sum_{t=-t^a}^{T} S_t^{a^*} / (1+d)^t}{\sum_{t=-t^a}^{T} I_t^{a^*} / (1+d)^t}. \tag{9.16}$$

As was the case for the PVNB and PVNS measures, use the BCR if the decision maker anticipates revenues from the investment, and use the SIR if the decision maker anticipates costs to be reduced. The SIR measure is documented in ASTM Standard E964-06 (2010) (ASTM 2012b).

9.5.2.3 Adjusted Internal Rate of Return

The adjusted internal rate of return (AIRR) is the annual yield from a project over the study period, taking into account reinvestment of interim receipts. Because the AIRR calculation explicitly includes the reinvestment of all net cash flows, it is instructive to introduce a new term, terminal value (TV). The terminal value of an investment, *a**, is the future value (i.e., the value at the end of the study period) of reinvested net cash flows excluding all investment costs. The terminal value for an investment *a**, is denoted as TV^{a^*}.

The reinvestment rate in the AIRR calculation is equal to the minimum attractive rate of return (MARR), which is assumed to equal the discount rate, *d*, a constant. When the reinvestment rate is made explicit, all investment costs are easily expressible as a time equivalent initial outlay (i.e., a value at the beginning of the study period) and all noninvestment cash flows (e.g., benefits, noninvestment costs, savings) as a time equivalent terminal amount. This allows a straightforward comparison of the amount of money that comes out of the investment (i.e., the terminal value) with the amount of money put into the investment (i.e., the time equivalent initial outlay).

The AIRR is defined as the interest rate, *r**, applied to the terminal value, TV^{a^*}, which equates (i.e., discounts) it to the time equivalent value of the initial outlay of investment costs. It is important to note that all investment costs are discounted to a time equivalent initial outlay (i.e., to the beginning of the study period) using the discount rate, *d*.

Several procedures exist for calculating the AIRR. These procedures are derived and described in detail in the report by Chapman and Fuller (1996). The most convenient procedure for calculating the AIRR is based on its relationship to the BCR (SIR). This procedure results in a closed-form solution for *r**. The AIRR—expressed as a decimal—is that value of *r** for which:

$$\begin{aligned} r^* &= (1+d)(\text{BCR}^{a^*})^{1/L} - 1 \\ &= (1+d)(\text{SIR}^{a^*})^{1/L} - 1. \end{aligned} \tag{9.17}$$

The AIRR measure is documented in ASTM Standard E1057-06 (2010) (ASTM 2012a).

9.5.3 Diffusion of Additive Manufacturing

In addition to examining the individual adoption of additive manufacturing, it is important to understand the trend in cumulative adoptions by all manufacturers. The diffusion of new technologies or innovations has a significant impact on the success of an industry. The diffusion process is studied in several disciplines: economics, communications, sociology, and marketing. Diffusion, for the purpose of this text, is defined as, "the spread of an innovation throughout a social system" (Koebel et al. 2004).

Rogers (1995) proposes a logistic S-curve model of diffusion, where at the early stage of diffusion there is an increasing rate, as seen in Figure 9.10. Toward the end of the diffusion, there is a decreasing rate. A widely accepted model of technology diffusion is presented by Mansfield (1995):

$$p(t) = \frac{1}{1 + e^{\alpha - \beta t}}, \tag{9.18}$$

where

$p(t)$ = the proportion of potential users who have adopted the new technology by time t;
α = location parameter; and
β = Shape parameter ($\beta > 0$).

A number of factors can affect how a new technology propagates through a business community. The communication structure, for example, affects how people hear about a new technology. The average size of firms in an industry affects their ability to adopt new technologies, as they might not have the resources to invest in it. Rogers (1995) proposed several variables that affect the diffusion of a new technology:

1. Perceived attributes of innovations
   a. Relative advantage to the adopter
   b. Compatibility with other currently used products and processes
   c. Complexity for the adopter

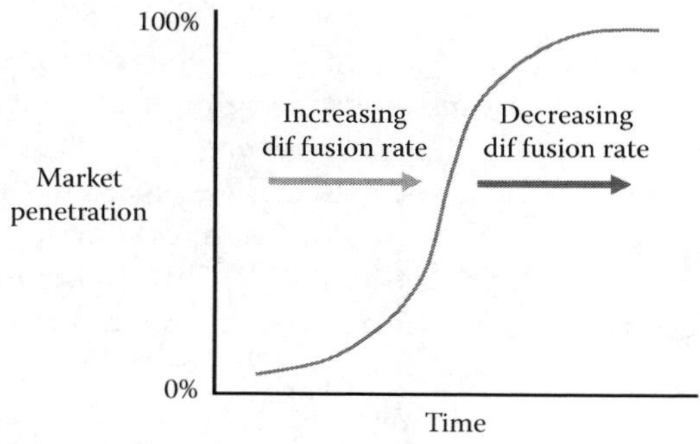

Figure 9.10 Logistic S-curve model of diffusion. (Modified from Rogers, E.M., *Diffusion of Innovations*, 4th edition, The Free Press, New York, 258, 1995.)

   d. Trialability of the new technology
   e. Observability of the results of an innovation
2. Information dissemination
3. Nature of the social system (e.g., attitudes, beliefs, etc.)
4. Extent of change agents' promotion efforts
5. Producer ability/profitability of adoption

These items should be considered when examining the adoption of a new technology within a business community.

Example 4

Thomas (2013) uses domestic unit sales to estimate future adoptions of additive manufacturing in the United States. Using the number of domestic unit sales, the growth in sales is fitted using least squares criterion to an exponential curve that represents the traditional logistic S-curve of technology diffusion. In order to examine additive manufacturing, it is assumed that the proportion of potential units sold by time t follows a similar path as the proportion of potential users who have adopted the new technology by time t. In order to examine shipments in the industry, it is assumed that an additive manufacturing unit represents a fixed proportion of the total revenue; thus, revenue will grow similarly to unit sales. The proportion used was calculated from 2011 data. The parameters α and β are estimated using regression on the cumulative annual sales of additive manufacturing systems in the United States between 1988 and 2011. The US system sales are estimated as a proportion of global sales. This method provides some insight into the current trend in the adoption of additive manufacturing technology. Unfortunately, there is little insight into the total market saturation level for additive manufacturing; that is, there is not a good sense of what percent of the relevant manufacturing industries (shown in Table 9.2) will produce parts using additive manufacturing technologies versus conventional technologies. In order to address this issue, a modified version of Mansfield's model is adopted from Chapman (2001):

$$p(t) = \frac{\eta}{1 + e^{\alpha - \beta t}}, \tag{9.19}$$

where

η = market saturation level in percent

The market saturation level is the percent of the output that will be produced using additive manufacturing after the technology has been adopted by all those who will adopt it. Because η is unknown, it was varied between 0.15% and 100% of the relevant manufacturing shipments, which means that Thomas (2013) calculates multiple diffusion curves, as shown in Figure 9.11. The 0.15% is derived from Wohlers' estimate that the 2011 sales revenue represents 8% market penetration, which equates to \$3.1 billion in market opportunity and 0.15% market saturation (not shown in Figure 9.11). At this level, additive manufacturing is forecasted to reach 50% market potential in 2018 and 100% in 2045. A potential scenario is that additive manufacturing would have between 5% and 35% market saturation. At these levels, additive manufacturing would reach 50% of market potential between 2031 and 2038, while reaching 100% between 2058 and 2065. The industry would reach \$50 billion between 2029 and 2031, while reaching \$100 billion between 2031 and 2044. As illustrated in the figure, it is likely that additive manufacturing is at the far left tail of the diffusion curve, making it difficult to forecast the future trends; thus, some caution should be used when interpreting this forecast. The figure also provides a small graph of the annual number of AM systems sold, which is increasing rapidly.

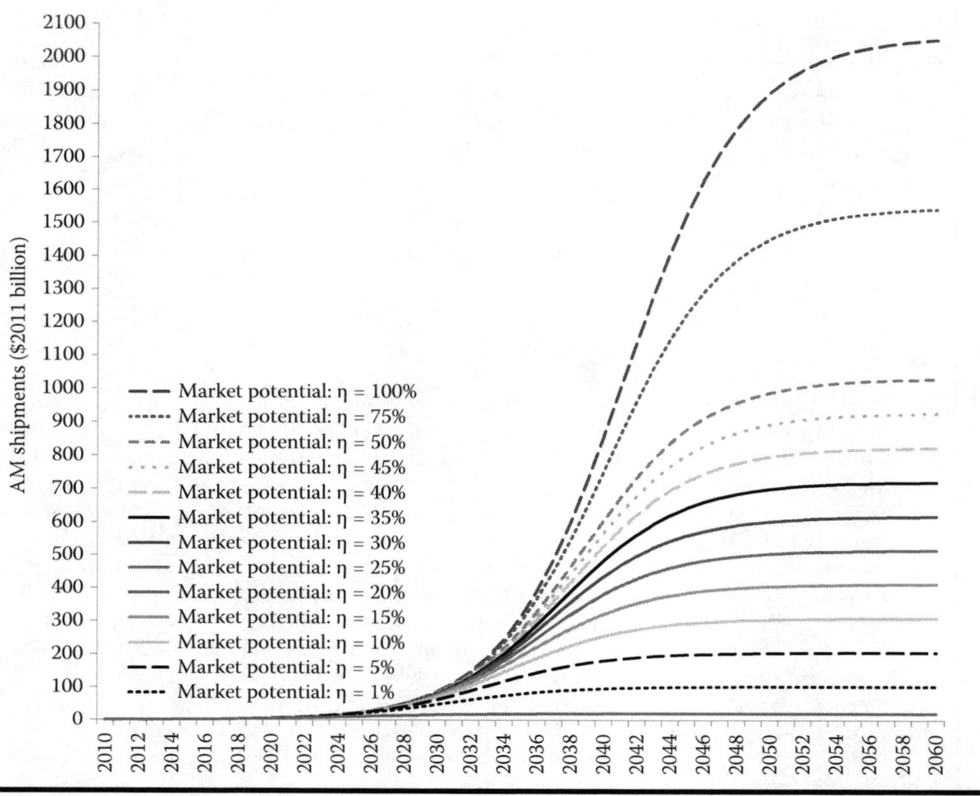

Figure 9.11 Forecasts of US additive manufacturing shipments, by varying market saturation levels.

9.6 Summary

This chapter introduced three aspects on the economics of additive manufacturing. The first involved measuring the value of goods produced using additive manufacturing, including output, value added, and industry classification. The second aspect of additive manufacturing economics involved measuring the costs and benefits of using this technology, including understanding of ill-structured costs and well-structured costs along with the various cost models. Two primary cost models were discussed, one that estimates a constant cost per part and the other that considers changes depending on the number of units being produced. The last aspect of additive manufacturing economics discussed is the adoption and diffusion of this technology. Additive manufacturing is significantly different from traditional methods; thus, determining when and how to take advantage of the benefits of additive manufacturing is a challenge. Manufacturers have to invest resources into understanding whether additive manufacturing makes sense for their business. If they adopt this technology, they will have to train or hire employees who are able to utilize it. This chapter discussed a firm's chain of capabilities along with the three capabilities of a firm: controllability, integration, and flexibility. Also discussed were five methods for measuring economic performance: present value of net benefits, present value of net savings, benefit-to-cost ratio, savings-to-investment ratio, and the adjusted internal rate of return. The chapter concluded with a discussion on the cumulative adoption of new technologies, such as additive manufacturing.

9.7 Review Questions

1. How is value added different from output?
2. How is industry activity categorized in the United States?
3. What are the seven categories of waste in lean manufacturing?
4. What are the two primary cost models for additive manufacturing?
5. What are the primary elements needed to produce a manufactured product?
6. What are the three capabilities of a firm?
7. Name five methods for measuring economic performance.
8. According to Everett Rogers, what are the variables that affect the diffusion of a new technology?

9.8 Review Problems

1. The following table provides economic information on the production of motor vehicles and parts in the United States for the year 2013. What is the value added for this industry activity?
2. The Digit's Widgets Company can produce 30 units per build using additive manufacturing technologies. The material cost for this build (i.e., all 30 units) is $50/kg and requires 5 kg of material. The postprocessing requires 0.75 h of labor per unit plus 0.50 h of labor for packaging. The hourly cost for postprocessing is $28/h while the labor cost for packaging is $15/h. Utilizing Hopkinson and Dickens cost model, what is the cost per unit of production?

	Motor Vehicles and Parts ($millions)
Total intermediate purchases	462,865
Compensation of employees	59,843
Taxes	3,062
Gross operating surplus	75,104
Total industry output	600,875

3. In Example 3, Atzeni and Salmi showed the cost of producing a landing gear using high-pressure die-casting and selective laser sintering. The Wacky Landing Gear Company in the United States can produce the same part with the following cost data:

	High-Pressure Die-Cast Part (US Dollars)	*Selective Laser Sintering Part (US Dollars)*
Material cost per part	1.28	22.34
Preprocessing cost per part	—	9.00
Processing cost per part[a]	0.42 + 24,000/*N*	442.15
Postprocessing cost per part	15.25	22.00
Assembly	0.75	—

[a] Includes the mold for high-pressure die-casting.
N, number of parts produced.

What is the total cost per part using each method? For what number of production runs is it cost effective to produce landing gears using selective laser sintering?

4. The Wacky Landing Gear Company from the previous question produces landing gear assemblies using high-pressure die-casting. However, the production cost of using selective laser sintering has dropped significantly, as seen in the following new cost data. Also included in the table is the company's projected production.

Year	*High-Pressure Die-Cast Part (US Dollars)*	*Selective Laser Sintering Part (US Dollars)*	*Projected Number of Parts Produced*
1	17.7 + 24,000/*N*	267.86	55.00
2	17.7 + 24,000/*N*	243.75	60.00
3	17.7 + 24,000/*N*	221.81	62.00
4	17.7 + 24,000/*N*	201.85	64.00
5	17.7 + 24,000/*N*	183.68	70.00

N, number of parts produced.

To switch to selective laser sintering there is an employee training cost of $15,000 in year one and a refresher course cost in year 3 of $10,000. Using a 3% discount rate and a 5-year study period, calculate the following items for switching to selective laser sintering:

- Present value net savings
- Savings-to-investment ratio
- Adjusted internal rate of return

5. The Fidgety Digit Company wants to assess what the prevalence of additive manufacturing will be in their industry in 10 years. Using the Rogers' logistic S-curve model of diffusion, they estimate a location parameter of 2 and a shape parameter of 0.01. What proportion of the potential users will have adopted the new technology in year 10?

Disclaimer

Certain trade names and company products are mentioned in the text in order to adequately specify the technical procedures and equipment used. In no case does such identification imply recommendation or endorsement by the National Institute of Standards and Technology, nor does it imply that the products are necessarily the best available for the purpose.

References

ASTM (2012a). ASTM 1057-6 (2010): Standard practice for measuring internal rate of return and adjusted internal rate of return for investments in buildings and building systems. In Bailey, S. J., N. C. Baldini, S. Emery, J. Ermigiotti, D. Gallagher, K. Hanratty, E. Moore, E. A. Olcese, K. A. Peters, J. L. Rosiak, D. A. Terruso, and J. Wright. *Annual Book of ASTM Standards*. West Conshohocken, PA: ASTM International, 357–364.

ASTM (2012b). ASTM E964-06 (2010): Practice for measuring benefit-to-cost and savings-to-investment ratios for buildings and building systems. In Bailey, S. J., N. C. Baldini, S. Emery, J. Ermigiotti, D. Gallagher, K. Hanratty, E. Moore, E. A. Olcese, K. A. Peters, J. L. Rosiak, D. A. Terruso, and J. Wright. *Annual Book of ASTM Standards*. West Conshohocken, PA: ASTM International, 306–315.

ASTM (2012c). ASTM E1074-09: Practice for measuring net benefits and net savings for investments in buildings and building systems. In Bailey, S. J., N. C. Baldini, S. Emery, J. Ermigiotti, D. Gallagher, K. Hanratty, E. Moore, E. A. Olcese, K. A. Peters, J. L. Rosiak, D. A. Terruso, and J. Wright. *Annual Book of ASTM Standards*. West Conshohocken, PA: ASTM International, 365–374.

ASTM (2012d). *ASTM Standards on Building Economics*. West Conshohocken, PA: ASTM International.

Atzeni, E. and A. Salmi (2012). Economics of additive manufacturing for end-usable metal parts. *International Journal of Advanced Manufacturing Technology* 62: 1147–1155.

Baumers, M. (2012). Economic aspects of additive manufacturing: Benefits, costs, and energy consumption, PhD thesis, Loughborough University.

Baumers, M., C. Tuck, R. Wildman, I. Ashcroft, and R. Hague (2011). Energy inputs to additive manufacturing: Does capacity utilization matter? In *22nd Annual International Solid Freeform Fabrication Symposium—An Additive Manufacturing Conference*, Laboratory for Freeform Fabrication and University of Texas, Austin, TX.

Campbell, I., J. Combrinck, D. De Beer, and L. Barnard (2008). Stereolithography build time estimation based on volumetric calculations. *Rapid Prototyping Journal* 14(5): 271–279.

Chapman, R. E. (2001). Benefits and costs of research: A case study of construction systems integration and automation technologies in commercial buildings, NISTIR 6763. Gaithersburg, MD: National Institute of Standards and Technology.

Chapman, R. E., D. T. Butry, A. L. Huang, and D. S. Thomas (2009). Benefits and costs of research: A case study of improved service life prediction, NIST Technical Note 1650. Gaithersburg, MD: National Institute of Standards and Technology.

Chapman, R. E. and S. K. Fuller (1996). Two case studies in building technology, NISTIR 5840. Gaithersburg, MD: National Institute of Standards and Technology, 27–37.

Di Angelo, L. and P. Di Stefano (2011). A neural network-based build time estimator for layer manufactured objects. *International Journal of Advanced Manufacturing Technology* 57(1–4): 215–224.

Dornbusch, R., S. Fischer, and R. Startz (2000). *Macroeconomics*. London, UK: McGraw-Hill.

Doubrovski, Z., J. C. Verlinden, and J. M. P. Geraedts (2011). Optimal design for additive manufacturing: Opportunities and challenges. *Proceedings of the ASME 2011 International Design Engineering Technical Conferences and Computers and Information in Engineering Conference*. Washington DC: ASME.

Holmström, J., J. Partanen, J. Tuomi, and M. Walter (2010). Rapid manufacturing in the spare parts supply chain: Alternative approaches to capacity deployment. *Journal of Manufacturing Technology* 21(6): 687–697.

Hopkinson, N. and P. M. Dickens (2003). Analysis of rapid manufacturing—Using layer manufacturing processes for production. *Proceedings of the Institution of Mechanical Engineers, Part C: Journal of Mechanical Engineering Science* 217(1): 31–39.

Huang, S. H., P. Liu, and A. Mokasdar (2013). Additive manufacturing and its societal impact: A literature review. *International Journal of Advanced Manufacturing Technology* 67: 1191–2103.

Khajavi, S. H., J. Partanen, and J. Holmstrom (2014). Additive manufacturing in the spare parts supply chain. *Computers in Industry* 65: 50–63.

Kim, B. (2015). Supply chain management: A learning perspective. Lecture 1–2, Coursera Lecture. Korea Advanced Institute of Science and Technology.

Kim, B. and C. Park (2013). Firms' integrating efforts to mitigate the tradeoff between controllability and flexibility. *International Journal of Production Research* 51(4): 1258–1278.

Koebel, C. T., M. Papadakis, E. Hudson, and M. Cavell (2004). *The Diffusion of Innovation in the Residential Building Industry*. Upper Marlboro, MD: Center for Housing Research, Virginia Polytechnic Institute and State University and NAHB Research Center.

Lindemann C., U. Jahnke, M. Moi, and R. Koch (2012). Analyzing product lifecycle costs for a better understanding of cost drivers in additive manufacturing. *Proceedings of the 2012 Solid Freeform Fabrication Symposium—An Additive Manufacturing Conference*, Austin, TX.

Mansfield, E. (1995). *Innovation, Technology and the Economy: Selected Essays of Edwin Mansfield*. Brookfield, VT: Edward Elgar.

Mognol, P., D. Lepicart, and N. Perry (2006). Rapid prototyping: Energy and environment in the spotlight. *Rapid Prototyping Journal* 12(1): 26–34. doi:10.1108/13552540610637246.

Morrow, W. R., H. Qi, I. Kim, J. Mazumder, and S. J. Skerlos (2007). Environmental aspects of laser-based and conventional tool and die manufacturing. *Journal of Cleaner Production* 15(10): 932–943.

Neef, A., K. Burmeister, and S. Krempl (2005). *Vom Personal Computer zum Personal Fabricator (From Personal Computer to Personal Fabricator)*. Hamburg, Germany: Murmann.

Reeves, P. (2008). How the socioeconomic benefits of rapid manufacturing can offset technological limitations. *RAPID 2008 Conference and Exposition*, Lake Buena Vista, FL.

Rogers, E. M. (1995). *Diffusion of Innovations*, 4th edition. New York: The Free Press.

Ruegg, R. T. and H. E. Marshall (1990). *Building Economics: Theory and Practice*. New York: Chapman and Hall.

Ruffo, M. and R. Hague (2007). Cost estimation for rapid manufacturing simultaneous production of mixed components using laser sintering. *Proceedings of the Institution of Mechanical Engineers, Part B: Journal of Engineering Manufacture* 221(11): 1585–1591.

Ruffo, M., C. Tuck, and R. Hague (2006a). Cost estimation for rapid manufacturing-laser sintering production for low to medium volumes. *Proceedings of the Institution of Mechanical Engineers, Part B: Journal of Engineering Manufacture* 220(9): 1417–1427.

Ruffo, M., C. Tuck, and R. Hague (2006b). Empirical laser sintering time estimator for Duraform PA. *International Journal of Production Research* 44(23): 5131–5146.

Rushing, A. S. and R. E. Chapman (2008). Cost-effectiveness tool for capital asset protection, National Institute of Standards and Technology, Gaithersburg, MD. https://www.nist.gov/services-resources/software/cost-effectiveness-tool-capital-asset-protection.

Stoneman, P. (2002). *The Economics of Technological Diffusion*. Oxford, UK: Wiley-Blackwell.

Thomas, D. (2013). *The US Additive Manufacturing Industry*. NIST Special Publication 1163. Gaithersburg, MD: National Institute of Standards and Technology.

Thomas, D. and A. M. Kandaswamy (2015). *Tracking Industry Operations Activity: A Case Study of US Automotive Manufacturing*. NIST Special Publication 1601. Gaithersburg, MD: National Institute of Standards and Technology.

UNSD (2015). National Accounts Main Aggregates Database, United Nations Statistics Division. https://unstats.un.org/unsd/snaama/Introduction.asp.

Walter, M., J. Holmstrom, and H. Yrjola (2004). Rapid manufacturing and its impact on supply chain management. *Logistics Research Network Annual Conference*, Dublin, Ireland.

Wohlers, T. (2012). *Wohlers Report 2012: Additive Manufacturing and 3D Printing State of the Industry*. Fort Collins, CO: Wohlers Associates, Inc.

Wohlers, T. (2015). *Wohlers Report 2015: Additive Manufacturing and 3D Printing State of the Industry*. Fort Collins, CO: Wohlers Associates, Inc.

Young, S. K. (1991). A cost estimation model for advanced manufacturing systems. *International Journal of Production Research* 29(3): 441–452.

Index

Note: Page numbers followed by *f* and *t* refer to figures and tables.

B

C

D

O

P

U

V

W